SELF-HEALING COMPOSITES

SELF-HEALING COMPOSITES

SHAPE MEMORY POLYMER-BASED STRUCTURES

Guoqiang Li
Louisiana State University, USA

WILEY

This edition first published 2015

© 2015 John Wiley & Sons Ltd

Registered office
John Wiley & Sons Ltd, The Atrium, Southern Gate, Chichester, West Sussex, PO19 8SQ, United Kingdom

For details of our global editorial offices, for customer services and for information about how to apply for permission to reuse the copyright material in this book please see our website at www.wiley.com.

Library of Congress Cataloging-in-Publication Data

Li, Guoqiang, author.
   Self-healing composites : shape memory polymer based structures / Guoqiang Li.
       pages cm
   Includes bibliographical references and index.
   Summary: "We hope this book will provide some background information for readers who are interested in using SMPs for self-healing"– Provided by publisher.
   ISBN 978-1-118-45242-4 (hardback)
   1. Composite materials. 2. Self-healing materials. I. Title.
   TA418.9.S62L53 2015
   620.1′18–dc23

                                                                                          2014021355

A catalogue record for this book is available from the British Library.

Set in 10/12pt TimesLTStd-Roman by Thomson Digital, Noida, India
Printed and bound in Malaysia by Vivar Printing Sdn Bhd

1   2015

To my wife, Deying, and daughters, Yiyao and Sarah

Contents

Preface

The history of human beings and advancement of civilization are generally categorized by the ability to produce and manipulate materials to meet the needs of the society, such as the Stone Age, Bronze Age, Iron Age, and Silicon Age. It is reasonable to ask what will be next. For many practical engineering applications, materials need to be lighter, stronger, tougher, smarter, and cost-effective. While lighter, stronger, and tougher materials have been a major theme in modern structure design for many years, smarter materials and structures are a comparatively new paradigm. Although it is difficult to define the exact meaning of smarter, because it evolves with time, the big picture is clear. We desire the synthetic materials and engineering structures to possess intelligence that is commensurate with or better than their biological counterparts. We are optimistic that synthetic materials can be better than biological counterparts because we have more types of elements to use while biomaterials build on a few types of atoms. Also, we have more knowledge of various biological systems and we have the ability to analyze, compare, combine, create, design, optimize, and manufacture materials and structures. With the advancement in the understanding of micro, nano, and molecular structures, as well as high fidelity computation technology, the concept of materials by design seems to be opening up endless opportunities for discovering and synthesizing new materials. However, no matter how powerful is the computation technology, scientists and engineers must have targets or pictures in mind, that is, visualizing and imagining what they want before they can start designing the materials by computation. We believe that the motivation should come from observation of nature.

Historically, bio-inspiration has been a driving force for the advancement of science and technology. The reason is simple. Mother nature has evolved millions and millions years and developed sophisticated natural materials or assemblies with multifunctions and properties. In today's technological understanding, most of the natural materials are smart. They sense, adapt, and react to external stimuli and, of course, self-heal internal damage. There are three levels of learning from nature, namely, bioinspiration, biomimicry, and bioreplication, with increasing complexity.

The purpose of bioinspiration is to recreate biological functions but not necessarily biological structures. An aircraft is an excellent example of bioinspiration of birds. The purpose of biomimicry is to recreate the biological functions by approximately replicating the biological structures. Dry adhesive by mimicking a gecko foot may be an example of biomimicry. The purpose of bioreplication is the direct reproduction of biological functions by replicating the biological structures. Because most biological structures are hierarchical and very complex, bioreplication is in its infancy. Based on current science and technology, we believe that the next generation of synthetic materials should be biomimetic smart materials by learning from mother nature, if not bioreplication. We must admit that, as compared to natural materials, the smart synthetic materials available are still in their infancy. We still have a long way to go and many challenges to overcome before we can mimic the simplest natural materials.

In the past several decades, the desire for lighter, tougher, and stronger materials in vehicles (car, aircraft, ship, bike, etc.), energy production, storage, and transport (windmill blade, piping, offshore oil platform, pressure vessel, on-board and off-board energy storage tank, etc.), and infrastructure (bridge, harbor, building, etc.) has driven the development of fiber reinforced polymer composite materials. Fiber reinforced polymer composite materials, while having high specific strength, stiffness, and corrosion resistance, are vulnerable to foreign object impact, due to stress concentration at various interfaces and brittleness of thermosetting polymer matrix, such as epoxy and synthetic fibers like glass fiber. In addition to the well-known ballistic impact, which can be visually detected by operators, low velocity impact is more dangerous in the sense that it usually avoids visual inspection. For example, a low velocity impact on a laminated composite may leave a barely visible indentation at the point of impact, but macroscopic cracks are likely to have been created at the back surface or inside the laminate. Low velocity impact is not uncommon. For instance, for aircraft, hail, bird strike during taking off or landing, debris strike in the runway, or even dropping a tool during routine inspection represent typical low velocity impact events. Therefore, for fiber reinforced polymer composite structures, while damage healing at the microscopic level such as healing a fatigue crack (micrometer or submicrometer scale) is essential, it is important that the composite has the ability to heal wide-opened macroscopic cracks such as delamination in the millimeter scale. There is no doubt that in fiber reinforced polymer composite materials, fibers are the primary load bearing component. Therefore, it is natural to think that we need to heal synthetic fibers such as carbon fibers and glass fibers, similar to healing the bone in a human body. While this is a legitimate attempt, it is extremely difficult based on current knowledge and technology. Of course, it is an interesting topic for materials scientists to pursue.

Actually, in fiber reinforced polymer composites, matrix dominated damage represents the major damage mode up to a certain loading level. Therefore, healing damage in the polymer matrix is not trivial. Because of the role played by the matrix in composites, healing matrix damage serves to recover the load carrying capacity, protect the fiber from damage, and extend the service life of the composite. Therefore, similar to studies by others, this book will focus on healing macroscopic damage in the polymer matrix. Another argument may be that, if the crack is wide-opened, it is visible with the naked eye, and thus directly injecting adhesive into the crack would heal the composite. This seems a legitimate argument and may be the case for some structures. However, it is not the case for many others, such as delamination and matrix cracking in a laminated composite and cracking in the foam core and debonding at the core/face sheet interface in a sandwich composite, which are inside the structures and inaccessible. Therefore, self-healing of a macroscopic crack in a polymer matrix is a genuine problem that deserves investigation. Unfortunately, healing structural-length scale damage represents a grand challenge for both intrinsic and extrinsic self-healing schemes. The reason is that almost all of them need external help to bring the fractured surfaces in proximity before the self-healing mechanisms can take effect.

Since 2006, our group has focused on healing of structure-length scale crack in shape memory polymer (SMP) based composite. Our first paper was published in 2008 in *Composites Science and Technology*. In 2010, we reported the bio-inspired two-step self-healing scheme, that is, close-then-heal (CTH), in several publications and with more and more details and depth. In CTH, we realized that for a wide-opened crack, we need to close/narrow the crack before we heal it molecularly, similar to self-healing of human skin. We demonstrated that CTH can heal low velocity impact induced damage in various types of composite structures, repeatedly, efficiently, molecularly, and timely. Most recently we have extended this concept to healing conventional thermosetting polymer composite by embedding SMP fibers. We have also conducted modeling work to understand the constitutive behavior of SMPs and SMP based composite better, as well as damage–healing mechanisms.

This book grew out of the work done in our group in the past several years. It is by no means a comprehensive representation of the entire self-healing picture. As self-healing has become a popular topic in both academia and industry, we hope this book will provide some background information for readers who are interested in using SMPs for self-healing. We are optimistic that self-healing materials and structures will become an engineering reality in the coming years rather than scientific fiction.

This book is arranged as follows. Chapter 1 will introduce some basics on damage in various composite structures and classification of self-healing schemes. Chapter 2 will focus on the introduction of self-healing in biological systems, which inspired the self-healing scheme in this book. Chapter 3 will focus on the thermomechanical behavior of thermosetting SMPs and SMP based syntactic foam. The effect of programming temperature (hot programming and cold programming) as well as programming stress condition (1-D, 2-D, and 3-D stress conditions) on the shape memory effect will be presented. Functional stability under environmental attacks will also be briefly discussed. Chapter 4 will provide some basics on solid mechanics modeling and discuss the two popular strategies to model SMPs – thermodynamic based and stress/structural relaxation based approaches. Modeling on both thermosetting SMP and SMP based syntactic foam will be reported. Chapter 5 will focus on testing and modeling of polyurethane SMP fibers, including physical, mechanical, thermomechanical, damping, and microstructure characterization. Viscoplasticity theory will be presented to model the effect of strain hardening by cold-drawing programming on stress recovery. The underlying mechanisms controlling the difference in strain memory and stress memory will be discussed. New definitions of stress fixity and stress recovery ratios will be given. Chapter 6 will deal with self-healing of SMP matrix based composite structures. Self-healing of notched beam, sandwich, grid stiffened, and 3-D woven fabric reinforced SMP polymer composite structures will be discussed in detail. Chapter 7 will focus on self-healing of conventional thermosetting polymer by embedding SMP fibers and thermoplastic particles. SMP fiber reinforcement in the form of 1-D, 2-D, and 3-D will be presented. Chapter 8 will discuss modeling of the healing process and evaluation of healing efficiency. The reptation model for a chain confined to a tube will be modified to describe the diffusion of molten thermoplastic molecules into a polymer matrix by incorporating the recovery stress due to the shape memory effect. Healing efficiency in terms of Mode I, Mode II, and Mixed Mode I&II fracture toughness through adhesively bonded joint configurations will be formulated. Chapter 9 will discuss some future perspectives on this topic.

Here, I would like to take this opportunity to express my sincere thanks to the funding agencies for supporting our research, particularly National Science Foundation (NSF) (CMMI 0900064, CMMI 0946740, and CMMI 1333997). My deepest thanks go to my previous and current graduate students either at Louisiana State University or at Southern University – Dr. Manu John, Dr. Wei Xu, Dr. Gefu Ji, Dr. Jones Nji, Dr. Amir Shojaei, Mr. Damon Nettles, Mr. Abe King, Mr. Naveen Uppu, Mr. Pengfei Zhang, Mr. Oludayo Ajisafe, Mr. Anqi Wang, and Ms. Qianxi Yang – and my previous and current research associates – Dr. Harper Meng, Dr. Tao Xu, Dr. Jinquan Cheng, and Dr. Zhenyu Ouyang – because much of the work I have written about in this book has been performed in collaboration with them. My special thanks go to Dr. Harper Meng. His assistance in obtaining all copyright permissions and proof-reading all chapters is greatly appreciated. Specifically, he collected all the raw materials for Chapter 2. Dr. Amir Shojaei also assisted in proofreading Chapters 4 and 5. My sincere thanks and appreciation also go to the John Wiley & Sons team for assistance provided during the course of writing this book. Finally, my sincere thanks go to my wife, Deying, and my daughters, Yiyao and Sarah. It is with their love and encouragement that this book was possible.

Guoqiang Li
Baton Rouge, Louisiana, USA

1

Introduction

Driven by the need for lighter and stronger engineering structures, fiber reinforced polymer composite materials have emerged as a new class of engineering materials for load bearing structures. From the sky to under the sea, from household products to infrastructure, from transportation vehicles to energy production, fiber reinforced polymer composites are gradually finding their way to replace traditional materials such as metal, concrete, wood, etc. This is due to the high specific strength, stiffness, and corrosion resistance of fiber reinforced polymer composites. Another advantage exists in its tailorability. Its anisotropic nature makes it work on demand, that is, optimize a design by aligning fibers along the direction that needs the highest strength and stiffness. Among the various types of fiber reinforced polymer composites, thermosetting polymer composites predominate. This is because thermosetting polymer composites are easy to manufacture using various techniques, such as resin transfer molding (RTM), vacuum assisted resin infusion molding (VARIM), filament winding, pultrusion, prepreg, or simply hand lay-up. Also, thermosetting polymer composites have better thermal and dimensional stability. However, thermosetting polymers are generally brittle and vulnerable to certain loadings, such as impact loading. Under impact loading, even under low velocity impact, severe damage such as macroscopic cracking may occur inside the composite structures, making repair a challenging task. In other words, there is an urgent need for thermosetting polymers to have self-healing capabilities. In this introductory chapter, we will briefly discuss thermosetting polymers. We will then illustrate typical fiber reinforced composite structures that use thermosetting polymer as the matrix. Then our focus will be on reviewing typical failure modes in thermosetting polymer composite structures under low velocity impact, followed by brief discussions on the repair strategies currently used in industry. Classification of self-healing systems will be the focus of the next section. Finally, we will present the organization of this book.

1.1 Thermosetting Polymers

It is well known that thermoset polymers such as epoxy, vinyl ester, polyester, phenolic, etc., are chemically or physically cross-linked polymers (chemical bonds between polymer chains, intermolecular van der Waals bonds, dipole–dipole interactions, and molecular entanglement). These cross-links serve as molecular anchorages, which prevent molecular motion of the

Self-Healing Composites: Shape Memory Polymer-Based Structures, First Edition. Guoqiang Li.
© 2015 John Wiley & Sons, Ltd. Published 2015 by John Wiley & Sons, Ltd.

polymer chains. This is how a thermoset obtains its strength, stiffness, and thermal stability and why it behaves in a brittle manner under mechanical loading. Once one chain factures, the force is transferred to its neighbors through the cross-linked network, leading to crazing, cracking, and ultimate macroscopic fracture at a relatively small strain.

Like any polymer, the physical/mechanical properties of thermosetting polymers depend on loading time (rate) and temperature. Owing to the cross-linked nature, thermosetting polymers are generally difficult to form into crystals because the cross-links prevent segments from motion and folding into lamella or spherulite. In other words, most thermosetting polymers are amorphous. Similar to thermoplastic polymers, amorphous thermosetting polymers show distinctively different behaviors at different temperatures. At temperatures below the so-called glass transition temperature (T_g), the network structure is frozen. The mobility of the segments is very low. Under external loading, stretching and rotation of the chemical bonds is the dominant deformation mode, and thus the polymer is rigid and very brittle. When the temperature is increased to the glass transition zone, we see drastic change in the mechanical properties. The stiffness may show one to two orders of decrease within a small temperature window. In this temperature range, the frozen molecules are gradually defrosted and their mobility increases dramatically so that the molecules can deform easily along the direction of loading. As a result, under the same load, the deformation in the glass transition region is significantly higher than that below T_g, leading to a considerable reduction in stiffness. Also, within this region, the viscous deformation (time dependent deformation) under dynamic loading absorbs a significant amount of energy. The deformation shows a delayed response to the dynamic loading, leading to viscoelastic behavior and vibration damping. When the temperature is above the glass transition zone, the polymer is rubbery but does not melt. Due to the cross-links, however, the ductility of the polymer is still comparatively small, even in the rubbery status. A further increase in temperature may result in decomposition and burning of the polymer. This shows a distinctive departure from thermoplastic, which becomes liquid and flows above the melting temperature and can be remolded.

The glass transition of thermosetting polymers can be explained by the free volume theory [1]. The volume in polymers consists of three parts – occupied volume, interstitial free volume, and hole free volume. The occupied volume does not change with temperature. The interstitial free volume may change linearly with temperature as it reflects the change in the interatomic distance. The hole free volume increases nonlinearly with temperature. When the temperature is increased to the glass transition region, the frozen hole free volume expands significantly, giving a free space for segmental motion of the cross-linked molecules. In other words, the mobility of the molecules increases drastically and large coordinated motion of the segments along the loading direction becomes possible. The significant increase in deformation leads to a dramatic reduction in stiffness. This is exactly what we see in the glass transition region.

Thermosetting polymers usually consist of two parts, the liquid resin and curing agent. Before mixing together, these two parts can be safely stored for months or years. When they are mixed together, the polymerization process starts, which is generally exothermic. Before it is cured into a solid, there is a time window from minutes to hours for us to work on it, such as using resin transfer molding to wet through the fibers in order to form a fiber reinforced polymer composite. The working window can be adjusted by controlling the amount of curing agent and the curing temperature. Clearly, reducing the amount of curing agent or lowering the curing temperature increases the time period before the resin hardens.

While thermoplastic polymers can also find applications in load bearing structures, particularly for thermoplastic polymers that have very high melting temperatures, thermosetting polymers have dominated the application in load bearing, fiber reinforced polymer composite structures. Therefore, this book will focus on thermosetting polymer composites.

1.2 Thermosetting Polymer Composites in Structure Applications

In recent years, advanced lightweight materials have become a technological and economic driver for societies. For example, advanced lightweight materials are essential for reducing vehicle weight to boost fuel economy of modern automobiles, while maintaining safety and performance. Replacing cast iron and traditional steel components with lightweight materials such as polymer composites allows vehicles to carry advanced emission-control equipment, safety devices, and integrated electronic systems, without an associated weight penalty. Using lighter materials also reduces fuel consumption of vehicles because it takes less energy to accelerate or decelerate a lighter object [2]. Another typical example is in the aviation industry. In the Dreamliner Boeing 787, it is estimated that about 50% of the materials used was carbon fiber reinforced polymer composites. In fact, fiber reinforced thermosetting polymer composites have been widely used in almost all man-made structures, particularly in high tech and high value structures, including, but not limited to, aerospace (fixed wing aircrafts, helicopters, etc.), defense (tank, armor, etc.), energy production, storage, and transportation (wind turbine blade, pipe, on-board and off-board storage tanks for natural gas or hydrogen, etc.), vehicles (car, truck, train, etc.), electrical and electronic (rods, tubes, molded parts, electrical housings, etc.), construction (bathtubs, decks, swimming pools, utility poles, bridge decks, railings, and repair, rehabilitation, reinforcement, and reconstruction of concrete structures, etc.), marine (ship hulls, decks, bulkheads, railings, offshore oil platforms, etc.), and consumer products (golf clubs, bicycles, fishing rods, skis, tennis rackets, snowmobiles, mobile campers, etc.).

The highly cross-linked nature provides thermosetting polymers with high strength, high stiffness, high thermal stability, and good chemical resistance. The trade-off for this gain in mechanical strength is a loss in toughness and ductility, and they are prone to developing microcracks under external loading such as cyclic fatigue loading. Because microcracks are not easily detectable, the propagation and coalescence of microcracks result in macrocracks and ultimate structural failure. For some types of loading such as impact loading, cracks may be created on a macroscopic scale, leading to imminent structural failure if immediate care is not taken. Owing to various types of loadings, such as fatigue, impact, vibration, creep, earthquake, hurricane, etc., and environmental conditioning, such as corrosion, hygrothermal effects, ultraviolet radiation, fire, etc., no engineering structures last forever. Therefore, self-healing is a highly desired feature for engineering structures in general and for fiber reinforced thermosetting polymer composites in particular.

1.3 Damage in Fiber Reinforced Thermosetting Polymer Composite Structures

Fiber reinforced thermosetting polymer composites have been used in the form of laminate, sandwich, grid stiffened, stitched, Z-pinned, three-dimensional (3-D) woven fabric, and hybrid structures. In addition to carrying the designed static/dynamic loads, most composite structures experience some kind of low or high velocity impact incidents during their life cycle. A low velocity impact is not uncommon. For example, dropping a tool on to a composite structure during a routine inspection characterizes a low velocity impact incident, not to mention incidents during manufacturing, transportation, installation, and service. For armor-grade composite structures, a low to high strain rate impact or blast is the primary criterion in structural design. Although both low and high velocity impacts are of concern, a low velocity impact is more dangerous because it is often missed by visual inspection. For example, after a low velocity impact on a laminated composite, only a barely visible indentation may be identified on the impacted surface. However, significant damage may have been induced inside the laminate and on the back surface, which cannot be detected with the naked eye. As a result of the damage, the residual load carrying capacity of the structure may be considerably reduced, leading to premature and catastrophic structural failure. Therefore, low velocity impact is the focus of this section.

Low velocity impact on composite structures (laminated, sandwich, grid stiffened, 3-D woven fabric reinforced) has been a topic of research interest for years all over the world. Many researchers have experimentally and theoretically investigated the low velocity impact response and residual strength of composite structures, including instrumented low velocity impact testing, analytical modeling based on the modified Hertz contact law or conservation of energy, and finite element modeling using commercial software packages like LS-DYNA. There is no shortage of literature in this research area. These studies have greatly enhanced understanding of impact behavior, damage, the energy dissipation mechanism, and residual structural performance. As a result, more and more impact tolerance/resistance composite structures are being designed and manufactured with confidence.

1.3.1 Damage in Laminated Composites

A low velocity impact induces various types of damage in fiber reinforced laminated composite structures. In addition to the visible or barely visible indentation on the impacted surface and cracking on the back surface, the most prevalent damage inside a laminated composite includes delamination, matrix cracking, fiber/matrix interfacial debonding, and fiber fracture (see Figure 1.1) [3].

After a low/high velocity impact, the residual load carrying capacity, particularly the in-plane load carrying capacity, reduces drastically. Figure 1.2 shows a laminated composite beam during an in-plane compression test after being impacted by a projectile with a mass of 7.45 g at a velocity of 390 m/s. It is clear that the damaged laminate shows global buckling and local buckling of the delaminated sublaminate. Because of the buckling, the compressive strength of the damaged laminate is reduced to below 50% of the undamaged control laminate.

1.3.2 Damage in Sandwich Composites

Catastrophic structure failure due to impact has been well documented in the literature; a well-known incident is the loss of the Space Shuttle *Columbia* [4]. A fiber reinforced polymer composite sandwich, while optimal for carrying a transverse load with a minimal weight penalty, is also vulnerable to impact damage [5–11]. Typical damage modes include indentation, face sheet cracking, core fracture, face sheet/core interfacial debonding, etc. In a sandwich construction, the two face sheets are responsible for carrying the transverse load and in-plane load, while the sandwich core is primarily responsible for fixing the skin and absorbing impact energy. Therefore, the key is to improve the core design in order to enhance impact tolerance of the sandwich. Various types of core materials have been studied such as foam core (polymeric

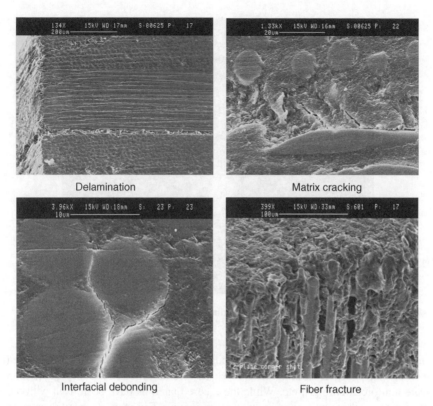

Delamination

Matrix cracking

Interfacial debonding

Fiber fracture

Figure 1.1 Low velocity impact damages inside a laminate. *Source:* [3] Reproduced with permission from Elsevier

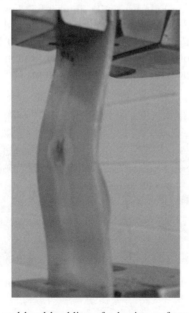

Figure 1.2 Mixed global and local buckling of a laminate after a moderate velocity impact.

Figure 1.3 Low velocity impact induced indentation on the sandwich skin

foam, metallic foam, ceramic foam, balsa wood, syntactic foam, etc.) [12,13], web core (truss, honeycomb, etc.) [14], 3-D integrated core [10], foam filled web core [15], laminated composite reinforced core [9], etc. While these core materials have been used with a certain amount of success, they are limited in one way or another. For example, brittle syntactic foam cores absorb impact energy primarily through macroscopic damage, significantly sacrificing residual strength [16–20], and web cores lack bonding with the skin and also have an internal open space for easy perforation by a projectile (impact windows) [16].

Figure 1.3 shows the indentation on a sandwich beam and Figure 1.4 shows top skin delamination after a low velocity impact by a DynaTup 8250HV machine with a hammer weight of 33 kg and velocity of 3.83 m/s, which translates to an impact energy of 242 J. The tup nose has a semispherical shape with a diameter of 12.7 mm. The sandwich beam has a carbon

Figure 1.4 Low velocity impact induced delamination, debonding, and shear fracture in the laminated skin and balsa wood core

fiber reinforced epoxy skin with a thickness of 2.54 mm and a balsa wood core with a thickness of 60.0 mm. During impact, the sandwich beam of 50.8 mm wide and 254.0 mm long was simply supported and the impact was at the center of the beam on the top skin. While only a visible indentation is seen on the top skin in Figure 1.3, Figure 1.4 shows significant delamination in the top skin, debonding between the top skin and the balsa wood core, as well as shear fracture at the core. Therefore, even under a low velocity impact, damage within the sandwich beam is dramatic and urgent repair is needed.

1.3.3 Damage in 3-D Woven Fabric Reinforced Composites

Delamination has been a major form of damage on a laminated composite. Therefore, enhancement in the thickness direction through stitching, Z-pinning, etc., has been widely used in laminated structures. Along the same lines is polymeric composite material reinforced with three-dimensional (3-D) fabric architecture, which has good impact tolerance [21–25]. Polymeric composite reinforced with 3-D woven fabric is an attractive candidate for use in weight sensitive industries such as aerospace, auto, and maritime where, in addition to carrying static and cyclic loads, structural components are expected to perform well under impact.

A number of studies has been conducted to understand the impact response of 3-D woven composites [26–33]. Recently, Baucom, Zikry, and Rajendran [34] investigated the effects of fabric architecture on damage propagation, perforation resistance, strength, and failure mechanisms in composite systems of comparable areal densities and fiber volume fractions, subjected to repeated impact. They found that 3-D systems survived more strikes before being perforated and absorbed more total energy compared to other systems. They reported transverse matrix cracking, fiber debonding from the matrix, fiber fracture, and fracture of Z direction fiber tows as failure modes in the 3-D systems. Most recent developments in the area of 3-D woven fabric composites include modeling the impact penetration of 3-D woven composite at the unit cell level [35], studying the transverse impact damage and energy absorption [36], investigating the compressive responses and energy absorption [37], and studying the effect of Z-yarns on the stiffness and strength [38]. While 3-D woven fabric reinforced composite has demonstrated considerable enhancement in terms of impact tolerance, it is still vulnerable to impact damage, in particular under repeated impact incidents. It is desired to add self-healing capacities to 3-D woven composites to further increase their impact tolerance and service life.

In order to demonstrate the damage mode in 3-D woven fabric reinforced polymer composite, a repeated impact test was conducted on a 3-D woven glass fabric reinforced thermosetting shape memory polymer composite foam [39]. The impact was conducted by the same DynaTup 8250HV impact machine with a hammer weight of 6.44 kg and velocity of 3.0 m/s. Figure 1.5 shows perforation of the 3-D woven fabric reinforced polymer composite and Figure 1.6 shows fracture of the reinforcing fiber tows, both after the ninth impact [39].

As compared to laminated composite, there is no doubt that the impact tolerance of the 3-D woven fabric reinforced polymer composite has been enhanced dramatically due to the 3-D network structure and the improved transverse shear resistance. However, it is clear that even repeated low velocity impact can cause significant damage. Therefore, like laminated composite and sandwich composite, 3-D woven fabric reinforced polymer composite also calls for self-healing capabilities.

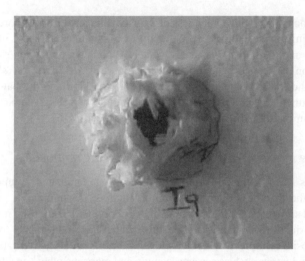

Figure 1.5 Perforation in a 3-D woven fabric reinforced polymer composite after repeated low velocity impact. *Source:* [39] Reproduced with permission from IOP Publishing

1.3.4 Damage in Grid Stiffened Composites

A basic grid structure is a latticework of rigid, interconnecting beams in two, three, or four groups and directions [40,41]. Figure 1.7 demonstrates an orthogrid structure and the terminology used to describe it. Nodes, ribs, beams, and cells are the grid structural elements. *Nodes* are the crossover points, *ribs* are the linear segments that span adjacent nodes, and *beams* are a collection of aligned ribs and nodes. *Cells* or *bays* are the spaces enclosed between ribs. Structurally related terms are center-to-center, in-plane, and out-of-plane.

Figure 1.6 Fracture of fiber tows after repeated low velocity impact. *Source:* [39] Reproduced with permission from IOP Publishing

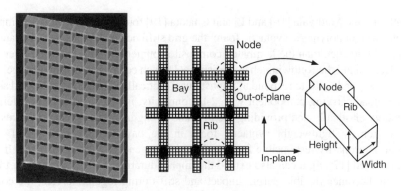

Figure 1.7 Schematic of an orthogrid.

Center-to-center indicates the distance between the centers of adjacent parallel beams. *In-plane* actions take place within the plane of the grid. *Out-of-plane* actions occur orthogonally or transversely to the plane of the grid. Element-level terms describe the rib cross-sectional dimensions where width is an in-plane measurement and depth (thickness) is out-of-plane.

The displayed grid segment in Figure 1.7 consists of beams placed in a bidirectional pattern, giving rise to the reference term *bi-grid* [41]. A special case of the bi-grid is one in which the beams intersect orthogonally with equal spacing. In this configuration, there are two identical mechanical directions and the term orthogrid is applied. Tri-grids are formed with three beam groups and directions [41]. A special case of tri-grids is the isogrid. The isogrid has three identical mechanical directions through the uniform distribution of beams at $0°/\pm60°$ to form equilateral bays. Similar to laminated composites, the in-plane stiffness of the isogrid is quasi-isotropic.

Another grid configuration, a *quadri-grid*, uses four beam groups [41]. When equally distributed at $0°/\pm45°/90°$, this grid has four equivalent directions. It is also quasi-isotropic in terms of in-plane stiffness, similar to an isogrid. Other grid patterns can be formed by varying these basic grid configurations. For example, in order to reduce the nodal build-up for an isogrid (three ribs cross over each other at the same point), one beam can be displaced a little so that the intersection point becomes a small triangle. Consequently, the 2-D *Kagome* grid is formed. Compared with bi-grids, the tri-grid or quadri-grid usually provides a higher in-plane shear resistance.

A number of studies have been conducted to investigate the structural behavior of grid stiffened composite structures experimentally and theoretically [40–56]. These tests and analyses show that grid stiffened structures are inherently strong and a resilient arrangement for composite materials. They are inherently resistant to impact damage, delamination and crack propagation because unidirectional composite ribs have no material mismatch. By having separate ribs, cracks do not propagate to the next ribs and may promote damage tolerance. They can achieve better performance in multiple directions by running the ribs in several directions and by finding optimal rib orientation. Also, the grids carry loads collaboratively and the overall load carrying capacity can be fully utilized because grid failure proceeds along the direction of the greatest strength. Furthermore, for the same amount of materials, grid panels are always thicker than their respective laminated composites and thus have higher flexural rigidity. Finally, other typical benefits of composite materials, such as light weight, high specific stiffness and strength, high corrosion resistance, tailorability, etc., are retained.

Recently, Li and Mathyala [16] and Li and Chakka [19] found that by filling the empty bays with lighter weight polymeric syntactic foam, the grid stiffened composite demonstrated much better impact tolerance than the laminated composite counterpart with the same fiber volume fraction. They found the following characteristics. (1) Each cell or bay is a small panel with an elastic boundary and tends to respond to impact quasi-statically. (2) The periodic grid skeleton, the primary load carrying component with 2-D continuity, is responsible for transferring the impact energy elastically and providing the in-plane strength and transverse shear resistance. (3) The extremely lightweight syntactic foam in the bay, the secondary load carrying component, is primarily responsible for absorbing impact energy through damage. If rubberized foam is used [17,18], it can also extend the impact duration. Of course, in order to ensure that the core becomes flexible during impact and stiff during regular service, the core must experience phase change such as a shape memory polymer [57]. (4) The grid skeleton and the foam develops a positive composite action; that is, the grid skeleton confines the foam to increase its strength and the foam provides lateral support to resist rib local buckling and crippling. In addition, the foam also provides additional in-plane shear strength for bi-grids such as the orthogrid. (5) The core and skin are fully bonded because the bay is fully filled, without the limitation of a honeycomb core or truss core.

However, grid stiffened composite also suffers from impact damage. Figure 1.8 shows the low velocity impact damage on an isogrid stiffened composite with impacts at different locations (rig, bay, and node) [19]. Figure 1.9 shows impact damage inside an orthogrid stiffened composite as determined by C-scan. In Figure 1.9, the pulse-echo transmission method was used to capture the signal and the color of the image changes with the strength of signal that is received by the transducer. Red color (gray color) represents an excess of 80% of the signal returning to the receiver, whereas blue color (dark color) represents the condition where 50–80% of the signal is being received by the receiver and green color (white color) indicates that less than 50% of the signal is being received. Therefore, red color (gray color) suggests composite without damage and green color (white color) means composite with some types of damage. As expected, the damage is localized primarily within the bay directly under impact. The 2-D grid skeleton does not suffer from significant damage, suggesting that the grid

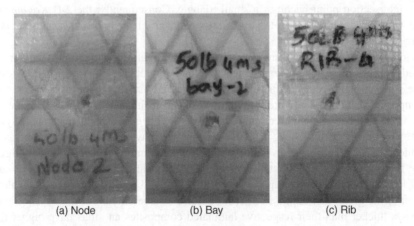

(a) Node   (b) Bay   (c) Rib

Figure 1.8 Localized damage in an isogrid stiffened composite subjected to impact energy of 193 J. *Source:* [19] Reproduced with permission from Elsevier

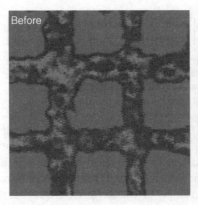 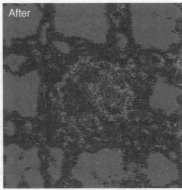

Figure 1.9 Pre- and post-impact of an orthogrid stiffened composite subjected to an impact energy of 48.8 J at the center of the bay. *Source:* [16] Reproduced with permission from Elsevier

skeleton is responsible for transferring the impact energy away from the impact point to the entire panel.

Most recently, Ji and Li [20] proved that if the brittle synthetic fibers are replaced by ductile millimeter metallic tubes, such as steel tubes or aluminum tubes, the impact tolerance of the grid stiffened composite panel can be further enhanced without a weight penalty. This is because crushing of the metallic tubes and plastic deformation of the tube wall absorb a significant amount of impact energy. Also, because the tubes are hollow, the increase in density is insignificant. It is believed that, while traditionally metals and polymer composites are competitors, this new composite construction utilizes the advantage of both polymer and metal, and may provide new opportunities for hybrid and lightweight metal/polymer composite structures.

1.4   Repair of Damage in Thermosetting Polymer Composite Structures

When polymer composites used as structural materials are damaged, there are only a few methods available to extend their functional lifetime. Ideal repair methods are those that can be executed quickly, electively, and directly on a damaged site, eliminating the need for removing a component for repair. However, the mode of damage must also be taken into consideration as repair strategies that work well for one mode might be completely useless for another. For example, matrix cracking can be repaired by sealing the crack with resin, while fiber breakage would require new fiber replacement or a fabric patch to achieve the recovery of strength.

The purpose of repair is to provide a smooth path for load transfer and/or reestablish the material's continuity, ideally not significantly changing the original stress distribution. For matrix damage, if the damage is accessible, injecting adhesive directly can restore the load carrying capacity. For fiber fracture, it is very difficult to re-establish continuity in fibers based on the current available technology. Therefore, external repair by bonding additional fiber reinforced polymer layers is essential. Currently, two techniques are available for repairing laminated composites, along with two types of repair materials [58,59]. The two techniques are scarf repair (including stepped repair) and lap (including a single or double lap) repair (see the schematics in Figure 1.10). In scarf repair, the damaged materials are removed and form a

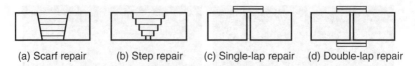

(a) Scarf repair   (b) Step repair   (c) Single-lap repair   (d) Double-lap repair

Figure 1.10 Schematic of various patch repair approaches

V-shaped space. The new repairing material is laid up layer by layer until the designed level is reached. After curing, a scarf repair is formed. The step repair is a revised version of the scarf repair. Each ply is sanded so that a flat surface is exposed, leading to a stepped finish. The advantage of scarf repair is that it provides a straighter load transfer path without increasing the structure weight. The limitation is that it needs a longer time and effort to prepare for the repair. In lap repair, the repairing layers are directly bonded to the cleaned and roughed surface of the damaged area, either on one side (single lap) or on both sides (double lap). As compared to scarf repair, the stress transfer path is disturbed and the repaired structure sees an increase in weight.

The two repair materials are a wet lay-up material, which is cured at room temperature, and a prepreg material, which is cured at an elevated temperature. However, the currently used repair materials have limitations. The wet lay-up material usually requires three to seven days for complete curing of the resin. In many applications, however, a seven-day wait is unacceptable. A heat activated curing prepreg can reduce the repair time to several hours. However, these materials generally require freezer storage and have a limited shelf life. Heat and pressure are required to cure the adhesive and patch material in order to obtain a uniform, nonporous adhesive layer. The most common heating method for field repairs requires heat blankets that are controlled by a programmable temperature controller. The heat blankets are a series of electrical resistance wires embedded in silicone rubber. There are several disadvantages to this heating method. First, prepreg curing requires a curing temperature with a narrow tolerance. However, due to thermally complex structures, achieving curing temperatures within the required range is often difficult. Second, heating large areas using heat blankets requires large amounts of energy that can easily exceed available power sources. Third, for structures with complicated geometries, the required curing pressure is generally difficult to apply. Finally, the time required for field level repair is still too long. This is because the curing cycle includes heating the blanket to the curing temperature, followed by soaking and cooling down to the ambient temperature, which may take several hours. Li *et al.* [60] used fiber reinforced ultraviolet (UV) curing resin composite to repair a low velocity impact damaged laminated composite beam. Hybrid scarf repair and patch (single lap) repair were used. Because of the fast curing of the UV curing resin, it has been found that fiber reinforced UV curing resin is a fast, strong, durable, and cost effective method to repair low velocity impact damaged composite laminates. This fast curing resin has also been used to repair damaged concrete structures [61,62] and to join composite pipes [63]. However, one limitation of such a repair strategy is the large volume change during the curing process of UV curing resins.

It should be pointed out that none of these methods of repair is an ideal solution to damage in load carrying structures. These methods are temporary solutions to extend the service time of the material, and each of these repair strategies requires monitoring of the damage and manual intervention to enact the repair. Also, all these repair methods require an interruption to the service. This is unacceptable for some structures, such as aircraft during flight, where immediate and in-service care is needed to avoid a catastrophic incident. Therefore, self-healing strategies are urgently needed.

1.5 Classification of Self-Healing Schemes

Because of the need for self-healing in thermosetting polymer composite materials, various self-healing approaches have been explored and have been published in the form of papers, books, and patents. Therefore, it is time to classify self-healing approaches so that engineers can utilize the results in design and researchers can plan new research topics. It is noted that, due to the vast body of literature in the area of self-healing, this book is by no means a comprehensive review of all the existing literature on self-healing. Rather, this section aims to provide a brief overview of the literature and to focus on the classifications. Detailed reviews can be found in recent review papers [64–93] and books [94–100].

Based on the characteristics of healing – healing by the polymer itself or healing by an externally added healing agent – self-healing can be broadly divided into two categories: intrinsic self-healing and extrinsic self-healing. Intrinsic self-healing can be further categorized into physical healing or chemical healing. For physical healing, molecules with high mobility at temperature above the glass transition or melting temperature entangle each other at the fractured surface by segmental interdiffusion or reptation, which is driven by the thermo-dynamic force and forms physical cross-linking, such as healing of thermoplastic polymers [101]. For chemical healing, chemical bonds such as the covalent bond, ionic bond, hydrogen bond, etc., are reestablished at the interface through various means, such as a thermosetting epoxy with unreacted epoxide [102], ionic attraction in ionomers [103], polymers with thermally reversible cross-links [104], supramolecule chemistry (hydrogen bonds, metal–ligand coordination, π–π stacking interactions) [90,105], dynamic covalent bond exchange [106], etc. Extrinsic self-healing can also be divided into chemical or physical healing. For a liquid healing agent contained in microcapsules [107], hollow fibers or hollow channels [108,109], and microvascular networks [110], the healing is through in situ polymerization of the contained monomers when contacting with the embedded catalyst. This healing mechanism is chemical, with the formation of covalent bonds at the interface. For a solid healing agent such as thermoplastic healing agents, the healing is through diffusion of molten thermoplastic molecules into the fractured matrix, leading to molecule entanglement and thus physical healing [111,112]. The bio-inspired close-then-heal (CTH) self-healing scheme, which aims at healing wide-opened, millimeter scale cracks and is the focus of this book, belongs to this category of extrinsic physical healing. Figure 1.11 shows a schematic of the categorization of self-healing schemes. Classification of self-healing can also be based on other functional or compositional criteria. For example, based on the repeatability of healing, some self-healing can be categorized as healing only one time, while others can heal more than once. For instance, most of covalent bonded healing chemistry can only heal one time, unless a reversible covalent bond or dynamic covalent bond exchange is used, while most physical based healing is repeatable.

It deserves mentioning that most of the existing self-healing schemes can only heal microscopic cracks because almost all the healing mechanisms need to bring the fractured surfaces in contact before healing occurs. Of course, this is not an issue at all in lab testing as the fractured pieces can be manually brought into contact easily. However, in real world applications, this simple operation represents a grand challenge for application of the

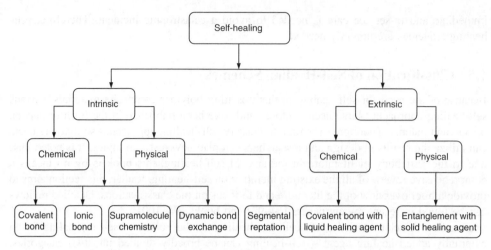

Figure 1.11 Schematic classification of self-healing schemes

self-healing schemes in engineering structures. In full-scale structures, almost all the structural components are subjected to a certain constraint by their neighboring components. In other words, the boundary conditions of the structural components are not free. The cracks cannot be brought into contact manually. This is why we propose to use the shape memory effect to bring the fractured matrix into contact before the self-healing scheme takes effect.

1.6 Organization of This Book

With the fast advancement in materials science and engineering, as well as an understanding of self-healing mechanisms and modeling capability, self-healing is evolving towards engineering practice instead of engineering dream. Self-healing is gradually changing the paradigm of structural design. While the literature related to self-healing of thermosetting polymer composites is growing quickly and in vast volume, most examples are based on either a liquid healing agent encased in microcapsules, hollow fibers, and microvascular networks or a solid healing agent such as thermoplastic particles dispersed in a thermosetting polymer matrix. The challenge facing liquid healing and solid healing agents is the difficulty in healing macroscopic or structural length scale damage. The author's research group has been working on shape memory polymer (SMP) based healing systems, which has demonstrated the ability to heal structural length scale damage, such as impact damage, repeatedly, efficiently, and molecularly. Therefore, the purpose of this book is to summarize systematically the experimental and theoretical advancement made in our lab and discuss future endeavors needed to enhance this area of study.

The book is organized as follows. Chapter 1 has introduced some basics on damage in various composite structures and classifications for self-healing. Chapter 2 will focus on the introduction of self-healing in biological systems, which inspired the self-healing scheme in this book. Chapter 3 will focus on the thermomechanical behavior of thermosetting SMPs and SMP based syntactic foam. Chapter 4 will provide some basics on solid mechanics modeling and discuss the two popular strategies to model SMPs – thermodynamic based and stress/structural relaxation based approaches. Modeling on both thermosetting SMP and SMP based syntactic foam will be reported. Chapter 5 will focus on testing and modeling of polyurethane SMP fibers, including physical, mechanical, thermomechanical, damping, and microstructure characterization. Viscoplasticity theory will be presented to model the effect of strain hardening by cold-drawing programming on stress recovery. The underlying mechanisms controlling the difference in strain memory and stress memory will be discussed and new definitions of stress fixity and stress recovery ratios will be given. Chapter 6 will deal with the self-healing of SMP matrix based composite structures. Self-healing of notched beam, sandwich, grid stiffened, and 3-D woven fabric reinforced SMP polymer composite structures will be discussed in detail. Chapter 7 will focus on self-healing of conventional thermosetting polymer by embedding SMP fibers and thermoplastic particles. SMP fiber reinforcement in the form of 1-D, 2-D, and 3-D will be presented. Chapter 8 will briefly discuss modeling of damage–healing by revising the reptation model. An evaluation of healing efficiency using the adhesively bonded configuration will also be discussed. Healing efficiency in terms of Mode I, Mode II, and Mixed Mode I&II fracture modes will be formulated. Chapter 9 will give some future perspectives on this topic.

References

1. Cohen, M.H. and Turnbull, D. (1959) Molecular transport in liquids and gasses. *Journal of Chemical Physics*, **31**, 1164–1169.
2. Tolouei, R. and Titheridge, H. (2009) Vehicle mass as a determinant of fuel consumption and secondary safety performance. *Transportation Research, Part D: Transport and Environment*, **14**, 385–399.
3. Pang, S.S., Li, G., Helms, J.E., and Ibekwe, S.I. (2001) Influence of ultraviolet radiation on the low velocity impact response of laminated beams. *Composites Part B: Engineering*, **32**, 521–528.
4. NASA (2003) *Columbia Accident Investigation Board (CAIB) Final Report*, NASA.
5. National Research Council (2011) *Opportunities in Protection Materials Science and Technology for Future Army Applications*, The National Academies Press, Washington, DC.
6. Ishai, O. and Hiel, C. (1992) Damage tolerance of a composite sandwich with interleaved foam core. *Journal of Composite Technology and Research*, **14**, 155–168.
7. Reddy, T.Y., Wen, H.M., Reid, S.R., and Soden, P.D. (1998) Penetration and perforation of composite sandwich panels by hemispherical and conical projectiles. *ASME Journal of Pressure Vessel Technology*, **120**, 186–194.
8. Evans, A.G., Hutchinson, J.W., and Ashby, M.F. (1998) Multifunctionality of cellular metal systems. *Progress in Materials Science*, **43**, 171–221.
9. Hasebe, R.S. and Sun, C.T. (2000) Performance of sandwich structures with composite reinforced core. *Journal of Sandwich Structures and Materials*, **2**, 75–100.
10. Hosur, M.V., Abdullah, M., and Jeelani, S. (2005) Manufacturing and low-velocity impact characterization of foam filled 3-D integrated core sandwich composites with hybrid face sheets. *Composite Structures*, **69**, 167–181.
11. Hou, W.H., Zhu, F., Lu, G.X., and Fang, D.N. (2010) Ballistic impact experiments of metallic sandwich panels with aluminum foam core. *International Journal of Impact Engineering*, **37**, 1045–1055.
12. Shutov, F.A. (1991) Syntactic polymer foams, in *Handbook of Polymer Foams and Foam Technology* (eds D. Klempner and K.C. Frisch), Hanser Publishers, pp. 355–374.
13. Bart-Smith, H., Hutchinson, J.W., Fleck, N.A., and Evans, A.G. (2002) Influence of imperfections on the performance of metal foam core sandwich panels. *International Journal of Solids and Structures*, **39**, 4999–5012.
14. Wicks, N. and Hutchinson, J.W. (2004) Performance of sandwich plates with truss cores. *Mechanics of Materials*, **36**, 739–751.
15. Bardella, L. and Genna, F. (2001) On the elastic behavior of syntactic foams. *International Journal of Solids and Structures*, **38**, 7235–7260.
16. Li, G. and Muthyala, V.D. (2008) Impact characterization of sandwich structures with an integrated orthogrid stiffened syntactic foam core. *Composites Science and Technology*, **68**, 2078–2084.
17. Li, G. and Jones, N. (2007) Development of rubberized syntactic foam. *Composites, Part A: Applied Science and Manufacturing*, **38**, 1483–1492.

18. Li, G. and John, M. (2008) A crumb rubber modified syntactic foam. *Materials Science and Engineering A*, **474**, 390–399.
19. Li, G. and Chakka, V.S. (2010) Isogrid stiffened syntactic foam cored sandwich structure under low velocity impact. *Composites Part A: Applied Science and Manufacturing*, **41**, 177–184.
20. Ji, G. and Li, G. (2013) Impact tolerant and healable aluminum millitube reinforced shape memory polymer composite sandwich core. *Materials and Design*, **51**, 79–87.
21. Ko, F. and Hartman, D. (1986) Impact behaviour of 2D and 3D glass–epoxy composites. *SAMPE Journal*, **22**, 26–30.
22. Chou, S., Chen, H.C., and Wu, C.C. (1992) BMI resin composites reinforced with 3D carbon-fibre fabrics. *Composites Science and Technology*, **43**, 117–128.
23. Voss, S., Fahmy, A., and West, H. (1993) Impact tolerance of laminated and 3-dimensionally reinforced graphite–epoxy panels, in *Advanced Composites 93: International Conference on Advanced Composite Materials* (eds T. Chandra and A.K. Dhingra), The Minerals, Metals and Materials Society, pp. 591–596.
24. Billaut, F. and Roussel, O. (1995) Impact resistance of 3-D graphite/epoxy composites, in *Proceedings of the Tenth International Conference on Composite Materials, ICCM-10* (eds A. Portsartip and K. Street), Woodhead, 14–18 August 1995, pp. V551–V558.
25. Chou, S., Chen, H.C., and Chen, H.E. (1992) Effect of weave structure on mechanical fracture behavior of three-dimensional carbon fiber fabric reinforced epoxy resin composites. *Composites Science and Technology*, **45**, 23–35.
26. Siow, Y.P. and Shim, V.P.W. (1998) An experimental study of low velocity impact damage in woven fiber composites. *Journal of Composite Materials*, **32**, 1178–1202.
27. Davies, G.A.O., Hitchings, D., and Zhou, G. (1996) Impact damage and residual strengths of woven fabric glass/ polyester laminates. *Composites Part A*, **27**, 1147–1156.
28. Kim, J.K. and Sham, M.L. (2000) Impact and delamination failure of woven fabric composites. *Composites Science and Technology*, **60**, 745–763.
29. Park, S.J., Park, W.B., and Lee, J.R. (2000) Characterization of the impact properties of three-dimensional glass fabric-reinforced vinyl ester matrix composites. *Journal of Materials Science*, **35**, 6151–6154.
30. Naik, N.K., Sekher, Y.C., and Meduri, S. (2000) Damage in woven-fabric composites subjected to low-velocity impact. *Composites Science and Technology*, **60**, 731–744.
31. Naik, N.K., Sekher, Y.C., and Meduri, S. (2000) Polymer matrix woven fabric composites subjected to low velocity impact: Part I – Damage initiation studies. *Journal of Reinforced Plastics and Composites*, **19**, 912–954.
32. Naik, N.K., Sekher, Y.C., and Meduri, S. (2000) Polymer matrix woven fabric composites subjected to low velocity impact: Part II – Effect of plate thickness. *Journal of Reinforced Plastics and Composites*, **19**, 1031–1055.
33. Baucom, J.N. and Zikry, M.A. (2005) Low-velocity impact damage progression in woven E-glass composite systems. *Composites Part A: Applied Science and Manufacturing*, **36**, 658–664.
34. Baucom, J.N., Zikry, M.A., and Rajendran, A.M. (2006) Low-velocity impact damage accumulation in S2-glass composite systems. *Composites Science and Technology*, **66**, 1229–1238.
35. Sun, B., Liu, Y., and Gu, B. (2009) A unit cell approach of finite element calculation of ballistic impact damage of 3D orthogonal woven composites. *Composites Part B: Engineering*, **40**, 552–560.
36. Sun, B., Hu, D., and Gu, B. (2009) Transverse impact damage and energy absorption of 3-D multi-structured knitted composite. *Composites Part B: Engineering*, **40**, 572–583.
37. Huang, H.J. and Waas, A.M. (2009) Modeling and predicting the compression strength limiting mechanisms in Z-pinned textile composites. *Composites Part B: Engineering*, **40**, 530–539.
38. Rao, M.P., Sankar, B.V., and Subhash, G. (2009) Effect of Z-yarns on the stiffness and strength of three-dimensional woven composites. *Composites Part B: Engineering*, **40**, 540–551.
39. Nji, J. and Li, G. (2010) A self-healing 3D woven fabric reinforced shape memory polymer composite for impact mitigation. *Smart Materials and Structures*, **19**, paper 035007 (9 pages).
40. Huybrechts, S. and Tsai, S.W. (1996) Analysis and behavior of grid structures. *Composite Science and Technology*, **56**, 1001–1015.
41. Dutta, P.K., Bailey, D.M., Tsai, S.W., Jensen, D.W., Hayes Jr, J.R., McDonald, W.E., Smart, C.W., Colwell, T., Earl, J.S., and Chen, H.J. (1998) *Composite Grid/Frame Reinforcement for Concrete Structures*, Construction Productivity Advancement Research (CPAR) Program, USACERL Technical Report 98/81.
42. Noor, A.K. (1988) Continuum modeling of repetitive lattice structures. *Applied Mechanics Review*, **41**, 285–296.

43. Jaunky, N. and Knight Jr, N.F. (1996) Formulation of an improved smeared stiffener theory for buckling analysis of grid-stiffened composite panels. *Composites Part B: Engineering*, **27B**, 519–526.
44. Evans, A.G., Hutchinson, J.W., and Ashby, M.F. (1998) Multifunctionality of cellular metal systems. *Progress in Materials Science*, **43**, 171–221.
45. Jaunky, N., Knight, Jr, N.F., and Amburb, D.R. (1998) Optimal design of general stiffened composite circular cylinders for global buckling with strength constraints. *Composite Structures*, **41**, 243–252.
46. Hohe, J., Beschorner, C., and Becker, W. (1999) Effective elastic properties of hexagonal and quadrilateral grid structures. *Composite Structures*, **46**, 73–89.
47. Lennon, R.F. and Das, P.K. (2000) Torsional buckling behaviour of stiffened cylinders under combined loading. *Thin-Walled Structures*, **38**, 229–245.
48. Deshpande, V.S., Ashby, M.F., and Fleck, N.A. (2001) Foam topology bending versus stretching dominated architectures. *Acta Materialia*, **49**, 1035–1040.
49. Evans, A.G., Hutchinson, J.W., Fleck, N.A., Ashby, M.F., and Wadley, H.N.G. (2001) The topological design of multifunctional cellular metals. *Progress in Materials Science*, **46**, 309–327.
50. Kidane, S., Li, G., Helms, J.E., Pang, S.S. and Woldesenbet, E. (2003) Analytical buckling load analysis of grid stiffened composite cylinders. *Composites Part B: Engineering*, **33**, 1–9.
51. Han, D.Y. and Tsai, S.W. (2003) Interlocked composite grids design and manufacturing. *Journal of Composite Materials*, **37**, 287–316.
52. Ambur, D.R., Jaunky, N., and Hilburger, M.W. (2004) Progressive failure studies of stiffened panels subjected to shear loading. *Composite Structures*, **65**, 129–142.
53. Rackliffe, M.E., Jensen, D.W., and Lucas, W.K. (2006) Local and global buckling of ultra-lightweight IsoTruss structures. *Composites Science and Technology*, **66**, 283–288.
54. Jadhav, P. and Mantena, P.R. (2007) Parametric optimization of grid-stiffened composite panels for maximizing their performance under transverse loading. *Composite Structures*, **77**, 353–363.
55. Fan, H.L., Meng, F.H., and Yang, W. (2007) Sandwich panels with Kagome lattice cores reinforced by carbon fibers. *Composite Structures*, **81**, 533–539.
56. Li, G. and Cheng, J.Q. (2007) A generalized analytical modeling of grid stiffened composite structures. *Journal of Composite Materials*, **41**, 2939–2969.
57. John, M. and Li, G. (2010) Self-healing of sandwich structures with grid stiffened shape memory polymer syntactic foam core. *Smart Materials and Structures*, **19**, paper 075013 (12 pages).
58. Ahn, S.H. and Springer, G.S. (1998) Repair of composite laminates – I: Test results. *Journal of Composite Materials*, **32**, 1036–1074.
59. Ahn, S.H. and Springer, G.S. (1998) Repair of composite laminates – II: Models. *Journal of Composite Materials*, **32**, 1076–1114.
60. Li, G., Pourmohamadian, N., Cygan, A., Peck, J., Helms, J.E., and Pang, S.S. (2003) Fast repair of laminated beams using UV curing composites. *Composite Structures*, **60**, 73–81.
61. Li, G., Hedlund, S., Pang, S.S., Alaywan, W., Eggers, J., and Abadie, C. (2003) Repair of damaged RC columns using fast curing FRP composites. *Composites Part B: Engineering*, **34**, 261–271.
62. Li, G. and Ghebreyesus, A. (2006) Fast repair of damaged RC beams using UV curing FRP composites. *Composite Structures*, **72**, 105–110.
63. Peck, J., Pang, S.S., Li, G., Jones, R., and Smith, B. (2007) Effect of UV-cured FRP joint thickness on coupled composite pipes. *Composite Structures*, **80**, 290–297.
64. Meng, H. and Li, G. (2013) A review of stimuli-responsive shape memory polymer composites. *Polymer*, **54**, 2199–2221.
65. Gould, P. (2003) Self-help for ailing structures. *Materials Today*, **6**, 44–49.
66. Billiet, S., Hillewaere, X.K.D., Teixeira, R.F.A., and Du Prez, F.E. (2013) Chemistry of crosslinking processes for self-healing polymers. *Macromolecular Rapid Communications*, **34**, 290–309.
67. Herbst, F., Döhler, D., Michael, P., and Binder, W.H. (2013) Self-healing polymers via supramolecular forces. *Macromolecule Rapid Communications*, **34**, 203–220.
68. Aïssa, B., Therriault, D., Haddad, E., and Jamroz, W. (2012) Self-healing materials systems: overview of major approaches and recent developed technologies. *Advances in Materials Science and Engineering*, Article ID 854203 (17 pages).
69. Zhang, M.Q. and Rong, M.Z. (2012) Theoretical consideration and modeling of self-healing polymers. *Journal of Polymer Science Part B: Polymer Physics*, **50**, 229–241.

70. Guimard, N.K., Oehlenschlaeger, K.K., Zhou, J., Hilf, S., Schmidt, F.G., and Barner-Kowollik, C. (2012) Current trends in the field of self-healing materials. *Macromolecular Chemistry and Physics*, **213**, 131–143.

71. Hu, J., Zhu, Y., Huang, H., and Lu, J. (2012) Recent advances in shape memory polymers: structure, mechanism, functionality, modeling and applications. *Progress in Polymer Science*, **37**, 1720–1763.

72. Zhang, M.Q. and Rong, M.Z. (2012) Design and synthesis of self-healing polymers. *Science China Chemistry*, **55**, 648–676.

73. Brochu, A.B.W., Craig, S.L., and Reichert, W.M. (2011) Self-healing biomaterials. *Journal of Biomedical Materials Research A*, **96**, 492–506.

74. Hager, M.D., Greil, P., Leyens, C., van der Zwaag, S., and Schubert, U.S. (2010) Self-healing materials. *Advanced Materials*, **22**, 5424–5430.

75. Mauldin, T.C. and Kessler, M.R. (2010) Self-healing polymers and composites. *International Materials Reviews*, **55**, 317–346.

76. Blaiszik, B.J., Kramer, S.L.B., Olugebefola, S.C., Moore, J.S., Sottos, N.R., and White, S.R. (2010) Self-healing polymers and composites. *Annual Review of Materials Research*, **40**, 179–211.

77. Murphy, E.B. and Wudl, F. (2010) The world of smart healable materials. *Progress in Polymer Science*, **35**, 223–251.

78. Meng, H. and Li, G. (2013) Reversible switching transitions of stimuli-responsive shape changing polymers. *Journal of Materials Chemistry A*, **1**, 7838–7865.

79. Hager, M.D., Greil, P., Leyens, C., van der Zwaag, S., and Schubert, U.S. (2010) Self-healing materials. *Advanced Materials*, **22**, 5424–5430.

80. Samadzadeha, M., Bouraa, S.H., Peikaria, M., Kasirihab, S.M., and Ashrafic, A. (2010) A review on self-healing coatings based on micro/nanocapsules. *Progress in Organic Coatings*, **68**, 159–164.

81. Burattini, S., Greenland, B.W., Chappell, D., Colquhoun, H.M., and Hayes, W. (2010) Healable polymeric materials: a tutorial review. *Chemical Society Reviews*, **39**, 1973–1985.

82. Syrett, J.A., Becer, C.R. and Haddleton, D.M. (2010) Self-healing and self-mendable polymers. *Polymer Chemistry*, **1**, 978–987.

83. Amendola, V. and Meneghetti, M. (2009) Self-healing at the nanoscale. *Nanoscale*, **1**, 74–88.

84. Meng, H. and Li, G. (2013) Reversible switching transitions of stimuli-responsive shape changing polymers. *Journal of Materials Chemistry A*, **1**, 7838–7865.

85. Wu, D.Y., Meure, S., and Solomon, D. (2008) Self-healing polymeric materials: a review of recent developments. *Progress in Polymer Science*, **33**, 479–522.

86. Yuan, Y.C., Yin, T., Rong, M.Z., Zhang, M.Q. (2008) Self healing in polymers and polymer composites. Concepts, realization and outlook: a review. *eXPRESS Polymer Letters*, **2**, 238–250.

87. Wool, R.P. (2008) Self-healing materials: a review. *Soft Matter*, **4**, 400–418.

88. Trask, R.S., Williams, H.R. and Bond, I.P. (2007) Self-healing polymer composites: mimicking nature to enhance performance. *Bioinspiration and Biomimetics*, **2**, 1–9.

89. Balazs, A.C. (2007) Modeling self-healing materials. *Materials Today*, **10**, 18–23.

90. Yang, Y. and Urban, M. (2013) Self-healing polymeric materials. *Chemical Society Reviews*, **42**, 7446–7467.

91. Pretsch, T. (2010) Review on the functional determinants and durability of shape memory polymers. *Polymers*, **2**, 120–158.

92. Hearon, K., Singhal, P., Horn, J., Small IV, W., Olsovsky, C., Maitland, K.C., Wilson, T.S., and Maitland, D.J. (2013) Porous shape-memory polymers. *Polymer Reviews*, **53**, 41–75.

93. Meng, H., Mohamadian, H., Stubblefield, M., Jerro, D., Ibekwe, S., Pang, S.S., and Li, G. (2013) Various shape memory effects of stimuli-responsive shape memory polymers. *Smart Materials and Structures*, **22**, paper 093001 (23 pages).

94. Zhang, M.Q. and Rong, M.Z. (2011) *Self-Healing Polymers and Polymer Composites*, ISBN 978-0-470-49712-8, Wiley-VCH.

95. van der Zwaag, S., Schmets, A.J.M., and Zaken, G. (eds) (2007) *Self-Healing Materials: An Alternative Approach to 20 Centuries of Materials Science*, ISBN-13: 978-1-402-06249-0, Springer.

96. Binder, W.H. (ed.) (2013) *Self-Healing Polymers*, ISBN: 978-3-527-33439-1, Wiley-VCH.

97. Ghosh, S.K. (ed.) (2008) *Self-healing Materials: Fundamentals, Design Strategies, and Applications*, ISBN-13: 978-3-527-31829-2, Wiley-VCH.

98. Amendola, V. and Meneghetti, M. (eds) (2011) *Self-Healing at the Nanoscale: Mechanisms and Key Concepts of Natural and Artificial Systems*, ISBN-13: 978-1-439-85473-0, CRC Press.

99. Nosonovsky, M. and Rohatgi, P.K. (2011) *Biomimetics in Materials Science: Self-Healing, Self-Lubricating, and Self-Cleaning Materials*, ISBN-13: 978-1-461-40925-0, Springer.
100. Verma, R. (2011) *Self Healing Composite Materials: The Smart Technology for Future Aerospace*, ISBN-13: 978-3-844-30942-3, LAP LAMBERT Academic Publishing.
101. Wool, R.P. and O'Connor, K.M. (1981) A theory crack healing in polymers. *Journal of Applied Physics*, **52**, 5953–5963.
102. Hemmelgarn, C.D., Margraf, T.W., Havens, D.E., Reed Jr, J.L., Snyder, L.W., Louderbaugh, A., and Dietsch, B.A. (2008) *Composites Self-Healing System*, US PCT No. PCT/US08/60055.
103. Varley, R. and van der Zwaag, S. (2008) Towards an understanding of thermally activated self-healing of an ionomer system during ballistic penetration. *Acta Materialia*, **56**, 5737–5750.
104. Chen, X., Dam, M.A., Ono, K., Mal, A., Shen, H., Nutt, S.R., Sheran, K., Wudl, F. (2002) A thermally remendable cross-linked polymeric material. *Science*, **295**, 1698–1702.
105. Sijbesma, R.P., Beijer, F.H., Brunsveld, L., Folmer, B.J.B., Hirschberg, J.H.K.K., Lange, R.F.M., Lowe, J.K.L., and Meijer, E.W. (1997) Reversible polymers formed from self-complementary monomers using quadruple hydrogen bonding. *Science*, **278**, 1601–1604.
106. Montarnal, D., Capelot, M., Tournilhac, F., and Leibler, L. (2011) Silica-like malleable materials from permanent organic networks. *Science*, **334**, 965–968.
107. White, S.R., Sottos, N.R., Geubelle, P.H., Moore, J.S., Kessler, M.R., Sriram, S.R., Brown, E.N., and Viswanathan, S. (2001) Autonomic healing of polymer. *Nature*, **409**, 794–797.
108. Pang, J.W.C., Bond, I.P. (2005) A hollow fibre reinforced polymer composite encompassing self-healing and enhanced damage visibility. *Composites Science and Technology*, **65**, 1791–1799.
109. Trask, R.S., Williams, G.J., Bond, I.P. (2007) Bioinspired self-healing of advanced composite structures using hollow glass fibres. *Journal of the Royal Society Interface*, **4**, 363–371.
110. Toohey, K.S., Sottos, N.R., Lewis, J.A., Moore, J.S., and White, S.R. (2007) Self-healing materials with microvascular networks. *Nature Materials*, **6**, 581–585.
111. Zako, M. and Takano, N. (1999) Intelligent material systems using epoxy particles to repair microcracks and delamination damage in GFRP. *Journal of Intelligent Material Systems and Structures*, **10**, 836–841.
112. Hayes, S.A., Jones, F.R., Marshiya, K., and Zhang, W. (2007) A self-healing thermosetting composite material. *Composites Part A: Applied Science and Manufacturing*, **38**, 1116–1120.

2

Self-Healing in Biological Systems

Self-healing in biological systems is a physiological and/or psychological recovery process that can occur without external intervention. With millions of years of evolution and natural selection (survival of the fittest), most biological systems, including humans, plants, animals, insects, bacteria, and even deoxyribonucleic acid (DNA), have developed remarkable self-healing capabilities. This chapter will introduce some typical examples of self-healing in plants, animals, and human beings, and will discuss their implications in developing potential biomimetic self-sealing engineering materials and engineering structures. Compared to plants, animals have developed more advanced self-healing strategies. The focus of this chapter will thus be on animal self-healing. Another highlight will be on human beings. Humans not only have a physiological self-healing capability but also possess psychological self-healing capabilities.

2.1 Self-Healing in Plants

Self-healing is popular in plants. One typical example is grass (see Figure 2.1). Grass has a strong self-healing capability and grows quickly after damage. No matter how often the grass is mowed in one's yard, it reproduces itself, of course with a sufficient energy supply – sunlight, water, and nutrients (minerals from the soil). Immediately after being mowed the grass has a very rough straight edge on the top of each grass blade. On the next day, these rough straight edges have disappeared. After one or two weeks, the grass may have fully recovered, following the generic code. In addition to a self-healing leaf, grass also possesses a regeneration capability. Some grasses have additional stems that grow sideways, either below ground or just above it. Both stems below and above ground have been used by grass to reach out of the ground and establish new grass culms. The stems nurture the new plant until it is strong enough to survive on its own.

2.2 Seal-Healing in Animals

Over the years, various strategies have been developed by animals to self-heal injuries or reproduce a part of the body. In the following, we will present several typical examples for animal self-healing.

Self-Healing Composites: Shape Memory Polymer-Based Structures, First Edition. Guoqiang Li.
© 2015 John Wiley & Sons, Ltd. Published 2015 by John Wiley & Sons, Ltd.

Figure 2.1   Grass can heal after cutting. *Source:* [1] Reproduced with permission from Elsevier

2.2.1 Self-Healing by Self-Medicine

Animals, especially wild animals, can survive after injury without access to treatments by humans. Through millions of years of evolution, animals have the capability to identify and utilize natural medicines. Zoologists and botanists have started to investigate how wild animals use plant medicines to prevent and cure illness [2]. In addition to the ability to find medicines in nature, they have developed more advanced self-healing capabilities. For example, some animals know how to use their saliva to heal wounds, such as dogs, cats, rodents, and primates. Saliva consists of many compounds that are antibacterial or able to promote healing [3]. For instance, the enzymes in saliva, including lysozyme and peroxidase [4], defensins [5], and IgA [6], are antibacterial and thrombospondin is antiviral [7,8]. Protease, a secretory leukocyte protease inhibitor, is both antibacterial and antiviral [9]. Nitrates, after breaking down into nitric oxide in saliva, can inhibit bacterial growth [10]. VEGF [11], TGF-β1 [12], leptin [13,14], IGF-I [15,16], lysophosphatidic acid [17,18], hyaluronan [19], and NGF [20–22] are all epidermal growth factors [23]. In saliva there are also analgesic and opiophin [24]. In addition to the medicinal effect, licking also has the function of deriding the wound and removing gross contamination from the affected area [3]. Figure 2.2 shows an adult male toque macaque licking the female's wound.

2.2.2 Self-Healing Lizard

Some animals can regenerate a body part if they either passively or actively lose the part due to either attack by predators or reproductions. One example is the lizard or gecko. Lizards consist of a large family with over 9766 species belonging to a widespread group of squamate reptiles [26]. The size of different species ranges from small lizards that are just a few centimeters long to much larger ones reaching a few meters. Figure 2.3 shows different species of lizards with long tails [27]. The tails of lizards often have different and vivid colors from the rest of their body parts. This feature is to attract potent predators to strike the tail, instead of other parts of the body. Many lizards can detach their tails upon attack to escape from predators. The detached tails can writhe and wiggle, making it look like a living body with continuous struggle, distracting the predator's attention. Over a period of several weeks, the lizard can regrow a new tail. The new tail is a little different from the original one in that it contains cartilage rather than bone, and the new tail is distinctly discolored compared with other parts of

Figure 2.2 An adult male licking the female's wound. *Source:* [25] Reproduced with permission from Springer

the body. This capability of lizards is called the pars pro toto principle, which means sacrificing part of the whole body to allow them to escape [28].

2.2.3 Self-Healing Starfish

Lizards are not the only animals with the capability to heal part of their bodies. Starfish can grow back their arms. Starfish are echinoderms belonging to the class Asteroidean, which has about 1500 species around the world [29] (see Figure 2.4). Starfish typically have a central disc surrounded by five arms. Some starfish may have four or six arms. Starfish may fragment for the purpose of reproduction. Starfish reproduce asexually by splitting its central disc into fragments. Fragmentation of starfish may also be due to an attack by a predator. Starfish can actively shed arms during escape. The loss of the arms is achieved by rapid softening of a special type of connective tissue in response to nervous signals. Starfish have amazing

Figure 2.3 Different species of lizards with long tails. *Source:* [27] Reproduced with permission from Elsevier

Figure 2.4 Starfish have self-healing capabilities. *Source:* [30] Reproduced with permission from Springer

self-healing capabilities. Some species of starfish can regrow lost arms and even entire limbs from a single arm. Some species of starfish need at least part of the central body disc to regenerate the whole starfish. The regrowth of starfish needs several months or even years.

2.2.4 Self-Healing of Sea Cucumbers

In addition to limbs and tails, certain animals can even self-heal their internal organs, such as sea cucumbers (see Figure 2.5). Sea cucumbers belong to *Holothurioide*, which has about 1250 species around the world [31,32]. They have been in existence for six hundred million years. After millions of years of evolution, sea cucumbers have developed several abilities to survive. They can use camouflage by changing their colors according to the environmental color and can change skin hardness upon attack or sleep. Furthermore, sea cucumbers have the capability to regenerate internal organs as well as limbs. Sea cucumbers may be attacked by crabs, various

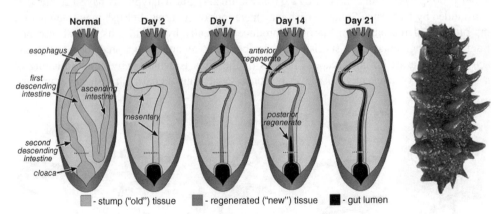

Figure 2.5 Sea cucumbers have the capability of internal organ regeneration. Anatomy of the regenerating digestive tube in noneviscerated individuals of the sea cucumber *H. glaberrima* at different time points of regeneration. *Source:* [34] Reproduced with permission from Elsevier

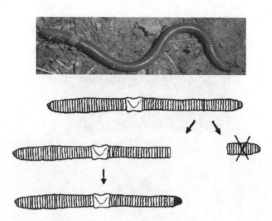

Figure 2.6 Earthworm has the ability to regenerate a lost segment. Schematic drawing of the amputation and regeneration of earthworm segments. *Source:* [37] Reproduced with permission from Springer

fish and crustaceans, sea turtles, and sea stars. When attacked by enemies, sea cucumbers instinctively spit out their internal organs containing toxins through their anus to confuse predators. The sea cucumbers can then take the opportunity to escape. If it does not die, after around 60 days they can reproduce their new internal organs [33]. Some sea cucumbers are also able to mutilate their own bodies by splitting into two pieces when attacked by predators. If each piece contains part of the cloaca, the fragments regenerate into a whole.

2.2.5 Self-Healing of Earthworms

There are about 2700 different types of earthworm on earth [35] (see Figure 2.6). They have tube and segmented shapes and live in soil. The digestive system of earthworms runs through the whole body. Research results on self-healing of earthworms are not conclusive. Although some research did not demonstrate the healing capability of certain earthworms, there are also studies that have shown the regeneration capability of earthworms. It is believed that the regeneration capability of earthworms varies depending on the species and the extent of the damage [36].

Most earthworms, such as the red wiggler, can regenerate several segments lost from their head and tail. The general rule is that the more the head segments are lost, the less the possibility they can reproduce. March-loving blackworm can always regenerate, regardless of the segment that is broken. Most earthworms can regenerate a new tail if the tail is dissected. If the breaking location is at the middle, the two halves may generate into two whole earthworms with the front part generating a new tail and the tail part growing a new head [38].

2.2.6 Self-Healing of Salamanders

Salamanders are 550 extant species of amphibians within the family of Caudata [39] (see Figure 2.7). The shape of their bodies is like a lizard, with slender bodies, short noses, and long tails. Salamanders live in or near water, or in wetland. Some salamanders are fully aquatic, some take to the water intermittently, and some are entirely terrestrial. Salamanders are often referred to as having a "perfect" regeneration capability, because they can heal any structure or cell type of their body entirely [40]. They are able to perform scar-free repair of deep tissue wounds after injury. Salamanders are capable of regenerating lost limbs, damaged lungs, sliced

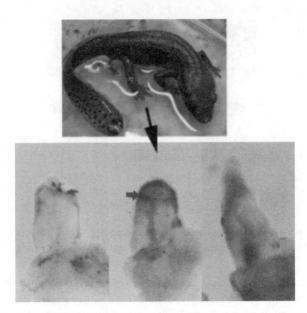

Figure 2.7 Salamanders have the perfect healing capability. *Source:* [42] Reproduced with permission from John Wiley & Sons, Ltd

spinal cord, eyes, and even part of lopped-off brain. At the beginning, the new body part may look pale in comparison to the rest of the salamander. Eventually, the color matches perfectly. It has been found that the immune cells called macrophages play a critical role in the early stages of healing. Removing these cells can prevent regeneration and lead to scarring [41].

2.3 Self-Healing in Human Beings

Self-healing of human beings is more complicated in that the self-healing of human beings is for a state of optimal well-being, which goes beyond the curing of wound and illness [43]. Self-healing of human beings mainly involves two aspects, which are psychological healing and physiological healing.

2.3.1 Psychological Self-Healing

When the psychological aspect of a person is hurt, the physical body is vulnerable to illness. Stress hormones impair physical functions. Muscle tension due to stress can worsen or cause pain in muscles. Emotional tension, depression, and anger can impair social relationships and functions, leading to increased psychological symptoms [44]. Several methods can help psychological healing and even promote physical healing. Examples of these methods include, but are not limited to, stress release, peace of mind, energizing, having enough sleep, music, and opening up one's feelings to another. Special guidance can be obtained from professional training to fully achieve this psychological healing [45,46].

2.3.2 Physiological Self-Healing

Human beings cannot regenerate limbs and any part of a body like other animals, although small children may grow a small piece of fingertip. The reason is that the human body does not

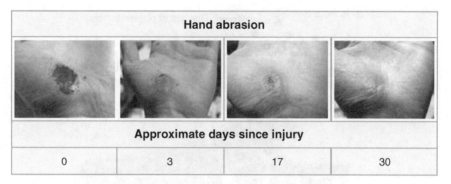

Figure 2.8 Healing of hand skin (http://en.wikipedia.org/wiki/Wound_healing, Wikimedia Commons, freely licensed)

have regeneration blastema [47]. However, human beings do possess amazing physical self-healing ability using blood. Without the self-healing capabilities, a small cut would cause bleeding to death of individuals [48]. If the wound is not extensive, the healing can be so perfect that no one can notice the wound after a few weeks. Figure 2.8 shows the healing process of a hand after being cut.

Human skin is a perfect example of self-healing. Human skin serves as a protective barrier and consists of two layers: an outer epidermal layer and an underlying dermal layer, as shown in Figure 2.9. The keratinized stratified epidermal layer continuously rebuilds the surface of the skin and the underlying thick layer of collagen-rich dermal connective tissue provides support and nourishment [49]. If a cut happens on the skin, blood flows from the vessels to the wound site. Human blood mainly consists of plasma, white blood cells, platelets and red blood cells. Plasma is a mixture of water, sugar, fat, protein, and potassium and calcium salts. The chemicals in plasma help form blood clots necessary to stop further bleeding. White blood cells produce antibodies to help fight infections caused by bacteria, viruses, and foreign proteins. Red blood cells carry oxygen to the living cells. Platelets are the most important member of blood when the human body is injured. They gather and stick to the edges of the wound. They

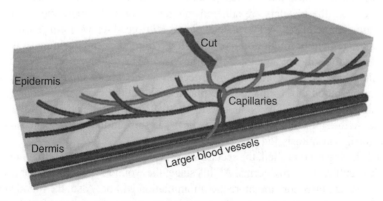

Figure 2.9 Schematic diagram of self-healing in the dermis layer of skin. *Source:* [51,52] Reproduced with permission from Nature Publishing Group and John Wiley & Sons, Ltd; copyright © 2010 John Wiley & Sons, Ltd, DOI: 10.1002/adma.201002561

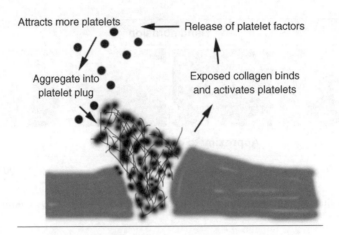

Figure 2.10 Schematic of clot formation. Adapted from [55,56]

release chemicals that help start the process of blood clotting so that further bleeding can be stopped [50].

The self-healing of human skin can be summarized in four general stages. The first stage of the healing process is called hemostasis. Hemostasis itself is a three-step process that includes vasoconstriction, platelet plug formation, and formation of a blood clot. Once the skin is cut, the blood vessel at the damaged area contracts to decrease the flow of blood to this area. At the same time, collagen helps platelets adhere to the site of the cut. In the second step, the platelets create a plug to block further bleeding, as shown in Figure 2.10. Platelets only stick to the vessel location where the cut is made because of their special coating made of a special protein. When the platelets adhere to the collagen fibers, they release granules that contain serotonin, adenosine diphosphate glucose pyrophosphorylase, and thromboxane [53]. Serotonin can cause further shrinkage; adenosine diphosphate glucose pyrophosphorylase can attract more platelets to the wounded vessel; and thromboxane promotes the release of more granules. Two types of main chemicals function during the clot formation. One is a coagulation factor; the other is an anticoagulant. At the same time, in the injured vessel, thirteen coagulation factors turn prothrombin into thrombin, which produces fibrin for the fibrin mesh. As a positive feedback, the more the platelets are adhered (bigger wound), the more the granules will be released to attract more platelets. The platelet plug is reinforced with a fibrin "mesh," which is a criss-cross arrangement of tough bodily tissue. In the third step, the fibrin mesh traps red and white blood cells to form a blood clot at the injured site. The red blood cells form a solid mass and the white blood cells aid in reducing infection at the site of injury [53]. About 30–60 minutes after the blood clot formation, the platelets within the clot contract due to actin and myosin contractile proteins. The contraction pulls the fibrin mesh and brings the cut wound closer inside. At the same time of contraction, the tissue surrounding the damaged site begins to divide and repair via mitosis [54].

Once the bleeding is controlled, the second stage is to stop infection by utilizing white blood cells, which engulf and destroy germs. At this stage, the wound swells and becomes inflamed. The more severe the infection, the more the inflammation will be. Also, the blood vessel starts to dilate to bring more white blood cells to the infection site. Two days after the injury, the fibrin mesh that holds the clot together is dissolved by a fibrinolysis process. The third stage is

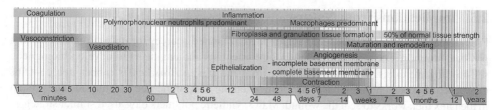

Figure 2.11 Wound healing stages (http://en.wikipedia.org/wiki/File:Wound_healing_phases.png, Wikimedia Commons, freely licensed; author Mikael Häggström)

the proliferation process. This stage rebuilds new granulation tissue, which is composed of collagen and an extracellular matrix, in which a new network of blood vessels develop. In stage four, maturation occurs once the wound is closed. This stage involves remodeling of collagen. The complete healing process may take months or years. For detailed phases or stages and time of the complicated healing process, refer to Figure 2.11.

2.4 Summary

After millions of years of evolution, biological systems have developed advanced self-healing capabilities. Some animals can regenerate tails, arms, or even any part of the body, including internal organs, for different purposes, such as reproduction or escaping from predators. With this self-healing capability, some animals can technically live forever. The self-healing process in biological systems are complex with sophisticated chemical processes. The underlying mechanism of self-healing has not been fully revealed and understood until now. Uncovering the mystery behind self-healing is expected to have a tremendous impact on many areas of human life.

Not only animals, but also human beings, possess amazing self-healing abilities. Although we cannot regenerate limbs, the human body can heal skin or bone wounds effectively if the wound is not extensive. Without this healing ability, a small cut would lead to death by continuous bleeding. The self-healing of human skin is a multilength and multitime scale process. The self-healing of a skin wound involves more than thirteen chemicals and reactions in blood. The flow of these chemicals can be inherently and instinctively controlled for optimal self-healing functions. Moreover, biological systems like human skin are smart in that wound often leads to pain from the nervous system, which indicates the location and the severity of the wound. In other words, biological systems have a damage sensing capability besides self-healing functionality. The nervous system transmits instant signals to retract the biological body from the danger situation. The nerve or sensing system can also recover after the self-healing process. For example, human skin after injury healing can regain the ability of feeling pain, temperature, moisture, etc.

Over the years, biological systems have developed sophisticated strategies to heal injuries. Researchers are actively working on understanding and mimicking the self-healing mechanisms found in nature and applying them to synthetic materials. Today we are still at the early stage of discovery and there is still a long way to go. However, by taking inspiration from nature, we believe that we will be able to develop synthetic materials with remarkable self-healing performances in the future.

2.5   Implications from Nature

From the above discussion, nature has provided unlimited sources for mimicking self-healing physical wounds. These motivate and inspire scientists and engineers to develop biomimetic self-healing materials. For example, latex from the *Hevea Brasiliensis* rubber tree contains capsules of a special protein that break open when the bark of the tree is damaged. The protein in the latex then works to seal the wound. Many synthetic polymers have used similar strategies by holding liquid healing agent in encapsulated microcontainers [57–59] or hollow fibers [60,61]. By mimicking the blood supply system in human skin, a microvascular self-healing system has been developed [51,62].

Although the self-healing process of human skin has been cited as four stages, we believe that it can be simplified to two steps for biomimetic engineering materials, that is, close then heal (CTH). In human skin, the first thing is to stop bleeding by clotting (for a small wound) or by suturing (for a large wound) before new cells gradually grow. Therefore, in order to mimic the self-healing process of human skin, at least the two steps, close then heal, are needed. If a damage sensing capability is added, the engineering material will mimic the human skin better.

Recently, our group has developed self-healing of macroscopic cracks in polymer matrix using shape memory functionality, by means of the CTH scheme. At first glance, self-healing using the shape memory effect seems straightforward. This is because shape memory is defined as the ability for deformed polymer to return to its original shape upon external stimuli. Because a macroscopic crack is considered as deformation, shape recovery can restore the original shape and thus close the crack. It sounds so simple! This is actually true for some cases such as indentation or a small scratch on shape memory polymer coatings, which can recover upon heating [63]. Indeed, if a shape memory polymer rod is broken by tension, heating the broken pieces can bring them together, as long as the shape recovery ratio is 100% and the broken pieces are free to move. The reason is that the external tension load, which creates the damage, also programs the polymer in tension. Therefore, the broken two pieces shrink and approach each other upon shape recovery. Unfortunately, in real world structures, structural components are constrained by their surroundings and are thus not free to move. Therefore, for polymers programmed by the external load only, or ad hoc programming, the crack cannot be closed. In other words, we must provide additional programming to the polymer beforehand. We must understand the relationship between the crack width that needs to be closed and the recovery stress of the SMP. To this end, we must understand the thermomechanical constitutive behavior of the SMPs. All of these will be discussed in this book.

References

1. Naderia, J.R. and Ramanb, B. (2005) Capturing impressions of pedestrian landscapes used for healing purposes with decision tree learning. *Landscape and Urban Planning*, **73**, 155–166.
2. Animal self-medication: Do wild animals heal themselves? http://www.natural-wonder-pets.com/do-wild-animals-heal-themselves.html.
3. Wound licking. From Wikipedia, the free encyclopedia, http://en.wikipedia.org/wiki/Wound_licking.
4. Ihalin, R., Loimaranta, V., and Tenovuo, J. (2006) Origin, structure, and biological activities of peroxidases in human saliva. *Archives of Biochemistry and Biophysics*, **445**, 261–268.
5. Abiko, Y., Nishimura, M., and Kaku, T. (2003) Defensins in saliva and the salivary glands. *Medical Electron Microscopy*, **36**, 247–252.
6. Schenkels, L., Veerman, E., and Nieuw, A.A. (1995) Biochemical composition of human saliva in relation to other mucosal fluids. *Critical Reviews in Oral Biology and Medicine*, **6**, 161–175.

7. Campo, J., Perea, M., del Romero, J., Cano, J., Hernando, V., and Bascones, A. (2006) Oral transmission of HIV, reality or fiction? *An update. Oral Diseases*, **12**, 219–228.
8. Baron, S., Singh, I., Chopra, A., Coppenhaver, D., and Pan, J. (2000) Innate antiviral defenses in body fluids and tissues. *Antiviral Research*, **48**, 71–89.
9. Ashcroft, G., Lei, K., and Jin, W. (2000) Secretory leukocyte protease inhibitor mediates non-redundant functions necessary for normal wound healing. *Nature Medicine*, **6**, 1147–1153.
10. Benjamin, N., Pattullo, S., Weller, R., Smith, L., and Ormerod, A. (1997) Wound licking and nitric oxide. *The Lancet*, **349**, 1776.
11. Pammer, J., Weninger, W., Mildner, M., Burian, M., Wojta, J., and Tschachler, E. (1998) Vascular endothelial growth factor is constitutively expressed in normal human salivary glands and is secreted in the saliva of healthy individuals. *The Journal of Pathology*, **186**, 186–191.
12. Schrementi, M., Ferreira, A., Zender, C., and DiPietro, L. (2008) Site-specific production of TGF-beta in oral mucosal and cutaneous wounds. *Wound Repair and Regeneration*, **16**, 80–86.
13. Frank, S., Stallmeyer, B., Kämpfer, H., Kolb, N., and Pfeilschifter, J. (2000) Leptin enhances wound re-epithelialization and constitutes a direct function of leptin in skin repair. *Journal of Clinical Investigation*, **106**, 501–509.
14. Gröschl, M., Topf, H., Kratzsch, J., Dötsch, J., Rascher, W., and Rauh, M. (2005) Salivary leptin induces increased expression of growth factors in oral keratinocytes. *Journal of Molecular Endocrinology*, **34**, 353–366.
15. Costigan, D., Guyda, H., and Posner, B. (1988) Free insulin-like growth factor I (IGF-I) and IGF-II in human saliva. *The Journal of Clinical Endocrinology and Metabolism*, **66**, 1014–1018.
16. Todorović, V., Pesko, P., and Micev, M. (2008) Insulin-like growth factor-I in wound healing of rat skin. *Regulatory Peptides*, **150**, 7–13.
17. Sugiura, T., Nakane, S., Kishimoto, S., Waku, K., Yoshioka, Y., and Tokumura, A. (2002) Lysophosphatidic acid, a growth factor-like lipid, in the saliva. *The Journal of Lipid Research*, **43**, 2049–2055.
18. Balazs, L., Okolicany, J., Ferrebee, M., Tolley, B., and Tigyi, G. (2001) Topical application of the phospholipid growth factor lysophosphatidic acid promotes wound healing *in vivo*. *Regulatory, Integrative and Comparative Physiology*, **280**, R466–R472.
19. Pogrel, M., Low, M., and Stern, R. (2003) Hyaluronan (hyaluronic acid) and its regulation in human saliva by hyaluronidase and its inhibitors. *Journal of Oral Science*, **45**, 85–91.
20. Li, A., Koroly, M., Schattenkerk, M., Malt, R., and Young, M. (1980) Nerve growth factor: acceleration of the rate of wound healing in mice. *Proceedings of the National Academy of Sciences*, **77**, 4379–4381.
21. Kawamoto, K. and Matsuda, H. (2004) Nerve growth factor and wound healing. *Progress in Brain Research*, **146**, 369–384.
22. Nam, J., Chung, J., Kho, H., Chung, S., and Kim, Y. (2007) Nerve growth factor concentration in human saliva. *Oral Diseases*, **13**, 187–192.
23. Jahovic, N., Güzel, E., Arbak, S., and Yeğen, B. (2004) The healing–promoting effect of saliva on skin burn is mediated by epidermal growth factor (EGF): role of the neutrophils. *Burns*, **30**, 531–538.
24. Wisner, A., Dufour, E., and Messaoudi, M. (2006) Human opiorphin, a natural antinociceptive modulator of opioid-dependent pathways. *Proceedings of the National Academy of Sciences*, **103**, 17979–17984.
25. Dittus, W.P.J. and Ratnayeke, S.M. (1989) Individual and social behavioral responses to injury in wild toque macaques (*Macaca sinica*). *International Journal of Primatology*, **10**, 215–234.
26. Ochoa-Ochoa, L.M., Campbell, J.A., and Flores-Villela, O.A. (2014) Patterns of richness and endemism of the Mexican herpetofauna, a matter of spatial scale? *Biological Journal of the Linnean Society*, **111**, 305–316.
27. Recknagel, H., Elmer, K.R., Noonan, B.P., Raselimanan, A.P., Meyer, A., and Vencese, M. (2013) Multi-gene phylogeny of Madagascar's plated lizards, Zonosaurus and Tracheloptychus (Squamata: Gerrhosauridae). *Molecular Phylogenetics and Evolution*, **69**, 1215–1221.
28. Sipos, P., Győry, H., Hagymási, K., Ondrejka, P., and Blázovics, A. (2004) Special wound healing methods used in ancient Egypt and the mythological background. *World Journal of Surgery*, **28**, 211–216.
29. Zhang, G.Y., Ren, H.H., Zhang, Y.B., Ma, L.Q., Yang, Y.L., and Wang, S. (2013) Chemical constituents of the starfish *Asterias rollestoni* Bell. *Biochemical Systematics and Ecology*, **51**, 203–206.
30. Ramsay, K., Kaiser, M., and Richardson, C. (2001) Invest in arms: behavioural and energetic implications of multiple autotomy in starfish (*Asterias rubens*). *Behavioral Ecology and Sociobiology*, **50**, 360–365.
31. Chao, S.M., Chen, C.P., and Alexander, P.S. (1995) Reproductive cycles of tropical sea cucumbers (Echinodermata: Holothuroidea) in southern Taiwan. *Marine Biology*, **122**, 289–295.

32. Kang, J.H., Kim, Y.K., Kim, M.J., Park, J.Y., An, C.M., Kim, B.S., Jun, J.C., and Kim, S.K. (2011) Genetic differentiation among populations and color variants of sea cucumbers (*Stichopus Japonicus*) from Korea and China. *International Journal of Biological Sciences*, **7**, 323–332.

33. Moore, C.R. and Price, D. (1948) A study at high altitude of reproduction, growth, sexual maturity, and organ weights. *Journal of Experimental Zoology*, **108**, 171–216.

34. Mashanov, V.S., Zueva, O.R., and Garcia-Arraras, J.E. (2012) Expression of Wnt9, TCTP, and Bmp1/Tll in sea cucumber visceral regeneration. *Gene Expression Patterns*, **12**, 24–35.

35. Kalman, B. (2004) *The Life Cycle of an Earthworm*, Crabtree Publishing Co.

36. Earthworm, from Wikipedia, the free encyclopedia, http://en.wikipedia.org/wiki/Earthworm.

37. Varhalmi, E., Somogyi, I., Kiszler, G., Nemeth, J., Reglodi, D., Lubics, A., Kiss, P., Tamas, A., Pollak, E., and Molnar, L. (2008) Expression of PACAP-like compounds during the caudal regeneration of the earthworm *Eisenia fetida. Journal of Molecular Neuroscience*, **36**, 166–174.

38. Clark, P. (2013) Urban Jungle: the changing natural world at our doorsteps, making heads or tails out of severed earthworms, http://www.washingtonpost.com/wp-srv/special/metro/urban-jungle/pages/130604.html.

39. Pough, F.H. (2007) Amphibian biology and husbandry. *ILAR Journal*, **48**, 203–213.

40. Templeton, G. (2013) Salamander macrophages may hold key to "perfect" tissue regeneration, http://www.geek .com/science/salamander-macrophages-may-hold-key-to-perfect-tissue-regeneration-1556106/.

41. Godwin, J.W., Pinto, A.R., and Rosentha, N.A. (2013) Macrophages are required for adult salamander limb regeneration. *Proceedings of the National Academy of Sciences*, **110**, 9415–9420.

42. Ju, B.G. and Kim, W.S. (1998) Upregulation of cathepsin D expression in the dedifferentiating salamander limb regenerates and enhancement of its expression by retinoic acid. *Wound Repair and Regeneration*, **6**, S349–S357.

43. Shirley, Oriented to self care, Shirley's Wellness Cafe is an educational web site dedicated to promoting natural health for humans and their animals – See more at: http://www.shirleys-wellness-cafe.com/#sthash.UKRXV0ZQ .YpjmDdUF.dpuf.

44. Wikipedia, http://en.wikipedia.org/wiki/Self-healing.

45. Circle for Self Healing, http://www.circleforselfhealing.com/.

46. Self Healing Expressions, Guiding the self to healing, one lesson at a time, http://www.selfhealingexpressions. com/index.php.

47. Thiaoouba Prophecy, Believing is not enough you need to know, http://www.thiaoouba.com/heal.htm.

48. Huang, X., Zhang, Y., Zhang, X., Xu, L., Chen, X., and Wei, S. (2013) Influence of radiation crosslinked carboxymethyl-chitosan/gelatin hydrogel on cutaneous wound healing. *Materials Science and Engineering: C*, **33**, 4816–4824.

49. Martin, P. (1997) Wound healing – aiming for perfect skin regeneration. *Science*, **276**, 75–81.

50. Learn Genetics, Genetic Science Learning Center, What is blood? http://learn.genetics.utah.edu/content/begin/ traits/blood/blood.html.

51. Toohey, K.S., Sottos, N.R., Lewis, J.A., Mooer, J.S., and White, S.R. (2007) Self-healing materials with microvascular networks. *Nature Materials*, **6**, 581–585.

52. Hamilton, A.R., Sottos, N.R., and White, S.R. (2010) Self-healing of internal damage in synthetic vascular materials. *Advanced Materials*, **22**, 5159–5163.

53. The McGraw-Hill Companies, Inc., Hemostasis, http://www.mhhe.com/biosci/esp/2002_general/Esp/folder_ structure/tr/m1/s7/trm1s7_3.htm.

54. Kohn, C., Bleeding and healing – How injuries heal? www2.waterforduhs.k12.wi.us/staffweb/.../Bleeding_ and_Healing.pptx

55. Jennings, L. (2009) Mechanisms of platelet activation: need for new strategies to protect against platelet-mediated atherothrombosis. *Thromb Haemost*, **102**, 248–257.

56. Heemskerk, J.W.M., Bevers, E.M., and Lindhout, T. (2002) Platelet activation and blood coagulation. *Thromb Haemost*, **88**, 186–193.

57. Zhao, Y., Zhang, W., Liao, L., Wang, S., and Li, W.J. (2012) Self-healing coatings containing microcapsule. *Applied Surface Science*, **258**, 1915–1918.

58. Kouhi, M., Mohebbi, A., Mirzaei, M., and Peikari, M. (2013) Optimization of smart self-healing coatings based on micro/nanocapsules in heavy metals emission inhibition. *Progress in Organic Coatings*, **76**, 1006–1015.

59. Song, Y.K., Jo, Y.H., Lim, Y.J., Cho, S.Y., Yu, H.C., Ryu, B.C., Lee, S.I., and Chung, C.M. (2013) Sunlight-induced self-healing of a microcapsule-type protective coating. *ACS Applied Materials and Interfaces*, **5**, 1378–1384.

60. Kousourakis, A. and Mouritz1 A.P. (2010) The effect of self-healing hollow fibres on the mechanical properties of polymer composites. *Smart Materials and Structures*, **19**, 085021.

61. Escobar, M.M., Vago, S., and Vázquez, A. (2013) Self-healing mortars based on hollow glass tubes and epoxy–amine systems. *Composites Part B: Engineering*, **55**, 203–207.

62. Patrick, J.F., Sottos, N.R., and White, S.R. (2012) Microvascular based self-healing polymeric foam. *Polymer*, **53**, 4231–4240.

63. Luo, X. and Mather, P.T. (2013) Shape memory assisted self-healing coating. *ACS Macro Letters*, **2**, 152–156.

3

Thermoset Shape Memory Polymer and Its Syntactic Foam

With the advancement of knowledge, scientists and engineers are often reminded of the limitations of their understanding and of the resulting output from its application to design, process, and product. Overcoming these barriers has been essentially the main focus of science and technology. Scientists and engineers are tasked not only with the implementation of viable solutions to problems but also with the creation of new knowledge, processes, and technology that broaden the collective toolbox. Historically, the advancement of materials was accompanied by a significant step forward in civilization. Indeed, the assigned names to particular periods of civilization have been based on access to certain materials and the tools created from them (e.g., The Stone Age, The Bronze Age, The Iron Age, etc.). Material science, in particular, constantly seeks to push our capabilities forward through improvements to existing methods, combinations of materials, or by new materials and systems that can be used as building blocks for larger projects or higher goals. When scientists and engineers are engaged in these grander efforts, they require more from components, and the materials that they are made of. The idea that a material can serve more than one function is a relatively new concept, and one that often requires the cooperation of multiple disciplines to realize. When properly conceived and implemented, these advanced materials become powerful vehicles by which all manner of progress is propelled.

The discovery of the shape memory effect (SME) by Chang and Read in 1932, described in Reference [1], is one of the revolutionary steps in the field of active materials research. Materials are said to show the shape memory effect if they can be deformed and fixed into a temporary shape and recover their original permanent shape only on exposure to external stimuli such as heat, light, magnetic, pH, moisture, etc. [2–5]. In other words, shape memory is a process that enables the reversible storage and recovery of mechanical energy through a cyclical change in shape. The mechanical energy of deformation is trapped as internal energy or back stress in a temporary shape and subsequently released through a change of the material morphology or a change in the rate of molecular relaxation when triggered by an external stimulus. Technological uses include durable, shape recovery eye-glass frames, packaging, temperature sensitive switches, generation of stress to induce mechanical motion, heat-shrink

Self-Healing Composites: Shape Memory Polymer-Based Structures, First Edition. Guoqiang Li.
© 2015 John Wiley & Sons, Ltd. Published 2015 by John Wiley & Sons, Ltd.

tubing, deployable structures, microdevices, biomedical devices, etc. [1,6–11]. Among the various shape memory materials, such as shape memory alloy (SMA) (for instance Ni–Ti alloy), shape memory ceramic, and shape memory polymer (SMP), SMPs have drawn increasing attention because of their scientific and technological significance [1,12,13]. While it seems that the first mention of the shape memory effect (SME) dated back to a patent in the 1940s with the term "elastic memory" and the application of heat shrinking tape in the 1960s [13], it is widely accepted that SMP was first developed by the CDF Chimie Company (France) in 1984 under the trade name of polynorbornene [1]. The reason is that a fundamental difference between the heat shrinking tape and the shape memory polymer is that the shape memory polymer can be repeatedly programmed and recovered, while the shrinking tape can recover only one time. Although having a long history, the shape memory polymer was not widely recognized and studied until the discovery of the segmented polyurethane shape memory polymer by Mitsubishi Heavy Industries Ltd, due to the versatility of urethane chemistry that allows easy structural tuning and the industrial significance of polyurethane [13]. It was found that SMP offers deformation to a much higher degree and a wider scope of varying mechanical properties compared to SMAs or ceramics. In addition to their inherent advantages of being cheap, lightweight, and easy process ability, SMPs offer extra advantages due to the fact that they may be biocompatible, nontoxic, and biodegradable [1]. Driven by the biomedical applications as first demonstrated by Lendlein and Langer [11], which showed that SMP could be used as a self-tightening suture for minimum invasive surgery, SMP has experienced fast growth in the past decade. Since the 2000s, a number of excellent reviews have been published that cover the various aspects of SMPs and their composites, the most recent being References [1] and [12] to [22].

Not all polymers have the shape memory effect (SME). For example, rubber can be deformed elastically and returns to its original shape when the load is removed. However, this type of elastic deformation/recovery is not considered as shape memory because a temporary shape cannot be fixed. In order for a polymer to have SME, it must have a stable polymer network and a reversible switching transition (see Figure 3.1). The stable network of SMPs determines the original shape, which can be formed by molecule entanglement, crystalline phase, chemical cross-linking, or an interpenetrated network. The lock in the network represents the reversible switching transition responsible for fixing the temporary shape, which can be a crystallization/melting transition, vitrification/glass transition, liquid crystal anisotropic/isotropic transition, reversible molecule cross-linking reaction, and supramolecular association/disassociation transition. Typical reversible molecule cross-linking reactions as switching transitions include photodimerization, Diels–Alder reaction, and oxidation/redox reaction of the mercapto group. A typical switching transition using supramolecular association/disassociation includes hydrogen bonding, self-assembly metal–ligand coordination, and self-assembly of β-CD. In addition to the above reversible switches, other stimuli that can significantly change the mobility of the SMP may also trigger the SME, such as moisture, water/solvent, ions, pressure, light, pH, etc. [22].

Thermosetting shape memory polymers are chemically or physically cross-linked polymers. Thermosetting shape memory polymers are preferred in engineering structures because, like their conventional thermosetting polymer counterparts, they have several advantages, such as high strength, high stiffness, high thermostability, high corrosion resistance, and high dimensional stability, as compared to thermoplastic shape memory polymers. A number of conventional thermosetting polymers can become shape memory polymers if the density of

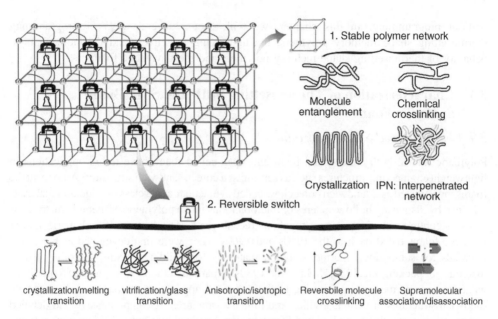

Figure 3.1 Various molecular structures of shape memory polymers (SMPs). A stable network and a reversible switching transition are the prerequisites for the polymers to show the shape memory effect (SME). The stable network can be molecule entanglement, chemical cross-linking, crystallization, and IPN; the reversible switching transition can be the crystallization/melting transition, vitrification/glass transition, anisotropic/isotropic transition, reversible chemical cross-linking, and association/dis-association of supramolecular structures. *Source:* [22] Reproduced with permission from Elsevier

cross-linking is appropriate. Compared to conventional thermosetting polymers, thermosetting shape memory polymers must have considerable ductility at temperatures above its glass transition temperature, the so-called rubbery or entropic elasticity. If the cross-linking density is too high, the "free" segmental length between two cross-linking points (net points) will be too short, and the rubbery elasticity will be too small to fix a temporary shape, and thus will not have considerable shape memory effect, such as conventional thermosetting epoxy, polyester, vinyl ester, etc.; on the other hand, if the cross-linking density is too low, the "free" segmental length between two cross-linking points will be too long, which may lead to long-range molecule chain slippage or elastic deformation till fracture. For polymers with long-range molecule chain slippage, while a temporary shape can be fixed, the fixed shape cannot be recovered, and thus no shape memory effect can be seen. For polymers with elastic deformation till fracture, such as rubber, again a temporary shape cannot be fixed, and the polymers will not have the shape memory effect. Therefore, the key for a thermosetting polymer to have the shape memory effect is the cross-linking density, which must be in between the density for conventional thermosetting polymers and conventional rubbers or elastomers. In other words, thermosetting shape memory polymers behave like "thermosetting elastomers," that is, glass-like conventional thermosetting polymers at temperatures below the glass transition temperature and rubber-like conventional elastomers at temperatures above the glass transition temperature. While an appropriate cross-linking density is the material basis for the shape memory effect, thermosetting shape memory polymers do not exhibit the shape memory effect without programming. In this chapter, we will discuss several ways of programming thermosetting shape memory polymers, the underlying principles, and thermomechanical behavior of thermosetting shape memory polymers and their syntactic foam.

3.1 Characterization of Thermosetting SMP and SMP Based Syntactic Foam

3.1.1 SMP Based Syntactic Foam

Polymeric foams, both close-celled foam and open-celled foam, have found applications in lightweight engineering structures such as serving as a core in sandwich structures. Among all the foams investigated as a sandwich core, close-celled syntactic foam, a class of structural materials obtained by dispersing hollow spheres or microballoons into a polymeric or metallic matrix, has enjoyed continuously increasing growth in recent years. The history of polymeric matrix syntactic foams dated back to the 1950s and 1960s. They were first developed as buoyancy materials for submarines, thermoinsulation materials for buildings, trains, aircraft, cushion materials for packing, etc. [23,24]. Since the 1980s, in particular the 1990s, syntactic foams have gained a new momentum, partly due to the increased application of foam cored composite sandwich structures in various civilian and military structures. A comprehensive and detailed survey on syntactic foams and related literature was provided by Shutov [25] on conventional polymer matrix syntactic foam and by Hearon et al. [26] on SMP based foam. Recent development in polymeric syntactic foams include static and dynamic modeling of the constitutive behavior of syntactic foams by using an analogy to geomaterials [27] or by using damage mechanics [28]; modeling of the effective elastic properties based on micromechanics, homogenization, and equivalent medium theorem [29–31]; making toughened syntactic foam by coating microballoons with rubber latex or adding crumb rubber particles into the polymer matrix [32–34]; and making functionally graded syntactic foam by changing the volume fraction of the microballoons or changing the density of the microballoons as the spatial location changes [35,36].

If microballoons are dispersed in a shape memory polymer matrix, smart syntactic foam is obtained. Like conventional polymer based syntactic foam, thermosetting shape memory polymer syntactic foam has all the advantages associated with polymeric syntactic foam, in addition to its shape memory properties and healing capabilities, as will be discussed in Chapter 6. In the following, the focus will be on a thermosetting polystyrene shape memory polymer and its syntactic foam.

3.1.2 Raw Materials and Syntactic Foam Preparation

The syntactic foam consists of hollow glass microspheres dispersed in a shape memory polymer matrix. The polymer is a styrene based thermoset SMP resin system sold commercially by CRG Industries under the name *Veriflex®* due to its availability and potentially adjustable glass transition temperature (T_g) (glass transition temperature: 62 °C, density: 920 kg/m^3, tensile strength: 23 MPa, flexural strength: 37.1 MPa, compressive strength: 32.4 MPa, and elastic modulus: 1.24 GPa, all at room temperature). The *Veriflex®* SMP is a two-part resin system. Part A is composed of styrene, divinyl benzene, and vinyl neo-decanoate. Part B is composed of benzoyl peroxide. The chemical structure for each component is shown in Figure 3.2.

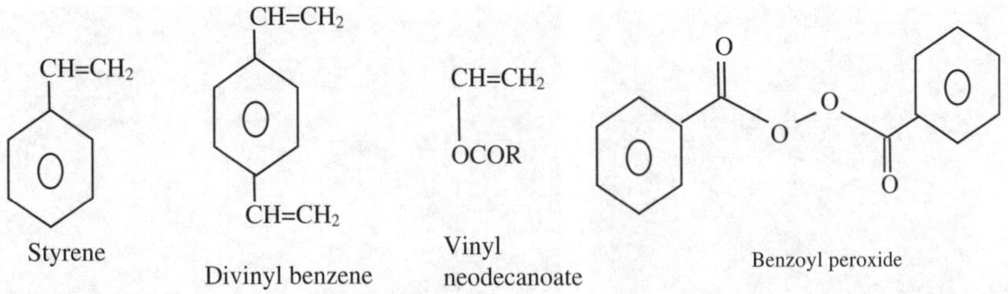

Figure 3.2 Chemical structure of each component in the polystyrene SMP. *Source:* [37] Reproduced with permission from Elsevier

The glass microballoons were obtained from Potters Industries (Q-CEL6014, average outer diameter: 85 μm, bulk density: 0.08 g/cm^3, effective density: 0.14 g/cm^3, maximum working pressure: 1.73 MPa, and wall thickness: 0.8 μm). These raw materials have been used previously for the smart syntactic foam [38] and conventional epoxy based syntactic foams [39,40].

The SMP based syntactic foam was fabricated by dispersing 40% by volume of the hollow glass microspheres into Part A of the SMP resin. Without the use of diluent, 40% by volume of microballoons was the maximum amount of fillers that can be incorporated without loss of workability. The microspheres were added to the resin while slowly stirring the mixture to minimize air bubbles in the resin. The microspheres were added in multiple steps to avoid agglomeration. After dispersion, Part B (hardening agent) was added and the solution mixed for 10 min in a weight ratio of Part A:Part B = 24:1. Unless a special composition is indicated, all the raw materials and compositions used in the experimental testing in this chapter are the same in order to keep the test consistent.

The mixture was then poured into a 229 mm × 229 mm × 12.7 mm steel mold. Teflon sheets were used on the inner surfaces of the steel mold to aid in demolding, and two-sided vacuum bag tape was used to seal areas to prevent loss of material from leaks. Hand clamps were used to hold the pieces together, keeping the face plates and tape in place. The open mold (without the top plate) was placed in a vacuum chamber (Abbess Instruments, USA) at 40 kPa for 20 minutes in order to remove any air pockets that may have been introduced during the mixing process. The mold was then closed and placed in a forced convection oven (Grieve model NB-350) for curing. The SMP resin manufacturer suggests using a closed mold to avoid material bake-off during curing. Based on previous studies [38,41,42], the curing cycle was 79 °C for 24 h, 107 °C for 3 h, and 121 °C for 9 h. After curing, the foam panel was demolded and was ready to be machined into required specimens for various tests. The fabrication process with the related equipment is shown in Figure 3.3. A typical SEM observation of the SMP based syntactic foam is shown in Figure 3.4.

3.1.3 DMA Testing

In order to perform programming and recovery effectively, the glass transition temperature (T_g) must be determined first as the shape memory effect of thermosetting shape memory polymers revolves around a temperature range centered at the T_g. Below this range the material is rigid or glassy and above it is in a rubbery-elastic state. Recovery must be conducted within or above

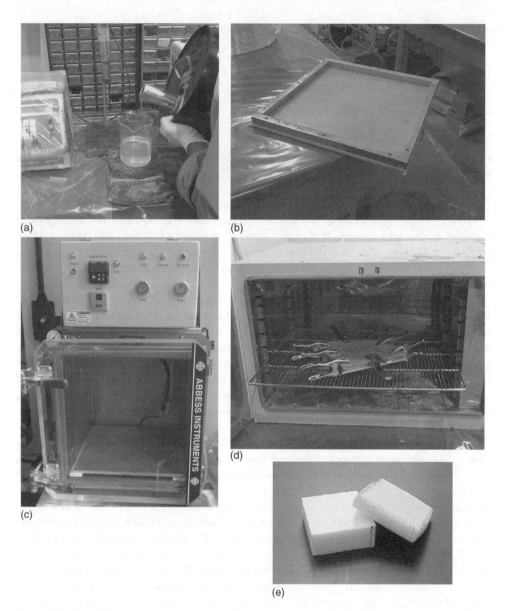

Figure 3.3 Fabrication process: (a) measuring the resin; (b) the mold; (c) the vacuum chamber; (d) the closed mold with clamps in the oven for curing; (e) block specimens (left – cut specimen; right – after compression programming)

the glass transition region in order to transform the material from the temporary shape to the permanent shape.

Dynamic mechanical analysis (DMA) is a technique used to study and characterize materials, proved to be extremely useful for studying viscoelasticity. With the complex modulus of the sample determined by the measurement of the strain response to a sinusoidal

Figure 3.4 SEM image of the SMP based syntactic foam sample: (left) overall view; (right) focused view. *Source:* [43] Reproduced with permission from Elsevier

stress input, the variable temperature leading to variations in the complex modulus can help locate the glass transition temperature of the material or identify transitions corresponding to other molecular motions. DMA was conducted on a DMA 2980 tester from TA instruments for both the polystyrene SMP and its syntactic foam. A rectangular bar with dimensions of 17.5 mm × 11.9 mm × 1.20 mm was placed into the DMA single cantilever clamping fixture. A small dynamic load at 1 Hz was applied to the platen and the temperature was ramped from room temperature to 120 °C at a rate of 3 °C/min. The amplitude was set to be 15 μm. The storage modulus and loss modulus of the pure SMP and the SMP based syntactic foam are shown in Figure 3.5 [41]. The glass transition region was determined using the storage modulus

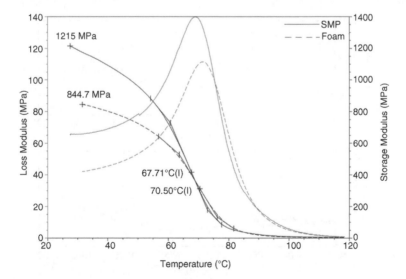

Figure 3.5 DMA results for the pure SMP and syntactic foam. The T_g regions were determined from the storage modulus curves. *Source:* [41] Reproduced with permission from Elsevier

curve, as suggested in ASTM E 1640-04. The intersection of tangent lines below the onset of the transition and to the inflection points denotes the upper and lower limits of the glass transition zone, with the average value in between determining the T_g value.

For the pure SMP, the glass transition zone is from 60.7 to 73.0 °C with $T_g = 67.7$ °C. The manufacturer lists a value of $T_g = 62$ °C as determined by differential scanning calorimetry (DSC). This is about 4 °C lower than the results obtained from their DMA tests, which is consistent with the current study. In the foam, it is observed that a shift up of about 2.8 °C in the glass transition region to 63.7–76.2 °C with $T_g = 70.5$ °C. This shift is caused by the addition of the glass microspheres, which may have increased the T_g of the SMP immediately surrounding them through physical adsorption or chemical reactions, restricting the mobility of the SMP molecules.

3.1.4 Fourier Transform Infrared (FTIR) Spectroscopy Analysis

Fourier transform infrared (FTIR) analysis works on the fact that chemical bonds and groups of chemical bonds vibrate at characteristic frequencies. During FTIR analysis, a modulated infrared (IR) beam is spotted on the specimen. The transmittance and reflectance of the infrared rays at different frequencies is then translated into an IR absorption plot consisting of reverse peaks, which after matching and identification provides information about the chemical bonding or molecular structure of materials, whether organic or inorganic.

FTIR spectra of both the pure SMP and the syntactic foam samples were recorded on a Bruker Tensor 27 single-beam instrument at 16 scans with a nominal resolution of $4 \, cm^{-1}$. Absorption spectra were saved from 4000 to $700 \, cm^{-1}$. FTIR spectra of the pure SMP and the SMP based syntactic foam are shown in Figure 3.6 [43]. Figure 3.6 indicates that the intensity ratio of the two peaks around 1746 and $1724 \, cm^{-1}$ (the peaks marked by circle) changed after

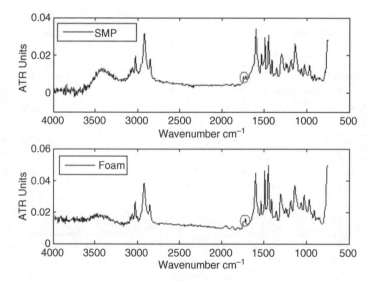

Figure 3.6 FTIR spectra of the pure SMP and the SMP based syntactic foam. *Source:* [43] Reproduced with permission from Elsevier

incorporation of the microballoons, which implies that some hydrogen bonds may have been formed between the silicon–hydroxyl (Si−OH) groups and the carbonyl (C=O) groups. This chemical reaction lowers the wavenumber of the carbonyl groups. Therefore, it can be concluded that an interfacial transition zone (ITZ) most likely exists in the syntactic foam and the foam can be perceived as a three-phase composite with ITZ coated microballoons dispersed in the SMP matrix. Although hydrogen bonds are not as strong as the primary chemical bonds, such as covalent bonds or ionic bonds, they are stronger than physical interaction such as absorption or interfacial friction. Because of the strong interfacial bonding, the mobility of the SMP molecules in the vicinity of the interface has been constrained and reduced, leading to an increase in glass transition temperature. This has been further evidenced by the DMA test results in Figure 3.5.

It is noted that incorporation of fillers does not always lead to an increase in T_g. It depends on the nature of the interaction between the filler and the polymer matrix. For example, incorporation of carbon nanotubes (CNTs) into the polymer matrix may see a decrease in T_g [44]. Previous tests on the same SMP based syntactic foam, but with an incorporation of 0.15% by weight of multiwalled carbon nanotubes, sees a T_g of 62 °C, which is almost the same as that of the pure SMP [38]. Obviously, this is as a result of competition between glass microballoons, which increase the T_g, and carbon nanotubes, which decrease the T_g. As a result of the competition, the T_g seems unchanged. The effect of carbon nanotubes on the reduction of T_g could be explained by the high degree of dispersion of CNTs, which form an interconnected space network and efficiently and rapidly transform the thermal energy to the entire polymer composite specimen. Usually, polymer is not a good thermal conductor. Therefore, the temperature distribution may not be uniform within the specimen. The test specimen needs to be overheated to above the real T_g. When the CNTs are introduced into the polymer, the thermal energy can be uniformly and quickly distributed to the specimen. Therefore, overheating may be minimal, which lowers the T_g and the T_g region. Another reason may be that incorporation of CNTs hinders the cross-linking of the polymer matrix, leading to a lowered T_g and T_g region.

3.1.5 X-Ray Photoelectron Spectroscopy

X-ray photoelectron spectroscopy (XPS) is a nondestructive quantitative spectroscopic analysis approach that measures the elemental composition, empirical formula, chemical state and electron state of the elements existing within the surface of a material. The principle behind this technique is the photoelectric effect. By calculating the binding energy, a characteristic of the electron configuration, determined as the attraction of the electrons to the nucleus, the number of electrons detected (sometimes per unit time) is plotted correspondingly in a typical XPS spectrum.

The XPS spectra of the pure SMP and the foam were collected on a Kratos AXIS165 high performance multitechnique surface analysis system with an information depth of 10 nm and a scan area of $700 \times 300 \, \mu m^2$. The purpose was to qualitatively evaluate the interface between the SMP matrix and the glass hollow microspheres. The XPS results shown in Figure 3.7 reveal that different binding energies exist in the pure SMP and the foam sample for the same emitted electrons (carbon in the 1s energy level, C (1s), and oxygen in the 1s energy level, O (1s)) [45]. This indicates that some chemical shifts may have occurred at the glass hollow microsphere/SMP matrix interface. The mobility of the SMP polymer chains in the vicinity of the interface

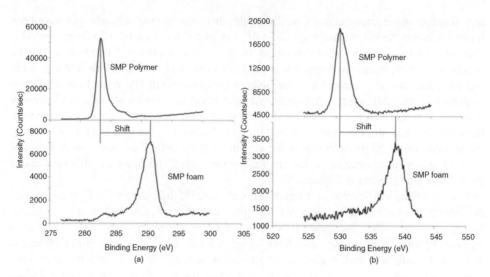

Figure 3.7 XPS spectra of the pure SMP and the SMP based syntactic foam for (a) the electron C (1s) and (b) the electron O (1s). *Source:* [45] Reproduced with permission from ASME

has probably been reduced, leading to an increase in the glass transition temperature of the foam, which echoes the DMA test results [41] and the FTIR test results [43].

3.1.6 Coefficient of Thermal Expansion Measurement

The coefficient of thermal expansion (CTE) was measured by using the Vishay BLH SR-4 general-purpose strain gages and a Yokagawa DC100 data acquisition system. The test procedure followed the Technical Note TN-513 (Vishay micromeasurement, 2007). The reference material was aluminum alloy. Both the test and reference sample dimensions were $30\,mm \times 12\,mm \times 5\,mm$. The temperature was ramped from room temperature to $90\,°C$ at an average heating rate of $1\,°C/min$ and then naturally cooled down to room temperature. The test result is shown in Figure 3.8 (a) [43] for the SMP. A cooling history for the SMP based syntactic foam is plotted as thermal deformation versus the temperature in Figure 3.8 (b) [45], where L_0 denotes the initial reference sample height. Because the cooling rate is extremely slow, averaging $0.17\,°C/min$, the thermal shrinkage can be perceived as the structural response. Linear CTEs in the rubbery state, α_r, and in the glassy state, α_g, were computed from the slopes at temperatures above and below T_g.

It is clear that the CTE can be divided into bilinear lines connected by a smooth curve, regardless of the pure SMP or the SMP based syntactic foam. The transition region corresponds to the glass transition zone. As compared to the pure SMP, the transition region becomes more gradual for the foam, possibly due to the existence of glass microballoons, which are less affected by the temperature change within the temperature range tested.

3.1.7 Isothermal Stress–Strain Behavior

Flat-wise compression was performed on an MTS QTEST/150 electromechanical frame outfitted with a moveable furnace as per the ASTM C 365 standard. Stress–strain responses

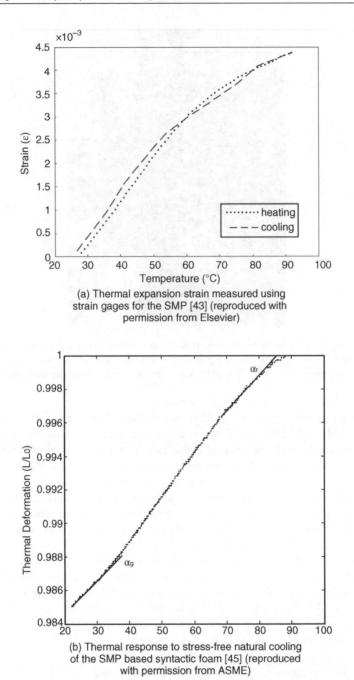

(a) Thermal expansion strain measured using
strain gages for the SMP [43] (reproduced with
permission from Elsevier)

(b) Thermal response to stress-free natural cooling
of the SMP based syntactic foam [45] (reproduced
with permission from ASME)

Figure 3.8 Measurement of thermal expansion behavior of (a) the SMP and (b) the SMP based syntactic foam

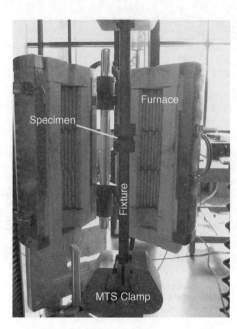

Figure 3.9 Setup used for isothermal stress–strain testing and constrained shape recovery testing. The fixture provided a 1-D external confinement and the furnace was used to trigger the shape memory effect by heating the specimen above its T_g. The MTS machine was used to record the resulting recovery stress

were generated for room temperature and two temperatures above the T_g (79 °C and 121 °C) for both the pure SMP and SMP based syntactic foam. Specimens of size 30 mm × 30 mm × 12.7 mm were placed in the preheated furnace and at least 30 minutes was allowed for temperature equilibration. Compression was then conducted in a strain-controlled manner at a constant rate of 1.3 mm/min to a strain level of 60%. Figure 3.9 shows the test setup.

Figure 3.10 shows the stress–strain response of the pure SMP and SMP based syntactic foam under uniaxial flat-wise compression at room temperature. The foam displays three distinct regions relating to microstructural changes. Region 1 (up to 5% strain) contains the linear elastic portion governed by microsphere cell wall bending and stretching. After the yield point we enter region 2 (5–40% strain). The curve plateaus as microsphere cell walls are damaged by crushing and fracture. In region 3 (>40% strain) the crushed microspheres are densified and the foam is consolidated, leading to a quick growth in stress [33]. These microstructural changes are shown in the SEM images taken after 5, 30, and 60% strain. Also, it is seen that both the yield strength and the modulus of elasticity of the foam is lower than those of the pure SMP because the microballoons serve as stress concentration centers. This observation agrees with conventional syntactic foams [34].

As the temperature increases, the polymer chains become more ductile, leading to reductions in the moduli and yield strength for both the pure SMP and foam. This behavior is particularly pronounced at temperatures higher than the T_g. Figure 3.11 shows the stress–strain responses at 79 °C, just above the T_g region. There is approximately a 97% reduction in the yield strength of the foam compared to room temperature and that of the pure SMP behaves like a rubber, obscuring the yield point. When the temperature is increased to 121 °C, even the foam tends to

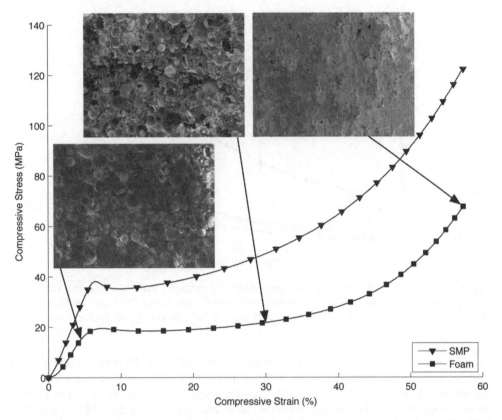

Figure 3.10 Stress–strain response of the pure SMP and syntactic foam at room temperature. The SEM images show the microstructure of the foam at compressive stain levels of 5, 30, and 60%. A progressive breaking of the microspheres occurs as the strain level is increased. *Source:* [41] Reproduced with permission from Elsevier

behave like a rubber, as evidenced by the near disappearance of the yield point. However, it was noticed that at a temperature higher than the T_g, the foam shows a higher yield strength (at 79 °C) and higher stiffness (at both 79 °C and 121 °C) than the pure SMP, which is opposite to the behavior at room temperature (below T_g). This is obviously due to the SMP's change from the glassy state below T_g to the rubbery state above T_g. Although the glass microballoons have a lower stiffness and lower compressive strength than the pure SMP at temperatures below T_g, they are less affected by the temperature rising in the temperature range investigated. Eventually, in the rubbery state, the glass microballoons seem stiffer and stronger than the SMP matrix. Hence, addition of stiffer particles (such as glass microspheres) increases the stiffness and yield strength of the composite foam at temperatures above T_g.

3.1.8 Summary

Polystyrene SMP and its syntactic foam can be fabricated by conventional manufacturing methods. No special equipment or device is needed. It is found that there is a certain physical/

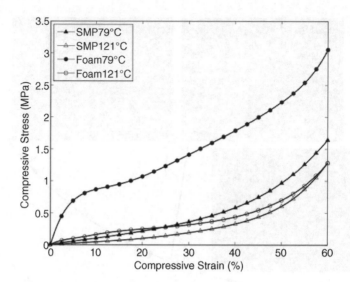

Figure 3.11 Stress–strain curves for the pure SMP and syntactic foam at 79 °C and 121 °C. At temperatures above T_g, the stiffness and strength were significantly lower. The foam was less affected due to the inclusion of microspheres. *Source:* [41] Reproduced with permission from Elsevier

chemical interaction between the glass hollow spheres and the SMP matrix, leading to an interfacial transition zone (ITZ) with reduced molecular mobility and an increased glass transition temperature. Therefore, the SMP based syntactic foam should be treated as a three-phase composite with ITZ coated glass microballoons dispersed in the SMP matrix. The uniaxial isothermal compression test shows that the SMP has a higher stiffness and yielding strength than the SMP based syntactic foam at temperatures below T_g; at temperatures above T_g, however, this tendency is reversed. Also, the SMP and its syntactic foam behave like rubber at temperatures well above T_g (such as 121 °C in this study), with the disappearance of the yielding point. The sharp decrease in stiffness from the glassy state to the rubbery state is a strong indication of good shape memory functionality of the polystyrene SMP and its syntactic foam.

3.2 Programming of Thermosetting SMPs

The shape memory behavior of SMPs derives from a combination of their molecular architecture and their deformation history. While the molecular architecture depends on the composition and polymerization process, which is difficult to alter once the polymerization process is completed, the deformation history can be actively altered by external forces, which is called programming or training. In this sense, shape memory is not an intrinsic material property, but a characteristic of an engineered system. In other words, shape memory polymers must be programmed or trained or educated in order to demonstrate the shape memory effect. This is unique for shape memory polymers because conventional thermosetting polymers, as the name indicates, cannot be physically or mechanically changed without damaging the polymers. Therefore, programming provides a way for engineers, rather than chemists, to change the behavior of polymers at will.

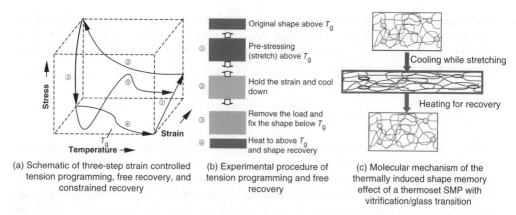

(a) Schematic of three-step strain controlled tension programming, free recovery, and constrained recovery

(b) Experimental procedure of tension programming and free recovery

(c) Molecular mechanism of the thermally induced shape memory effect of a thermoset SMP with vitrification/glass transition

Figure 3.12 Schematic of classical tension programming (steps 1 to 3), free shape recovery (step 4), fully constrained stress recovery (step 5), and schematic of molecular mechanisms for shape fixing and shape recovery

3.2.1 Classical Programming Methods

In order to give the polymer shape memory functionality, it usually needs to experience a typical three-step thermomechanical cycle called programming [12,46]; see Figure 3.12 (a) for a 1-D tensile stress (σ), tensile strain (ε), and temperature (T) relationship during the programming cycle. The programming starts at a temperature above the glass transition temperature (T_g) of the SMP. It involves a high strain deformation in the rubbery state, which is called "pre-deformation" or "pre-strain." Step 2 is a "strain storage" process by maintaining the pre-strain constant (strain controlled) or pre-stress constant (stress controlled) while cooling down to below T_g. Because of thermal contraction of the SMP during cooling, the tensile stress needed to maintain the pre-deformed shape increases as the temperature drops. The third step is a "low temperature unloading" process, which is defined as the removal of the stress in the glassy state. This completes the classical three-step thermomechanical programming or training. The low temperature unloading process may be accompanied by "springback," that is, some pre-strain may be rebounded. Because the thermosetting polymer in the glassy state is stiff and the stress applied at the rubbery state is small, the springback is usually very small. The amount of springback reflects the shape fixity capability of the polymer. Usually, in terms of strain, the shape fixity ratio is defined as [12]:

$$R_f(N) = \frac{\varepsilon_u(N)}{\varepsilon_m} \tag{3.1}$$

where N is the number of thermomechanical cycles, R_f is the shape fixity ratio, ε_m is the pre-deformation strain, and ε_u is the temporary strain fixed.

Step 4, which involves reheating to its starting temperature (above T_g) without applying any constraint, sometimes called the "free strain recovery" or "unconstrained recovery," brings the pre-strain back to zero (if the recovery ratio is 100%). Of course, if the strain is constrained when reheating (zero strain is allowed), constrained recovery occurs. The stress increases to a level, usually less than the programming stress, at the end of the constrained shape recovery; see step 5 in Figure 3.12 (a). The residual strain, as compared to the pre-deformation strain, defines the shape recovery capability of the polymer, which is represented by the shape recovery ratio as follows [12]:

$$R_r(N) = \frac{\varepsilon_m - \varepsilon_p(N)}{\varepsilon_m - \varepsilon_p(N-1)} \tag{3.2}$$

where R_r is the shape recovery ratio and ε_p is the permanent strain. Again, N and ε_m are the same as in Equation Equation (3.1).

A strain-controlled programming and free shape recovery are schematically shown in Figure 3.12 (b) for a polymer with the glass transition temperature (T_g) as its transition temperature (T_{trans}) and the molecular mechanisms are schematically shown in Figure 3.12 (c) [12]. The efficiency of a shape memory polymer is empirically controlled by its composition, as defined by the polymer's chemical structure, molecular weight, degree of cross-linking, and fraction of amorphous and crystalline domains [1,12,22]. The energy that is restored with shape recovery is a growing function of the energy supplied during the deformation at a high temperature [47].

As indicated by Behl and Lendlein [12], the thermally responsive shape memory effect can be understood from the point of view of entropy elasticity. During step 1 of programming, the deformation above T_g leads to molecular alignment along the loading direction; that is, the molecules transform from their original coiled configuration to a more straightened and ordered configuration, which leads to reduction in system entropy or nonequilibrium configuration. The work done by the external force leads to an increase in the internal energy in terms of back stress. In step 2, cooling freezes the aligned molecules because, as the temperature drops, the viscosity of the polymer increases and the molecular mobility decreases. As a result, at temperatures below T_g, the temporary configuration is frozen, leading to a temporary shape with reduced entropy or a nonequilibrium configuration. Step 3, removal of the external load, leads to springback of the material. Because the thermosetting shape memory polymer becomes stiff in the glassy state, the rebound corresponding to the removal of load is very small. The reason is that the load applied in step 1 is very small because the polymer is rubbery above T_g and is very flexible. As shown in Figure 3.12 (a), free shape recovery is represented by step 4. The driving force for shape recovery is the conformational entropy of the molecular segments in terms of micro-Brownian thermal motion. As the temperature rises above T_g, the viscosity reduces and molecule mobility increases, leading to recoil of molecules to their original configuration. Macroscopically, the polymer recovers to its original shape. Thermodynamically, the molecular segments experience a change from a temporary and ordered configuration to its random and coiled configuration during the shape recovery process. Since this process is accompanied by an increase in entropy, it is an autonomous process. For step 5 in Figure 3.12 (a), constrained shape recovery, tensile stress develops. The stress may first becomes negative (compressive stress) when the temperature rises due to thermal expansion of the SMP. Once the temperature comes to the start of the glass transition region, shape recovery occurs and the specimen starts to shrink. The compressive stress starts to reduce until it reaches zero. Further recovery leads to positive stress (tensile stress). The tensile stress increases until a peak tensile stress is achieved. As the temperature rises further, the stiffness of the polymer reduces and may be as high as two orders of reduction. Because the constrained recovery stress depends on both the stiffness and recovery tendency of the polymer, the competing effect of a gradual increase in the recovery tendency and a sharp decrease in stiffness leads to reduction of the tensile stress as the temperature rises. Once the temperature is above the glass transition region, the stiffness of the polymer is stabilized (rubbery state) and the recovery is complete, leading to a plateau in the constrained stress recovery, as shown in step 5 in Figure 3.12 (a). It is noted that the above description is based on tensile programming. For compression programming, the thermomechanical behavior is different from that of tensile programming. For instance, in step 2, the compressive stress will decrease as the temperature drops, instead of increasing, due to thermal contraction during temperature drops. On the other hand, the constrained recovery stress in step 5 will always be compressive and no tensile stress develops, because both thermal expansion and shape recovery make the specimen grow in length, instead of the opposite effect in the tensile programming case (thermal expansion makes the specimen longer, while shape memory makes the specimen shorter as the temperature rises). It is noted that in Figure 3.12 (c), the switching transition is vitrification/glass transition. This is true for the polystyrene SMP investigated in this chapter because no crystalline was found after programming, as evidenced by the X-ray diffraction (XRD) test [48].

The entropy change by programming can be quantified as follows. Consider the polymer in two statuses, one in the original status with entropy S_0 and the other in a deformed status with entropy S. According to Strobl [49], the change of entropy upon deformation is

$$\Delta S = S - S_0 = -\frac{1}{2} V c_p k \left(\lambda_1^2 + \lambda_2^2 + \lambda_3^2 - 3 \right) \qquad (3.3)$$

where ΔS is the entropy change upon molecule stretch, V is the volume considered, c_p is the chain density, and k is the Boltzmann constant; λ_i ($i = 1, 2, 3$) are the extension ratios in the three Cartesian directions. After stretching, λ_i ($i = 1, 2, 3$) are always equal to or greater than 1. Therefore, the term inside the parenthesis will always be positive. In other words, the entropy change will be negative, meaning a reduction in entropy after stretching or deformation.

3.2.2 Programming at Temperatures Below T_g – Cold Programming

While the above programming method has been widely accepted and used by many researchers, for practical applications, programming at very high temperatures is not a trivial task because it is a lengthy, labor-intensive, and energy-consuming process, in particular for large structures and for experimental investigations where special fixtures have to be designed to control heating–cooling cycles. There is a need to develop new programming approaches that depend on understanding the underlying shape memory mechanisms.

Instead of the classical entropy driven shape memory mechanism, Nguyen et al. [50] proposed a new concept that attributes the shape memory effects to structural and stress relaxation rather than the traditional phase transition hypothesis. They proposed that the dramatic change in the temperature dependence of the molecular chain mobility, which describes the ability for the polymer chain segments to rearrange locally to bring the macromolecular structure and stress response to equilibrium, determines the thermally activated shape memory phenomena of SMPs. The fact that the structure relaxes instantaneously to equilibrium at temperatures above T_g but responds sluggishly at temperatures below T_g suggests that cooling macroscopically freezes the structure in a nonequilibrium configuration below T_g, and thus allows the material to retain a temporary shape. Reheating to above T_g

reduces the viscosity, restores the mobility, and allows the structure to relax to its equilibrium configuration, which leads to shape recovery. As the temperature dependent phase transition assumption is abandoned, this theory implicitly indicates that acquiring the shape memory capacity may not necessarily require a temperature event.

Li and Xu [51] proposed that, as long as a nonequilibrium configuration can be created and maintained, SMPs will gain the shape memory capability, regardless of the temperature at which the nonequilibrium configuration is created. In other words, thermosetting SMPs can be programmed at temperatures below T_g. Instead of heating and then cooling steps, the programming can be conducted at a constant temperature, which was well below the T_g of the SMP. It is believed that this statement is true because it clearly indicates the pathways for the two most important requirements for programming and shape recovery: a way of fixing a temporary shape and a way of recovering the original shape. It is believed that the creation of the temporary shape should be in terms of plastic deformation. In order to distinguish from the traditional understanding of plastic deformation, that is, permanent deformation that cannot be reversibly recovered, we would like to define the plastic deformation by cold programming as "pseudo-plasticity." This effect is also termed by Xie [13] as "reversible plasticity shape memory effect (RP-SME)" to differentiate it from the classical shape memory cycles. Therefore, the key for cold programming of thermosetting SMPs is to create pseudo-plastic deformation. The nature of pseudo-plasticity is that, when the polymer is stressed to yielding, it overcomes the intermolecular rotation resistance barrier and initiates viscous flow. Plastic deformation and inhomogeneous shear banding are allowed during an adequate relaxation time. Once the load is released, high viscosity at low temperatures freezes the structure in a nonequilibrium state, resulting in a retained temporary shape or pseudo-plastic deformation. Upon heating, the viscosity decreases and molecule mobility increases. The increased chain mobility allows the structure to relax and to restore its equilibrium configuration. Hence, shape recovery is achieved.

It is noted that cold-drawing programming of thermoplastic SMPs has been conducted by several researchers. Lendlein and Kelch [52] indicated that the shape memory polymer (SMP) can be programmed by cold-drawing but did not give sufficient details. Ping et al. [53] investigated a thermoplastic poly(ε-caprolactone) (PCL) polyurethane for medical applications. In this polymer, PCL was the soft segment and can be stretched (tensioned) to several hundred percent at room temperature (15–20 °C below the melting temperature of the PCL segment). They found that the cold-drawing programmed SMP had a good shape memory capability. Rabani, Luftmann, and Kraft [54] also investigated the shape memory functionality of two shape memory polymers containing short aramid hard segments and poly(ε-caprolactone) (PCL) soft segments using cold-drawing programming. As compared to the study by Ping et al. [53], the hard segment was different but with the same soft segment PCL. Wang et al. [55] further studied the same SMP as that of Ping et al. [53]. They used FTIR to characterize the microstructure change during cold-drawing programming and shape recovery. They found that in cold-drawing programming, the amorphous PCL chains orient first at small extensions, whereas the hard segments and the crystalline PCL largely maintain their original state. When stretched further, the hard segments and the crystalline PCL chains start to align along the stretching direction and quickly reach a high degree of orientation; the hydrogen bonds between the urethane units along the stretching direction are weakened and the PCL undergoes stress-induced disaggregation and recrystallization while maintaining its overall crystallinity. When the SMP recovers, the microstructure evolves by reversing the sequence of

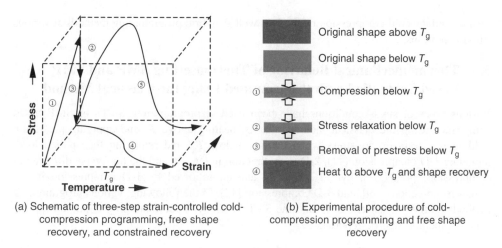

(a) Schematic of three-step strain-controlled cold-compression programming, free shape recovery, and constrained recovery

(b) Experimental procedure of cold-compression programming and free shape recovery

Figure 3.13 Schematic of cold-compression programming (steps 1 to 3), free shape recovery (step 4), and fully constrained stress recovery (step 5)

the microstructure change during programming. Zotzmann *et al.* [56] emphasized that a key requirement for materials suitable for cold-drawing programming is their ability to be deformed by cold-drawing. Based on their discussion, it seems that SMP with an elongation at break as high as 20% is not suitable for cold-drawing programming.

Because the ultimate tensile strain for most thermosetting SMPs is far below 20%, it is inferred that thermosetting SMPs cannot be programmed by cold-drawing. For the polystyrene SMP and the SMP based syntactic foam, it has been proved that the ultimate tensile strain is about 5% [57]. Also, the specimens deform all the way to fracture without appreciable pseudo-plastic deformation. Therefore, a temporary shape cannot be created under tension and the thermosetting SMP cannot be programmed by cold-drawing. However, it is worth exploring the potential for cold-compression programming. A schematic of cold-compression programming, free shape recovery, and constrained stress recovery is shown in Figure 3.13.

From Figure 3.13, the programming includes three steps: cold-compression (step 1, pre-stressing), holding the strain constant while the stress relaxes (step 2, stress relaxation), and removing the load (step 3, unloading). Step 2 can vary from almost zero to hours, which will affect the shape fixity ratio. Step 4 is a free shape recovery, the same as in Figure 3.12. Step 5 is a constrained stress recovery. It consists of two components: thermal stress due to thermal expansion during the heating process and stress due to shape recovery. The stress usually comes to a peak immediately before the start of the glass transition zone due to the high stiffness of the SMP in the glassy state. On entering the glass transition region, the stiffness drops sharply, leading to a decrease in recovery stress, although both thermal expansion and shape recovery contribute to the measured stress. Further rising temperatures above the glass transition region leads to stabilized recovery stress because the polymer becomes rubbery and its stiffness stabilizes.

In the following sections, the thermomechanical cycles of the thermosetting polystyrene SMP and its syntactic foam will be discussed, which are programmed using the classical method and the cold-compression method, as well as hybrid programming methods at various stress conditions.

3.3 Thermomechanical Behavior of Thermosetting SMP and SMP Based Syntactic Foam Programmed Using the Classical Method

Various types of stress conditions have been used in programming SMPs or SMP based composites using the classical method, that is, heating above T_g and applying pre-strain, holding pre-strain or pre-stress while cooling below T_g, and removing the applied load, including 1-D compression [41], 2-D (one direction in tension and the transverse direction in compression) [48,57,58], hybrid 2-D (1-D tension followed by 1-D compression in the transverse direction) [59], and 3-D compression [42]. In the following, various programming methods on the polystyrene SMP and the SMP based syntactic foam programmed using the classical method will be reported.

3.3.1 One-Dimensional Stress-Controlled Compression Programming and Shape Recovery

3.3.1.1 Compression Programming

In order to determine the thermomechancial behavior of the SMP and SMP based syntactic foam programmed using the classical method, the 1-D stress-controlled compression programming method (stress is constant) was conducted. Two stress levels were chosen: 47 kPa and 263 kPa. The experimental setup shown in Figure 3.14 was used to conduct the stress-controlled programming. Additional weight was added as necessary to reach the desired stress

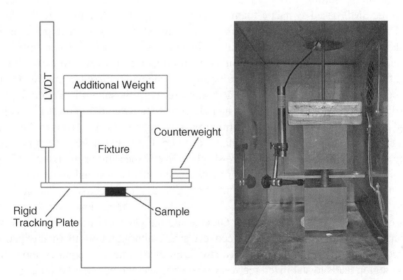

Figure 3.14 Setup used for the stress-controlled programming consisting of an LVDT, fixture, and weights. A static load was used to compress the specimen and the deformation measured using the LVDT. *Source:* [41] Reproduced with permission from Elsevier

level. The top half of the fixture and the load were suspended and preheated in an insulated forced convection environmental chamber to 79 °C (adopted from the Dynatup 8250 HV impact machine chamber). The specimen, with dimensions of 30 mm × 30 mm × 12.7 mm, was introduced into the system and 45 minutes were allowed for it to come to a uniform temperature. A rigid lightweight aluminum plate was situated on top of the specimen so that its motion could be tracked. A linear variable differential transducer (LVDT) (Cooper Instruments LDT 200) was independently mounted with an adjustable arm and magnetic base and used to record the displacement. The load was then gently lowered on to the specimen and the system was soaked in air at 79 °C for approximately 30 minutes in order for the deformation to stabilize, allowing for creep effects. This completed the first step (pre-strain) of the programming. In step 2 (cooling), the heating was first stopped and the system was allowed to cool naturally to room temperature while maintaining the stress level. The LVDT tracked the movement of the specimen during this cooling step. Once the room temperature was reached, the specimen was unloaded (step 3). This completed the three-step stress-controlled thermo-mechanical programming.

3.3.1.2 Free Shape Recovery

Several methods can be used to evaluate the shape memory behavior of programmed SMPs and their composites. Typically, three methods have been used. One is free shape recovery, that is, the specimen is free to move without any constraints. As discussed in Equation (3.2), the shape recovery ratio is determined by this test result. Fully constrained shape recovery, that is, the specimen is not allowed to move in the recovery direction, or zero strain, leads to an accumulation of recovery stress. This test is usually used to evaluate the capability for the SMPs to serve as actuators or in closing wide opened cracks in biomimetic self-healing systems [41,42]. The third method is partially constrained shape recovery, in which the specimen is subjected to a constant external load or constant stress during shape recovery. This test result is more useful to evaluate the capability of serving as an actuator or self-healing mechanism because partial constraints are more realistic in these applications. Therefore, three different external constraints were used in this study: fully constrained, partially constrained, and free. Due to limitations of the MTS machine used, partially constrained shape recovery was not determined through the thermomechanical recovery test; rather, it was conducted through direct observation under a scanning electron microscope (SEM).

Typical three-step stress-controlled programming at stress levels of 263 kPa and 47 kPa and free shape recovery for both the pure SMP and the SMP based syntactic foam are respectively shown in Figures 3.15 and 3.16. The programming consists of a high temperature loading (step 1), followed by cooling (step 2), during which time the stress is held constant, and unloading at room temperature (step 3). Step 4 is a free recovery step whereby the programmed specimen is reheated to 79 °C and soaked in air for about 30 minutes. These four steps complete a thermomechanical cycle. In step 1, the strain initially increases rapidly upon loading. Because the total load was applied to the specimen instantly at 79 °C and held, step 1 is represented by a stepped curve instead of a smooth curve, as described in Figure 3.12 (a). The applied pre-stress in step 1 is held for about 30 minutes at temperatures above T_g, which allows creep to develop. This is why the compressive strain at the end of step 1 is higher than the strain at the same stress level from the isothermal compressive stress–strain curve shown in Figure 3.11, which has no creep deformation. In step 2, there is also a gradual increase in strain caused by thermal

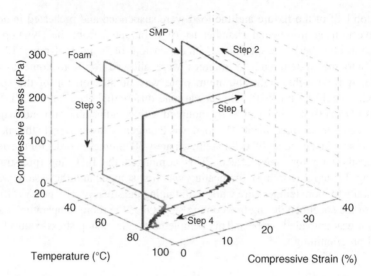

Figure 3.15 Four-step thermomechanical cycles (step 1: high temperature loading → step 2: cooling → step 3: room temperature unloading → step 4: free shape recovery) for the pure SMP and syntactic foam programmed under a stress-controlled condition with a pre-stress of 263 kPa at 79 °C followed by free recovery. *Source:* [41] Reproduced with permission from Elsevier

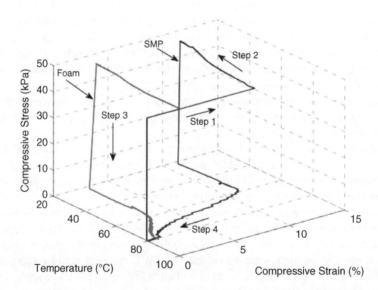

Figure 3.16 Four-step thermomechanical cycles (step 1: high temperature loading → step 2: cooling → step 3: room temperature unloading → step 4: free shape recovery) for the pure SMP and syntactic foam programmed under a stress-controlled condition with a pre-stress of 47 kPa at 79 °C followed by free recovery. *Source:* [41] Reproduced with permission from Elsevier

Table 3.1 Shape fixity and recovery ratios for the pure SMP and SMP based syntactic foam under stress-controlled programming and free shape recovery at 79 °C

Material	Pre-stress (kPa)	Pre-strain (%)	Fixity ratio (%)	Recovery ratio (%)
Pure SMP	47	11	98.5 ± 3.7	98.7 ± 3.8
	263	30	99.6 ± 1.4	98.4 ± 1.4
Foam	47	3	81.2 ± 11.8	87.6 ± 11.7
	263	12	96.7 ± 3.5	83.9 ± 3.0

Source: [41] Reproduced with permission from Elsevier.

contraction, viscoelasticity (creep) in the T_g region, and viscoplasticity below the T_g region. It is interesting to note that in the foam, there may be a thin layer of SMP that has been absorbed by the glass microballoons. Because of the reduced mobility of the SMP within this layer, the overall T_g of the foam is increased, as shown in Figure 3.5. Therefore, the actual T_g within this layer may be well above the average T_g shown because of its small volume. A similar behavior was found in SMP composites [60]. In step 3, the applied load is removed, leading to springback of the specimen or a reduction in the compressive strain. The strain maintained at the end of step 3 is the permanent strain stored due to programming. In step 4, which is a free recovery step driven by conformational entropy, most of the stored strain is released. From Figures 3.15 and 3.16 it is seen that the majority of the strain recovery occurs within the glass transition region in which the switching phase has been activated.

Shape fixity and shape recovery ratios were calculated based on the thermomechanical cycles of Equations (3.1) and (3.2), respectively. The results are summarized in Table 3.1. The shape fixity for both the foam and the pure SMP is close to 100%. This is due to the small stress applied during the programming and high stiffness of the material at room temperature when the stress is removed, leading to a very small springback. The 263 kPa pre-stress makes the specimen denser and stiffer than the 47 kPa pre-stress, leading to a smaller springback and higher shape fixity. For the shape recovery, the pure SMP is still able to recover almost all the stored strain. The shape recovery of the foam, however, has been reduced due to unrecoverable viscoplastic deformation below the T_g region. It may also include viscoplastic deformation within the T_g region for the interfacial transition zone confined by the glass microballoons. Also, the incorporation of microballoons may provide additional local intermolecular resistance to segmental rotation, and thus higher unrecoverable viscoplastic deformation, leading to a lower shape recovery than the pure SMP. Furthermore, some microballoons in the foam may have been crushed or damaged, and cannot be recovered.

In order to understand the rate of change of strain during the thermomechanical cycle better, changes of strain with time are plotted in Figure 3.17 at a pre-stress of 263 kPa and in Figure 3.18 at a pre-stress of 47 kPa. In step 1, the strain develops rapidly upon loading, gradually tapering off as creep progresses. The creep with time in step 1 is highlighted in the subplot in Figure 3.17. In step 2, the change of strain with time can be divided into two regions. For the region around T_g, the material continuously experiences creep, as represented by the curve with varying curvature. Below T_g, the material is glassy and the change of strain is primarily due to thermal contraction, as evidenced by an almost straight line with a constant slope. There is an increase in compressive stain of a few percent due to creep, viscoplastic deformation, and thermal contractions during cooling. In step 3, the unloading occurs

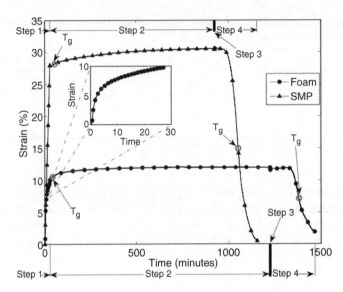

Figure 3.17 Strain versus time representation of the stress-controlled programming performed at 263 kPa followed by free recovery. The creep effect in the foam during step 1 is shown in the subplot. *Source:* [41] Reproduced with permission from Elsevier

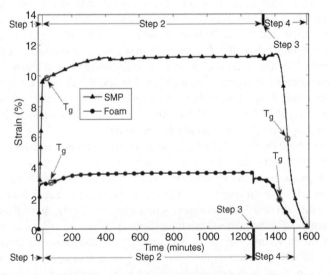

Figure 3.18 Strain versus time representation of the stress-controlled programming performed at 47 kPa followed by free recovery. *Source:* [41] Reproduced with permission from Elsevier

suddenly. This step is actually a point in Figures 3.17 and 3.18. Again, due to the relatively small stress applied and the large stiffness of the materials at room temperature, the springback is very small. In step 4, the strain recovery occurs primarily in the glass transition region.

3.3.1.3 Fully Constrained Recovery

Fully constrained recovery was performed using the MTS Q-TEST 150 machine and the associated furnace. Once the specimens were programmed under stress-controlled conditions, as described above, they were placed in the fixture shown in Figure 3.9 such that the strain was fixed and the stress was initially zero. Heating was performed at an average rate of 0.3 °C/min from room temperature until 79 °C and then held for approximately 20 minutes (some specimens were held for over 24 hours in order to investigate the stress relaxation behavior). The load cell of the MTS machine was used to record the recovered force as a function of time and temperature.

Since the programming is the same for the free recovery specimens and fully constrained specimens, the focus will be on step 4 of the thermomechanical cycles. The stress–temperature behavior under a fully constrained recovery condition is shown in Figure 3.19 for the two programming stresses (47 kPa and 263 kPa). The recovery stress–time behavior of the foam programmed at 47 kPa pre-stress is also highlighted by the inset in Figure 3.19. The recovery stress comes from two parts: thermal expansion stress and entropically stored stress or back stress. Since this is a 1-D fully constrained recovery, the thermal stress can be calculated as

$$\sigma = E\alpha\Delta T \tag{3.4}$$

where σ is the thermal stress, E is the average modulus over the temperature range, α is the average coefficient of thermal expansion, and ΔT is the temperature change.

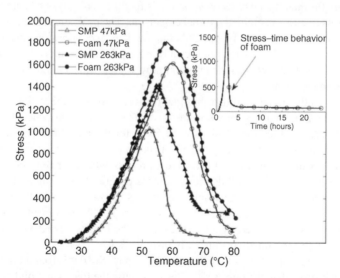

Figure 3.19 Recovery under a fully confined condition. The recovered stress for the pure SMP and foam after stress-controlled programming with two pre-stresses is shown. The inset shows the long-term stress relaxation of the foam programmed by a 47 kPa compressive stress. *Source:* [41] Reproduced with permission from Elsevier

When the temperature is below T_g, the modulus is high. As the temperature rises, the stress increases until the temperature reaches the start of the glass transition region. Here, the stress shows a peak. Once the temperature is within the T_g region, several competing factors take effect to control the recorded stress. First, the coefficient of thermal expansion may slightly decrease within the glass transition zone, as shown in Figure 3.8, which reduces the stress. Second, the SMP and the SMP based syntactic foam recover. Due to the constraint, however, this recovery (growth in the specimen length) is not allowed, leading to an accumulation of stress, which should increase the stress. Third, the E decreases sharply; for instance, as estimated from Figures 3.10 and 3.11, the modulus is about 600 MPa below T_g and about 10 MPa at 79 °C for the SMP, and about 300 MPa below T_g and about 20 MPa at 79 °C for the SMP based syntactic foam, leading to a continuous reduction in stress. The competition of the three factors leads to a continuous reduction in the recorded stress because the sharp decrease in modulus is the dominating factor. Once the temperature approaches the programming temperature (79 °C), the stress shows some relaxation initially (see the stress–time behavior inset for the foam programmed with 47 kPa in Figure 3.19). However, for a prolonged time period of about 24 hours, the stress does not appreciably decrease further. The reason is that, when holding the temperature constant at 79 °C, the material becomes rubbery, and stress relaxes quickly to equilibrium, leading to a stabilized recovery stress (the horizontal line in the stress–time plot). It is noted that this stabilized stress is the recovery stress that can be used in applications. For example, in self-healing applications [42], a certain period of time is required for the thermoplastic particles to melt and to diffuse into the fractured matrix. On the contrary, the peak stress cannot be used because the melting temperature of the thermoplastic particles is higher than the temperature corresponding to the peak stress. Therefore, only the sustained stress can be utilized for self-healing. Because of this, both the peak recovery stress and the stabilized recovery stress should be reported. Another reason that the stabilized recovery stress is more meaningful than the peak recovery stress is that the peak stress is mainly a buildup of the thermal stress due to constrained thermal expansion. The peak stress does not represent the shape memory capability of the polymer. Only the stabilized recovery stress in the rubbery state is a representation of the back stress stored during programming. Indeed, from Figure 3.19, the peak stress is much higher than the programming stress, which cannot be explained by the shape memory mechanism.

From the test results, it is found that the final recovery stress is 258.9 kPa for the SMP and 221.8 kPa for the foam, for a programming stress of 263 kPa. However, when the programming stress is 47 kPa, the final recovered stress is 51.5 kPa for the pure SMP and 69.5 kPa for the foam (higher than the programming stress). This abnormal behavior may be due to the expansion of the fixture during the shape recovery process, which applied a compressive stress to the specimen during heating. While the specimens programmed using a 263 kPa pre-stress experienced similar conditions, the results are not as sensitive as the 47 kPa programmed specimens because the 263 kPa pre-stress produced a strain in the specimens much higher than the small additional strain caused by the expansion of the fixture. Therefore, the effect on the 263 kPa pre-stress programmed specimens is insignificant.

Comparing the recovery stress of the foam with the pure SMP it is found that: (1) The foam has a higher peak stress than the pure SMP. This is because the foam has a higher modulus than the SMP at higher temperatures. (2) The temperature corresponding to the peak stress is higher for the foam than for the SMP. This is in agreement with the T_g test results. (3) The peak stress in specimens programmed by the 263 kPa pre-stress is higher than that in specimens programmed by the 47 kPa pre-stress. This is because the 263 kPa pre-stress makes the specimens denser and stiffer, producing a quicker accumulation of thermal stress during heating and ultimately a higher peak stress.

3.3.1.4 Partially Constrained Recovery

Stress-controlled (partially constrained) recovery was conducted using the setup shown in Figure 3.14, but without the LVDT. A specimen was first programmed under stress-controlled conditions. Then, its cross-section was exposed and an artificial crack was generated in a direction perpendicular to the programming axis using a sharp blade. An SEM (JEOL JSM – 6390) was used to catalog this damage. The specimen was then returned to the heating chamber and a constant uniaxial compressive stress equivalent (in both magnitude and direction) to that used during programming was applied. The temperature was set to 79 °C and the specimen was allowed to soak in air for 3 hours while maintaining the load. It was then slowly cooled to room temperature and unloaded. The cross-section of the specimen was again imaged under the SEM in order to evaluate the crack closing effect. SEM observations of the foam before and after recovery are presented in Figure 3.20 (a) and (b), respectively. During recovery there was an external stress constantly applied along the programming axis, which counteracts the internal entropically stored stress. Therefore, the specimen cannot recover to its original permanent shape. Instead, due to the external constraint, the recovered stress pushes the cracked surfaces into the empty space left by the artificial crack. This shows effective closing of the artificial crack, suggesting that partially constrained shape recovery of the SMP based syntactic foam is a way to close structural-scale damage. It is anticipated that a 3-D fully confined recovery condition would be even more effective at closing cracks regardless of their fracture modes or orientations since the 3-D confinement would provide a more stable frame for the recovered stress to act against.

3.3.1.5 Summary

In summary, thermosetting SMP and its syntactic foam are a type of material that can be taught or trained. Its shape recovery functionality depends on the programming and method of recovery used. For free recovery, higher programming stress leads to a higher shape fixity but a lower shape recovery ratio, suggesting that the pre-stress or pre-strain level must be within a certain limit during programming in order to maximize the shape recovery functionality of the SMP and the SMP based syntactic foam. Although the peak stress during the constrained recovery is determined by thermal expansion and is very high compared to the programming stress, the final stress recovered is dependent on the entropically stored stress and is close to the stress used for the stress-controlled programming. It is suggested that the stabilized recovery stress should be used to evaluate the shape memory capability of SMPs and their composites rather than the peak recovery stress. As compared to the pure SMP, the SMP based syntactic foam shifts the T_g upwards, possibly due to the existence of an interfacial transition zone with reduced molecular mobility, and increases the stiffness at higher temperatures. Even at a temperature of 79 °C, which is slightly above the T_g, the SMP and foam show a certain viscoelasticity, as evidenced by the creep during programming and stress relaxation during the constrained shape recovery. Viscoplasticity may also be present due to the higher T_g in the interfacial transition zone.

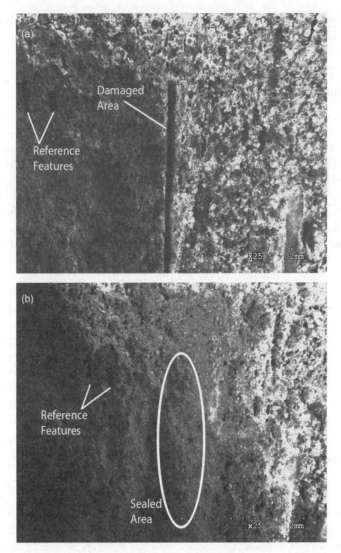

Figure 3.20 (a) SEM image of the cross-section of a stress-controlled programmed specimen after damage has been generated. Programming was performed in a direction perpendicular to the crack. (b) The same region after a partially constrained recovery showing the crack closing effect. Recovery occurred along the programming axis. Reference features already present in the specimen were chosen to aid in locating the damaged region. *Source:* [41] Reproduced with permission from Elsevier

3.3.2 Programming Using the 2-D Stress Condition and Free Shape Recovery

3.3.2.1 Biaxial Specimen and 2-D Programming

In order to create a 2-D stress condition in a uniaxial loading device (MTS Q-TEST 150 gear driven machine) special specimens were prepared. The specimens were machined into biaxial

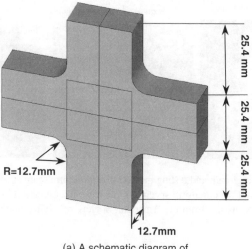

(a) A schematic diagram of
biaxial specimen dimensions

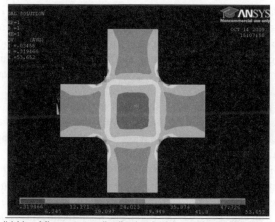

(b) Von Mises stress distribution in the specimen (there is a
piece of square material (dark color) at the center of the
specimen (about 25.4 mm by 25.4 mm) that is
subjected to a uniform 2-D stress condition)

Figure 3.21 Cruciform specimens for 2-D programming. *Source:* [57] Reproduced with permission from the American Society of Civil Engineers

cruciform specimens for programming and further testing. The cruciform specimen dimensions are given in Figure 3.21 (a). With such dimensions, the central square is roughly under a 2-D stress condition as validated by finite element analysis (see Figure 3.21 (b)).

After machining the cruciform foam specimen, two small steel tabs were fixed by using strong glue along the edge of the central square (the uniform 2-D stress condition as shown in Figure 3.21 (b)) on to one side surface of the specimen (one in the vertical and the other in the horizontal direction) to measure the displacement of the two directions (see Figure 3.22 (a)).

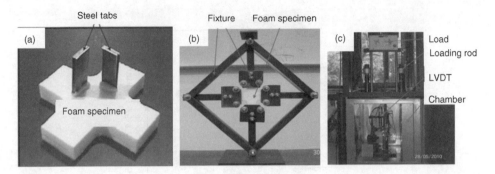

Figure 3.22 Programming setup and testing method of a cruciform foam specimen: (a) foam specimen attached with steel tabs and (b) cruciform specimen in a truss fixture, and (c) the test chamber. *Source:* [57] Reproduced with permission from the American Society of Civil Engineers

Then the cruciform specimen was introduced into a truss-form fixture and the bolts were tightened to grip the four arms of the specimen, with each arm length being 12.7 mm (as shown in Figure 3.22 (b)). The truss fixture can transform 1-D compression into 2-D loading (vertical compression and horizontal tension) by applying a compressive load directly to the top and bottom nodes of the truss fixture. The ratio of the vertical load and horizontal load can be adjusted by changing the length of the members of the truss fixture. Finally, the temperature was monitored with the help of thermocouples connected to a Yokagawa DC100 data acquisition system.

According to the previous study [41], the glass transition region of the foam is 63.69–76.19 °C with a glass transition temperature of $T_g = 70.50$ °C. In order to perform efficient programming and shape recovery tests, 79 °C was chosen as the testing temperature since it was just above the upper limit of the T_g region of the foam.

The device was put into a chamber, heated to 79 °C, held at 79 °C for about 25 minutes, and then the bolts were re-tightened to clamp the specimen tighter because it became soft at this temperature. After re-tightening, the setup was placed back into the chamber and two linear variable differential transducers (LVDTs) (Coopers Instrument ADT 200) were installed in conjunction with the two pre-attached tabs in the vertical and horizontal directions, respectively (see Figure 3.22 (c)). Also, the weight-bearing rod was installed on the top node of the truss fixture and the system was reheated to 79 °C for another 20 minutes to equilibrate the temperature. Immediately afterwards a weight was put on to the rod by hand. This compressive load led the specimen to compression in the vertical direction and tension in the horizontal direction. Approximately 20 min were allowed for the deformation to stabilize. This completed step 1 (pre-strain) of the programming. In the next step (cooling), the heating was first stopped, the chamber was kept closed, and the system was allowed to cool naturally for 8–10 h to room temperature while maintaining the applied load constant. The LVDTs tracked the deformation of the specimen during this cooling step. Because of creep and contraction during cooling, the strain was continuously increasing until the temperature reached room temperature. This completed step 2 of the programming. Then the specimen was unloaded (step 3). The entire process completed the three-step 2-D thermomechanical programming or shape fixity of the cruciform foam specimen.

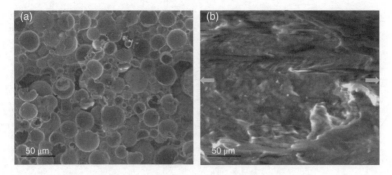

Figure 3.23 SEM pictures of (a) nonprogrammed and (b) 2-D programmed foam. *Source:* [57] Reproduced with permission from the American Society of Civil Engineers

In order to evaluate the effect of the pre-stress level on the thermomechanical behavior, four compressive pre-stress levels in the vertical direction were used. They were 168.3 kPa, 207.7 kPa, 247.6 kPa, and 300.7 kPa, respectively. Because of the symmetry of the truss fixture (the same length for each member) and the specimen, the tensile pre-stress in the horizontal direction was roughly the same as the corresponding vertical compressive pre-stress. In addition, some specimens with 300.7 kPa pre-stress experienced 10 thermomechanical cycles to evaluate the functional stability of the foam.

It is noted that 2-D programming changed the morphology of the foam. In addition to the likely molecular alignment along the loading directions, which cannot be visualized by direct observation, it seemed that the foam had been densified and the polymer flowed along the tension direction (see an SEM image in Figure 3.23). This change in morphology suggests that the foam transforms from isotropic to anisotropic [48].

3.3.2.2 Two-Dimensional Shape Recovery of Programmed Biaxial Specimens

Two-dimensional free shape recovery started immediately after completion of the programming by using the same heating chamber and LVDT system. According to a previous study [41], the chamber was heated up quickly from room temperature to 49 °C and allowed to soak in air for 20 minutes. Then the heating was continued again at an average heating rate of 0.3 °C/min until 77 °C. After soaking in air for 7 minutes, the heating was ramped quickly to 79 °C and the temperature was held for 30 minutes. This completed the 2-D free shape recovery.

Figure 3.24 shows the typical strain evolution during the entire thermomechanical cycle subjected to various pre-stress levels. Similar to previous 1-D stress controlled programming and free shape recovery [41], each direction (compression in the vertical direction and tension in the horizontal direction) shows a similar four-step thermomechanical behavior. From Figure 3.24, the thermomechanical cycle includes four steps: step1 (loading), step 2 (cooling while holding the load constant), step 3 (unloading), and step 4 (recovery). The small spikes are due to manual control of the testing setup.

Based on Figure 3.24 (a) and (b), the shape fixity ratio and shape recovery ratio were obtained. According to Equations (3.1) and (3.2), the shape fixity ratios are 95.6%, 96.7%, 98.6%, and 99.2% for the four prestress levels (168.3 kPa, 207.7 kPa, 247.6 kPa, and

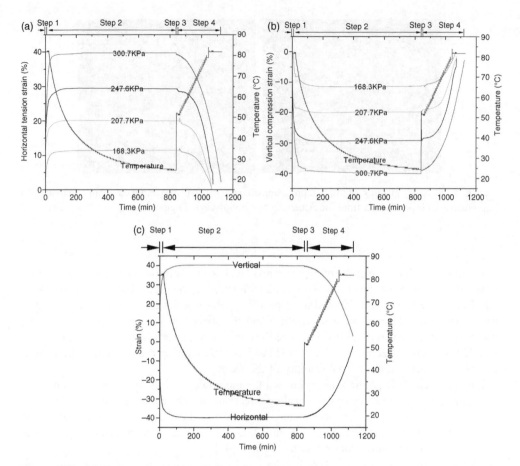

Figure 3.24 (a) Horizontal and (b) vertical strains of the cruciform sample under various stress levels during programming and recovery (steps 1 to 3 represent programming and step 4 represents free shape recovery) and (c) horizontal and vertical strains of the tenth thermo-mechanical cycle under the pre-stress level of 300.7 kPa (Note: tensile stress is treated as negative and compressive stress is treated as positive in (c)). *Source:* [57] Reproduced with permission from the American Society of Civil Engineers

300.7 kPa), respectively. The shape recovery ratios are 97.3%, 95.4%, 94.1%, and 91.6%, respectively. It is clear that as the pre-stress level increases, the shape recovery ratio decreases. This is understandable because a higher pre-stress may cause some damage to the foam (such as crushing of microspheres), which is not recoverable during shape recovery. For the shape fixity, with the increase in the pre-stress, the foam specimen becomes denser and stiffer, resulting in a smaller springback and higher shape fixity ratio.

Additionally, as shown in Figure 3.24 (c), the horizontal and vertical strains after 10 thermomechanical cycles are close to the strain evolution of the first thermomechanical cycle. The shape fixity ratio is 98.5% and the shape recovery ratio is 88.3%. It is noted that both the shape fixity ratio and shape recovery ratio are slightly lower than those in the first thermomechanical cycle (99.2% and 91.6%, respectively) under the same pre-stress level (300.7 kPa). This is because more unrecoverable damages have accumulated during each

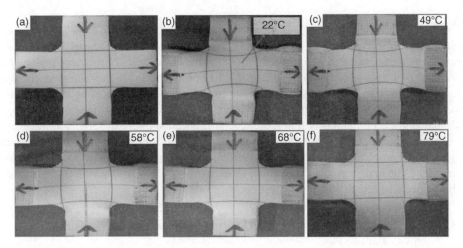

Figure 3.25 (a) Original and (b) programmed cruciform foam specimen and (c) to (f) its recovery process at various temperatures. *Source:* [57] Reproduced with permission from the American Society of Civil Engineers

additional thermomechanical cycle, leading to a decrease in the shape recovery ratio. Also the density and elasticity of the foam increases as the thermomechanical cycle increases, resulting in an increase in springback under the same pre-stress of 300.70 kPa and leading to a decrease in the shape fixity ratio. However, this change in the shape recovery ratio and shape fixity ratio is comparatively small. Therefore, the shape memory functionality of this foam is, generally speaking, stable.

To visualize the programming and recovery process, Figure 3.25 shows the original and programmed cruciform foam specimen and its recovery process under various temperatures. During step 1 and step 2 of the programming, the 2-D stress condition causes the orientation of the macromolecular chains of the SMP matrix to align along both vertical and horizontal directions. This leads to a decrease in entropy of the material system in the programming process. Meanwhile, some microspheres may be bent or buckled, and the deformed microspheres or crushed debris may align in the tension orientation due to compression in the transverse direction. Furthermore, a wrinkle-like pattern can be seen with the naked eye on the stretched specimen surface along both directions, as shown in Figure 3.25 (b). This reflects the 2-D stress conditions. It is believed that the central square of the cruciform foam specimen is changed from isotropy to anisotropy, as demonstrated by the negative Poisson's ratio after 2-D programming [48]. From Figure 3.25 (c) to (f), the recovery rate in horizontal and vertical directions are faster when the temperature is up to T_g, which is consistent with the strain evolution shown in Figure 3.24. Additionally, the recovery process occurs in both directions simultaneously for the following reasons. (1) The two directions were programmed simultaneously. As will be shown later in hybrid programming, the programming sequence has a certain effect on the recovery sequence. (2) The two pre-strain levels in the two directions were similar due to symmetry of the truss programming fixture. Again, it will be shown later that the pre-strain level plays a major role in determining the recovery sequence.

3.3.2.3 Summary

In summary, by the special 2-D training using cruciform specimens, the SMP based syntactic foam demonstrates shape memory in both programming directions. The recovery is along both directions simultaneously. It is found that the 2-D programming makes the syntactic foam transform from isotropy to anisotropy and the foam shows a negative Poisson ratio under a certain strain level. Because of the shape recovery behavior and the negative Poisson ratio, the SMP based syntactic foam may find some special applications, such as serving as a sealant in an expansion joint of a bridge deck or concrete pavement, which can overcome the sealant squeezing-out problem and concrete wall crushing problem when the bridge deck or pavement slab expand due to rising temperatures. The shape memory functionality of the foam is largely stable under repeated thermomechanical cycles.

3.3.3 Programming Using the 3-D Stress Condition and Constrained Shape Recovery

3.3.3.1 Three-Dimensional Confined Programming

The same thermosetting polystyrene SMP was used in this study. In addition to the 40% by volume of glass microballoons, 0.15% by volume of multiwalled carbon nanotubes was also added, which was the same as in Reference [38]. The multiwalled carbon nanotubes (Cheap Tubes Inc.) have a density of 2.1 g/cm^3, diameter of 20–30 nm, and length of 20–30 μm. Similar to Reference [38], a two-step procedure was used to prepare the syntactic foam. First, the carbon nanotubes were added to Part A of the polymer. The mixture was mixed with the assistance of an ultrasound mixer for 30 minutes at a frequency of 20 kHz (Sonics Vibracell VC 750W) and a three-roll mill for one pass (NETZSCH type 50) (see Figure 3.26). Second, microballoons and hardener (Part B of the polymer) were added to the carbon nanotube/Part A mixture and mixed with a spatula for 15 minutes. It was then poured into an aluminum mold of

Figure 3.26   Mixing of Part A with carbon nanotubes by a three-toll mill

304.8 mm × 304.8 mm × 12.7 mm for curing. The process started with 24 hours of room temperature curing, followed by post-curing in an oven at 75 °C for 24 hours, 90 °C for 3 hours, and 100 °C for 9 hours. The curing at 100 °C for 9 hours instead of 3 hours was based on the findings in Reference [38] to eliminate the post-curing effect. After curing, the SMP foam slab was brought to a drilling machine for machining. The drilling direction was perpendicular to the thickness direction of the panel. The specimens were cylinders with a diameter of 12.7 mm and a height of 25.4 mm.

Through testing of fiber reinforced polymer tube encased concrete cylinders, it has been demonstrated that uniaxial compression of the concrete core produces a 3-D compressive stress condition in the concrete core, and lateral confinement can be controlled by adjusting the confining tubes [61–67]. In this study, double-walled confining tubes were used. By adjusting the materials used for the inner tube or liner, varying lateral confinements were created. A systematic test program including a combination of two programming temperatures, three pre-strain magnitudes, three lateral confinement levels, and one recovery condition was implemented.

The confining tubes were double-walled tubes. The outer tube was made of medium carbon steel with a height of 25.4 mm, inner diameter of 25.4 mm, and wall thickness of 12.7 mm, so that it can serve as a "rigid" confining tube. Three types of inner tubes (or liners) were used. All of them had the same dimension: height of 25.4 mm, inner diameter of 12.7 mm, and outer diameter of 25.4 mm. Such a dimension ensures that the outer tube and the inner tube make a perfect double-walled tube. The difference between the three types of inner tubes was the materials used. The first one was medium carbon steel to provide "rigid" confinement; the second was nylon to provide "intermediate" confinement; and the third was rubber to provide "weak" confinement. The mechanical properties of the three types of liner materials are summarized in Table 3.2 and the double-walled confining tubes are shown in Figure 3.27 (a).

Strain-controlled programming was conducted. The advantage of using strain controlled programming instead of stress-controlled programming is that the geometry or dimensional stability of the specimen is maintained, which is critical for actual structures. In order to investigate the effect of programming temperature on the thermomechanical behavior of the foam, two temperatures, 71 °C and 79 °C, were used. Based on the DSC test results, the glass transition temperature (T_g) of the foam was about 62 °C [38], which is slightly lower than the foam without carbon nanotubes [41]. The reason may be: (1) the existence of carbon nanotubes makes the foam have a higher thermal conductivity, which may be easier for the specimen to come to equilibrium, that is, without overheating, and (2) the carbon nanotubes may hinder the polymerization process, leading to a lowered T_g. Therefore, 79 °C was well above the T_g; 71 °C was also used for the purposes of evaluating whether a programming temperature that was slightly above or close to the glass transition region could fix the shape and recover the stress or not. From the room temperature compressive stress–strain test, it is known that the foam

Table 3.2 Material properties of the confining tubes

Materials	Coefficient of thermal expansion (/°C)	Young's modulus (GPa)	Poisson's ratio
Medium carbon steel	1.08×10^{-5}	210.0	0.2
Nylon	8.20×10^{-5}	3.6	0.35
Rubber	6.90×10^{-4}	0.05	0.50

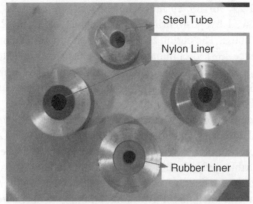

(a) Double-walled confining tubes

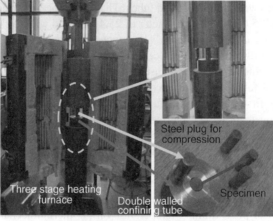

(b) Test setup [42] (reproduced with
permission from Elsevier)

Figure 3.27 Experimental setup

exhibited three regions: elastic region (strain between 0 and 7%), densification region (strain between 7 and 50%), and solidification region (strain above 50%) [38]. In order to evaluate the effect of the three regions on the thermomechanical performance, three pre-strain levels, corresponding to these three regions, were selected. They were 5%, 30%, and 60%. It is noted that the 60% compressive strain is extremely large for most polymers. For the foam, the deformation comes from two components, the SMP matrix and the microballoon. In order to investigate the maximum volume reduction of the SMP matrix without damaging the shape memory functionality, neat SMP specimens were compressed to failure under a "rigid" confinement condition at 79 °C. It was found that the maximum volume reduction of the SMP matrix was 17.0% at a failure axial strain of 18.3%.

For each pre-strain level and each temperature, the foam cylinder was first inserted into the double-walled confining tube. After that, the confined specimen was put into a temperature-controlled chamber (ATS heating chamber) in a gear-driven MTS Q-TEST 150 machine.

The temperature was gradually raised at a rate of 1 °C/min until the designed temperature (71 °C or 79 °C) was reached. Then the temperature was maintained for 30 minutes to achieve uniformity in the cylinder. After that, the programming started. It consisted of two steps. In step 1 (pre-stressing) the loading rig was brought into contact with a cylindrical steel plug (with a diameter of 12.7 mm and height of 25.4 mm) and started to load the specimen at a rate of 1.3 mm/min. Once the designed pre-strain level (5%, 30%, or 60%) was reached, the loading was stopped and the pre-strain was maintained constant for 20 minutes. In step 2 (cooling and unloading) the chamber was kept closed and natural cooling occurred. Depending on the pre-strain level and starting temperature, the cooling process took up to 6 hours. Because of stress relaxation and contraction during cooling, the stress was continuously reduced to zero, suggesting separation of the specimen from the loading rig. Once the stress became zero, the programming process or shape fixity process was completed. Figure 3.27 (b) shows the setup of the experiment. It is noted that in stress-controlled compression programming [41], the programming took three steps, that is, pre-stressing, cooling, and unloading. For strain controlled programming, due to the coupling between thermal contraction and stress relaxation, the stress reduced to zero during the cooling process. In other words, the cooling and unloading are coupled.

3.3.3.2 Confined Shape Recovery

A strain-controlled or fully confined recovery test was conducted. It started immediately after completion of the programming with the same loading device and environmental chamber. The chamber was heated at a ramp rate of 0.6 °C/min. The heating was continued until the temperature starting the programming (71 °C or 79 °C) was reached. After that, the temperature was maintained constant for several hours and the stress was continuously recorded. This process was stopped when further stress recovery was negligible.

The flowchart of the experimental design is schematically shown in Figure 3.28, which consists of 18 combinations (two temperatures (71 °C and 79 °C) × three inner tube materials

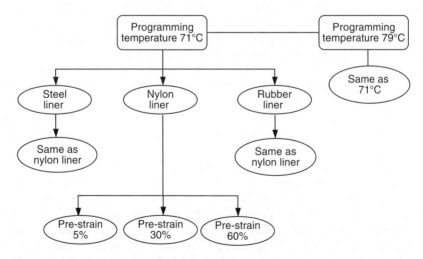

Figure 3.28 Flowchart of the experimental design. *Source:* [42] Reproduced with permission from Elsevier

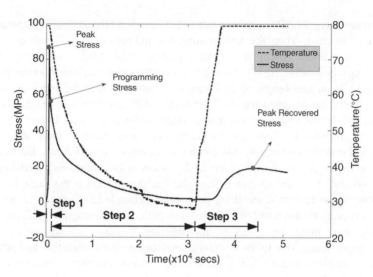

Figure 3.29 Axial stress–time and temperature–time behavior of the SMP foam with a nylon liner at a programming temperature of 79 °C and pre-strain level of 60%. The three steps (step 1: pre-stressing, step 2: cooling and unloading, and step 3: stress recovery) are shown by the three regions and the peak stress, programming stress, and peak recovered stress are indicated using black dots. *Source:* [42] Reproduced with permission from Elsevier

(steel, nylon, and rubber) × three pre-strain levels (5%, 30%, and 60%)). At least three effective specimens, that is, specimens that did not have any machine or procedural problems, were tested for each combination. The total number of effective specimens in this study was 54.

In order to understand the thermomechanical cycle of the syntactic foam under different test conditions better, the test results are presented in both 3-D and 2-D format. Typical 2-D axial stress–time and temperature–time curves for the foam confined by the nylon liner, programmed at 79 °C, and under 60% pre-strain level, and fully confined shape recovery are shown in Figure 3.29. Typical 3-D axial stress–axial strain–temperature thermomechanical cycles for the syntactic foam at a programming temperature of 71 °C, pre-strain level of 30%, and fully confined shape recovery are shown in Figure 3.30. Typical 3-D axial stress–axial strain–time behaviors at a programming temperature of 79 °C, pre-strain level of 30%, and fully confined shape recovery are shown in Figure 3.31.

Figure 3.29 gives one perspective of analyzing the thermomechanical behavior of the SMP based syntactic foam. Here we can observe step 1 loading, step 2 cooling and unloading, and step 3 stress recovery with respect to time. As shown in Figure 3.29, the programming stress, which corresponds to the start of the temperature drop, is highlighted using a red dot. From Figure 3.29, the stress experiences a significant change in step 1. The stress starts from zero until the programming strain is achieved, which leads to the peak stress. Because the machine needs to be stopped in a very short time period, this leads to an instant drop in the peak stress. As the holding time period proceeds, the stress is further reduced due to stress relaxation. In step 2, the stress still shows a certain relaxation when the temperature is within the T_g region. When the temperature drops below T_g, the specimen further shortens due to thermal contraction, leading to an almost linear curve. There were two competing tendencies in the specimen.

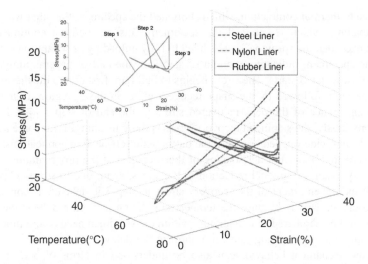

Figure 3.30 Axial stress–temperature–axial strain thermomechanical cycle at a programming temperature of 71 °C and pre-strain level of 30%. The subplot shows the three-step thermal mechanical cycle of a specimen confined by a steel liner (step 1 (pre-stressing) and step 2 (cooling and unloading) represent programming and step 3 represents stress recovery). *Source:* [42] Reproduced with permission from Elsevier

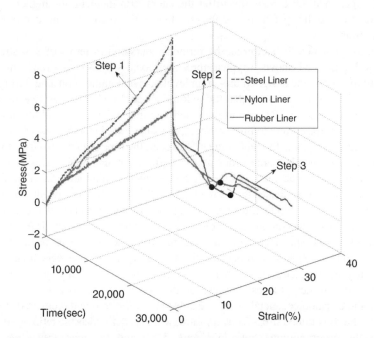

Figure 3.31 Axial stress–time–axial strain representation of the thermomechanical cycle of the foam at a programming temperature of 79 °C and pre-strain level of 30%. The black dots indicate the end of step 2 and start of step 3 (step 1 (pre-stressing), step 2 (cooling and unloading), and step 3 (stress recovery)). *Source:* [42] Reproduced with permission from Elsevier

One was due to thermal contraction, which shortened the specimen; the other was springback due to unloading, which lengthened the specimen. This competition continued until the contraction exceeded the springback, which led to fully unloading or separation of the loading rig from the specimen, indicating the end of step 2 or the end of programming. When the temperature rises, step 3 starts. The stress begins to build up, first due to the thermal stress by thermal expansion, followed by the shape recovery effect in the T_g region. In this study, we define the appearance of the peak recovered stress as the end of step 3 [42]. However, as discussed previously, the stabilized recovery stress is more meaningful for some applications, such as in closing wide-opened cracks in the biomimetic self-healing applications, which will be the focus of this book. Further soaking of the specimen at the programming temperature leads to a very small but gradual reduction in the stress, again due to stress relaxation. It is noticed that the development of the recovery stress in step 3 is delayed as compared to the temperature curve. This is because the temperature shown in Figure 3.29 is the "nominal" temperature of the chamber. The temperature within the specimen needs more time to become uniform. Using a slower heating rate can minimize the delay.

The thermomechanical behavior can also be understood in terms of 3-D stress–strain–temperature curves. During step 1, it can be visualized from Figure 3.30 that the specimen that was confined by the steel liner experienced the maximum stress, followed by the nylon liner and rubber liner. This is due to rigid confinement, intermediate confinement, and weak confinement provided by the steel liner, nylon liner, and rubber liner, respectively. Transverse confinement resists the lateral expansion of the foam specimen, leading to lateral compressive stress to the specimen. Of course, the stiffer the lateral confinement, the higher is the lateral compressive stress to the specimen, and thus the higher the axial stress needed to achieve the same axial strain.

The thermomechanical behavior can be further understood in terms of 3-D stress–strain–time curves. Figure 3.31 shows a typical three-step axial stress–axial strain–time behavior. Once the strain reaches the designed pre-strain level, it is maintained constant during the entire thermomechanical cycle. It is seen that the nylon liner confined specimen reaches the end of step 2 (black dot in Figure 3.31) earlier than that with the rubber liner, and the last is the one with the steel liner. Based on the liner stiffness, the steel liner is the highest, followed by the nylon liner, and the softest is the rubber liner. However, the time required to come to the end of step 2 does not follow the same order as the stiffness of the liners. This is because the coefficient of thermal expansion (CTE) of the rubber liner is about ten times that of the nylon liner and is also larger than that of the foam ($1.72 \times 10^{-4}/{}^\circ\text{C}$ [43]). As a result, the rubber liner contracts more than the nylon liner during cooling and always contacts with the foam cylinder, which applies a transverse compressive stress to the specimen and thus tends to take a little bit longer to separate the specimen from the loading rig. The steel liner needs a significant contraction to unload the specimen because the programming stress is very high, resulting in the longest time to complete step 2. Therefore, the confinement efficiency of the three liners depends not only on their stiffness but also on their CTE.

Based on the thermomechancial cycle, the stress recovery ratio and recovered stress can be determined. The constrained stress recovery ratio (which is defined as the ratio of the recovered stress over the programming stress in Figure 3.29) and the recovered stress for each combination of test parameters are summarized in Table 3.3.

For the programming temperature of 71 °C, it is seen that as the pre-strain level increases, the confined stress recovery ratio decreases. This is because as the pre-strain level increases, the

Table 3.3 Stress recovery ratio and recovered stress

Liner type	Pre-strain level (%)	Programming temperature 71 °C		Programming temperature 79 °C	
		Stress recovery ratio (%)	Recovered stress (MPa)	Stress recovery ratio (%)	Recovered stress (MPa)
Steel	5	84.32 ± 10.18	0.37 ± 0.15	58.26 ± 2.1	0.53 ± 0.01
	30	62.97 ± 11.22	2.01 ± 0.34	83.45 ± 3.57	1.75 ± 0.29
	60	0.48 ± 0.12	3.52 ± 1.29	1.05 ± 0.14	6.33 ± 1.35
Nylon	5	71.17 ± 8.74	0.17 ± 0.05	55.5 ± 12.94	0.28 ± 0.08
	30	58.17 ± 7.40	1.75 ± 0.72	70.45 ± 5.2	1.31 ± 0.15
	60	46.74 ± 1.73	26.57 ± 7.27	48.19 ± 13.1	17.52 ± 1.36
Rubber	5	87.42 ± 2.50	0.43 ± 0.15	55.55 ± 2.78	0.34 ± 0.03
	30	76.75 ± 3.89	1.04 ± 0.09	73.96 ± 2.54	1.20 ± 0.08
	60	71.53 ± 0.45	4.50 ± 0.35	56.9 ± 5.64	1.74 ± 0.24

Source: [42] Reproduced with permission from Elsevier.

programming stress also increases, which may cause more microballoon crushing. Because the crushed microballoons cannot be recovered, this leads to a reduction in the stress recovery ratio. As the confinement increases, the stress recovery ratio depends on the pre-strain level. At 5% and 30% pre-strain, the stress recovery ratio is the highest for the rubber liner, followed by the steel liner, and the least is for the nylon liner. This may be due to the combined effect of the programming stress level and the permanent dimensions of the specimen at the programming temperature, which depends on the stiffness and CTE of the liner. At the pre-strain level of 60%, the stress recovery ratio is still the highest for the rubber liner; however, it is followed by the nylon liner, instead of the steel liner. This is caused by the unrecovered damage because the steel liner created the largest amount of unrecoverable damage.

For the programming temperature of 79 °C, the stress recovery ratio is the highest for 30% pre-strain, followed by 5% pre-strain, and the least is by 60% pre-strain. It seems abnormal that the 30% pre-strain leads to a higher stress recovery ratio than that of the 5% pre-strain. The reason may be that, at 5% pre-strain, the pre-stress is very small because the specimen is very soft at 79 °C. As a result, when the temperature drops, the test specimen is unloaded in a very short time period and at a relatively higher temperature. Therefore, the entropic strain does not have sufficient time to be fixed, leading to a smaller stress recovery ratio. For the 60% pre-strain, the lower recovery ratio is due to the increased unrecoverable damage or loss of shape memory functionality. For the three liners, basically the steel liner leads to a slightly higher recovery ratio at 5% and 30% pre-strain than the nylon liner and rubber liner. The reason for this may be due to the larger diameter of the specimens with the steel liner immediately before programming at 79 °C. The larger diameter immediately before programming suggests that the foam specimen has a larger volume or less stiffness. Therefore, the programming stress by the steel liner is reduced, which is close to that by the nylon and rubber liners. At 60% pre-strain, the recovery ratio is the highest with the rubber liner, followed by the nylon liner, and the least by the steel liner. Again, this is related to the amount of unrecoverable damage produced by the three liners.

The recovered stress depends on both the stress recovery ratio and the programming stress. For the programming stress, the higher the pre-strain level, the higher is the programming stress, and the stiffer the liner, the higher the programming stress. However, the stress recovery ratio does not follow such a simple tendency, as discussed above. Therefore, the recovered stress has a more complex relationship with the test parameters. From Table 3.3, it is seen that the recovered stress increases as the pre-strain level increases, regardless of the type of liner and programming temperature. This is understandable because as the pre-strain level increases, the programming stress also increases. For the 60% pre-strain level, although the stress recovery ratio is very small for the steel liner, the very large programming stress leads to a recovered stress that makes the steel liner maintain the same tendency. It is noted that the recovered stress is the highest with the nylon liner, in particular at the 60% pre-strain level (26 MPa). Therefore, it is concluded that the recovered stress needs a proper combination of stiffness and CTE of the confining device (the stiffness and CTE of the nylon liner are between those of the steel liner and the rubber liner). Because the recovered stress by the nylon liner is as high as 26 MPa, it is recommended that SMP may be used as actuators if proper 3-D confined programming is conducted. This is not insignificant because SMP has been cited as being unable to serve as an actuator due to its low recovered stress.

The change in morphology of the foam after programming was observed with an SEM. In this study, Hitachi S-3600N VP-Scanning Electron Microscope was used to examine the microstructure change due to programming by different pre-strain levels and free shape recovery. The programming temperature and shape recovery temperature of the samples are 79 °C with a steel liner. Comparing Figure 3.32 (a) and (b), it is seen that programming by 5% pre-strain slightly increases the density of the foam, without much damage to the micro-balloons. Comparing Figure 3.32 (a) and (c), it is seen that the microstructure is fully recovered

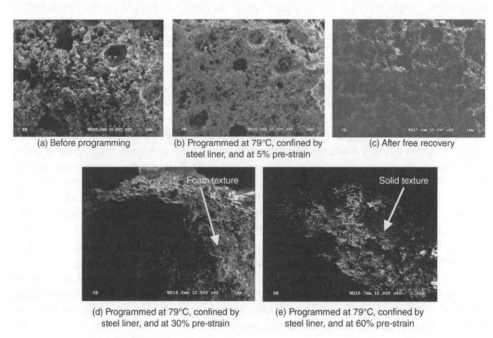

(a) Before programming   (b) Programmed at 79°C, confined by   (c) After free recovery
                                  steel liner, and at 5% pre-strain

(d) Programmed at 79°C, confined by   (e) Programmed at 79°C, confined by
steel liner, and at 30% pre-strain   steel liner, and at 60% pre-strain

Figure 3.32 SEM observation of the original microstructure of the foam (a), the effect of programming on changing the microstructure ((b), (d), and (e)), and the effect of shape memory on recovering the microstructure (c). *Source:* [42] Reproduced with permission from Elsevier

by free recovery, suggesting good shape memory functionality of the foam. Comparing Figure 3.32 (b) with Figure 3.32 (d), it is seen that densification of the foam occurs with microballoon crushing by the higher pre-strain level of 30%. At the 60% pre-strain level, Figure 3.32 (e), it is seen that the foam is solidified, with full crushing of all the microballoons. This supports the findings in the stress recovery test. The 60% pre-strain leads to significant damage and considerable loss of shape memory functionality.

3.3.3.3 Summary

In summary, under 3-D compressive confinement, strain-controlled programming can be represented typically by two steps, instead of three steps. The cooling and unloading are coupled into one step. The stress recovery ratio and recovered stress depend on the type of liner or the level of transverse confinement, the programming temperatures, and the pre-strain levels. For the parameters investigated in this study, the rubber liner leads to the highest stress recovery ratio and the nylon liner leads to the highest recovered stress. A higher pre-strain leads to a higher recovered stress, regardless of the liner types and programming temperatures. The foam programmed by the nylon liner has a potential to be used as a 1-D actuator due to its considerably higher stress (26 MPa) recovered. Because one weakness cited for SMPs and their composites is the lower recovery stress, which limits their applications as actuators, this study shows that 3-D confined programming is a way of increasing the recovery stress. This is because the 3-D confinement needs an appreciably higher energy input, which leads to a higher recovery stress as the energy output during shape recovery is positively correlated with the energy input during programming.

3.4 Thermomechanical Behavior of Thermosetting SMP and SMP Based Syntactic Foam Programmed by Cold Compression

3.4.1 Cold-Compression Programming of Thermosetting SMP

3.4.1.1 Programming by Isothermal Flat-wise Uniaxial Compression Test

For the thermosetting polystyrene SMP, cold compression programming was conducted. Based on the DMA test result (Figure 3.5), the T_g region of the SMP is between 60.7 and 73.0 °C. Therefore, cold programming suggests that the programming temperature should be below 60.7 °C. In this study, room temperature (20 °C) was adopted for programming. The programming was conducted by a uniaxial compression test. Uniaxial flat-wise compression was performed with an MTS QTEST150 electromechanical frame outfitted with a moveable furnace (ATS heating chamber) compliant with the ASTM C 365 standard. The loading rate was 1.3 mm/min all the way to the test pre-strain level. The specimen size was 30 mm × 30 mm × 12 mm. Temperature control and monitoring were achieved through a thermocouple (Omega XC-20-K-12) placed in the chamber near the SMP specimen, monitored by the Yokogawa DC 100 data acquisition system. Stress-strain responses were then generated for different pre-strain levels and stress relaxation times. In this study, three pre-strain levels (5%, 10%, and 30%), corresponding to the elastic zone (5%) and post-yielding zone (10% and 30%), respectively, were selected. The stress relaxation time was determined to be 0 min, 30 min, 120 min, and 260 min for the 5% pre-strain level, and 0 min, 5 min, 15 min, 30 min, and 120 min for the 10% and 30% pre-strain levels. At least three effective specimens were tested for each pre-strain level and stress relaxation time.

The strain evolution during the material programming process is presented in Figure 3.33. Obviously, the shape fixity greatly depends on the pre-strain levels. SMP specimens programmed by the 5% pre-strain level cannot fix a temporary shape regardless of the length of the stress relaxation time. Upon removal of the load, immediate full springback is observed. This is understandable because the temporary shape by cold programming depends on the pseudo-plastic deformation. Without yielding, no pseudo-plastic deformation can be created and a temporary shape cannot be fixed. For specimens programmed by the 30% pre-strain, however, a decent amount of strain is preserved, even when the load is instantly removed (zero relaxation time). With a zero stress relaxation time, the shape fixity ratio is still about 73%. Therefore, the pre-strain level plays a key role in programming at the glassy temperature. As documented in a previous study [41], the uniaxial compression yielding strain of the same thermosetting SMP is about 7% at the same glassy temperature. Obviously, 5% pre-strain falls in the elastic region of the SMP. Therefore, immediate full springback occurs regardless of the relaxation time held. At 30% pre-strain, the SMP specimen already yields and thus is able to maintain a decent temporary fixed strain even without stress relaxation. Therefore, a post-yielding pre-strain level determines the success of programming at the glassy temperature.

It can also be observed from Figure 3.33 that, with 30% pre-strain, a longer stress relaxation time in step 2 (Figure 3.13) tends to enhance the shape fixity ratio. As the relaxation time continuously increases, the shape fixity asymptotically approaches an upper bound, which is equal to the difference between the pre-strain and elastic springback (ratio of the relaxed stress over the relaxed modulus). A further increase in the relaxation time can hardly bring up any significant increase in the shape fixity ratio. This behavior can be understood from the point of view of stress relaxation because the mode of loading during cold-compression programming is stress relaxation (constant strain). For polymers under the stress relaxation condition, the relaxation is fastest when the loading time is around the stress relaxation time. When the loading time is much longer than the stress relaxation time, further stress relaxation becomes negligible, that is, the stress tends to plateau and so does the strain. Molecularly, the pseudo-plastic deformation is caused by overcoming intermolecular segmental rotation resistance. Because of the limited segmental length between the cross-linked net points, no molecular chain slippage is allowed and thus the available deformation is limited, or the deformation saturates with the stress relaxation time or programming time.

It is noted that, for the thermosetting polystyrene SMP used in this study, a compression deformation higher than 30% is also allowed at the glassy temperature. This shows a distinct departure from conventional thermosetting polymers, which break at a much smaller compressive strain level. The reason again persists in the special cross-linked structure with an appropriate cross-link density. For conventional thermosetting polymers, the cross-link density is very high and the segmental length between the cross-link net points is very short. There is little free space or degree of freedom for the segment to rotate or translate, leading to brittle behavior or elastic behavior till fracture. No appreciable plastic deformation can be formed and thus no temporary shape can be fixed. As a result, no shape memory effect can be observed. Conversely, thermosetting SMPs such as the polystyrene SMP in this study have a proper cross-link density and the segmental length is long enough to allow appreciable pseudo-plastic deformation without fracture. Therefore, a temporary shape can be fixed and the shape memory effect can be demonstrated.

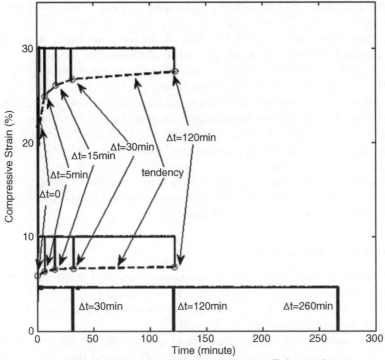

(a) Shape fixity results at temperatures below T_g for specimens programmed at different pre-strain levels (5%, 10%, and 30%) [51] (reproduced with permission from Elsevier)

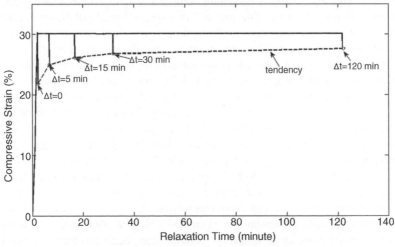

(b) Shape fixity results at temperatures below T_g for specimens programmed at 30% (adapted from (a) for clarity)

Figure 3.33 Cold-compression programming results

With 10% pre-strain, which is about 3% higher than the yield strain, a tendency similar to 30% pre-strain is observed. Therefore, as long as the pre-strain is over the yield strain, a certain amount of shape fixity can be realized. Of course, as the pre-strain increases, the shape fixity ratio also increases. For example, at the zero stress relaxation time, the shape fixity is about 62.5% for the 10% pre-strain level, which is lower than the corresponding shape fixity of 73% for the 30% pre-strain level. It is also observed that the shape fixity with 10% pre-strain plateaus earlier than that with 30% pre-strain as the stress relaxation time increases, possibly due to less viscoelastic and viscoplastic deformation with the lower pre-strain level.

In summary, the test results show that cold compression is an effective and efficient method for programming. It is found that the pre-strain level must be larger than the yielding strain of the SMP in order to fix a temporary shape at temperatures below T_g. It is also found that a longer stress relaxation time leads to a larger shape fixity ratio. The upper bound of the shape fixity is determined by the difference between the pre-strain and the springback, which is the ratio of the relaxed stress over the relaxed modulus.

3.4.1.2 Free Shape Recovery Test

Once the specimens were programmed, an unconstrained free shape recovery test was implemented, where the compressed SMP specimen was heated to $T_{high} = 79\,°C$ at an average heating rate of $q = 0.82\,°C/min$. The same LVDT system as given in Reference [41] was used to track the movement of the specimen during heating.

Figure 3.34 shows the entire thermomechanical cycles, including the three steps of cold-compression programming (steps 1 to 3 in Figure 3.13 (a)) and unconstrained free shape recovery during the heating process (step 4 in Figure 3.13 (a)). From Figure 3.34 (a), which is programmed by 30% pre-strain, it is observed that initially the programmed specimen only shows slight and gradual thermal expansion, while as the temperature approaches T_g the influence of the entropy change is becoming dominant, leading to rapid strain recovery. At temperatures well above T_g, most of the pre-strain has been released and the strain converges to a stabilized value. It is interesting to note that a similar sigmoidal-type strain recovery path is shared by all the specimens with different relaxation times during programming, implying that the strain release mechanism is generally independent of the holding time during programming. As discussed above, during programming, the holding time affects the stress relaxation of the SMP. However, it does not affect the nature of the pseudo-plastic deformation (rotation and translation of the segments between cross-link net points). Therefore, regardless of the holding time during cold-compression programming, the shape recovery behaves similarly because the temporary shape is created by the same mechanism. With 10% pre-strain (Figure 3.34 (b)), the shape recovery follows a tendency similar to that with 30% pre-strain. A noticeable difference exists in the shape recovery ratio. With the 10% pre-strain, the shape recovery ratio is about 100%, regardless of the stress relaxation time during programming; with the 30% pre-strain, there is a small amount of strain that cannot be recovered. A possible reason is that, with the 30% pre-strain, some damage may have been created within the SMP specimen, which cannot be recovered during free shape recovery. Overall, the shape memory capability of the thermosetting SMP programmed by cold-compression is considerable. In other words, the approach of programming at the glassy temperature is much simpler and easier with a considerable shape memory capability, and thus should be an alternative programming approach in practice.

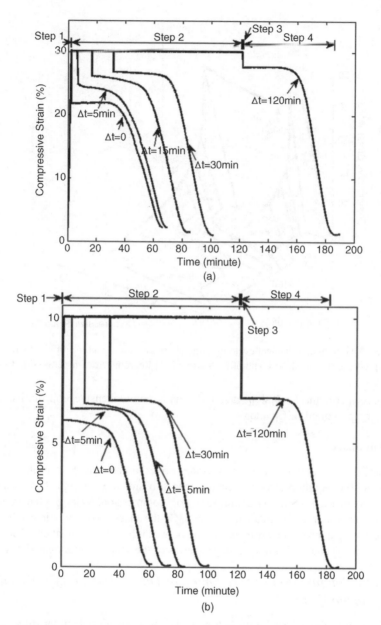

Figure 3.34 Strain–time response during the entire thermomechanical cycle for specimens programmed with (a) 30% and (b) 10% pre-strain (the four steps for the specimen with 120 min of stress relaxation time during programming are also shown). *Source:* [51] Reproduced with permission from Elsevier

The 3-D stress–strain–time behaviors for the entire themomechanical cycle, which include the three-step cold-compression programming process and the one-step heating recovery, are shown in Figure 3.35, for both the 10% and 30% pre-strain levels. An extremely nonlinear, and time and temperature dependent constitutive behavior is revealed. In-depth understanding of

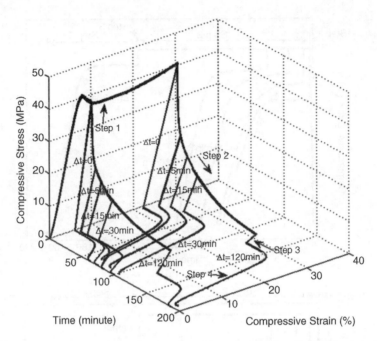

Figure 3.35 3-D thermomechanical cycle in terms of stress–strain–time for different stress relaxation times with pre-strain levels of 10% and 30%. *Source:* [51] Reproduced with permission from Elsevier

this complex thermomechanical behavior demands comprehensive constitutive modeling, which will be presented in Chapter 4.

3.4.1.3 Summary

In summary, the concept that the shape memory effect in nature is the transition between an equilibrium and nonequilibrium configuration of the SMP structure can explain the shape memory mechanism of a thermoset SMP programmed by cold compression. As long as the pre-strain level is appropriate so as not to cause damage in the SMP, the shape recovery ratio can be as high as 100%. There is no appreciable difference found between the classical method of programming (hot programming) and cold programming in terms of the shape memory capability.

3.4.2 Cold-Compression Programming of Thermosetting SMP Based Syntactic Foam

3.4.2.1 Programming of the Foam Below the Glass Transition Temperature

The isothermal uniaxial flat-wise cold-compression programming was performed on the MTS QTEST150 electromechanical frame outfitted with a moveable furnace (ATS heating chamber) compliant with the ASTM C 365 standard. Again, the block specimen size was 30 mm × 30 mm × 12 mm. The loading rate was set to be 1.3 mm/min. A thermocouple placed in the chamber near the SMP specimen was used to control the environmental temperature. As suggested by Li and Xu [51], successful shape fixity at glassy temperatures requires a post-yield pre-strain. Two pre-strain levels, 30% and 20%, which were above the yield strain of

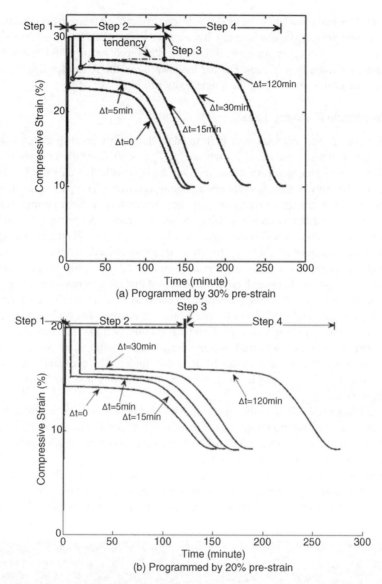

Figure 3.36 Strain–time response during the entire thermomechanical cycle for specimens programmed with (a) 30% pre-strain and (b) 20% pre-strain (the four steps shown in the figure are for the curve with 120 min of stress relaxation time). *Source:* [45] Reproduced with permission from ASME

7% for the same foam at room temperature [41], were selected with stress relaxation times of 0 min, 5 min, 15 min, 30 min, and 120 min, respectively. At least three effective specimens were tested for each stress relaxation time period.

The strain evolution with time during the material programming process can be observed in Figure 3.36. A decent shape fixity ratio (70.5% for 20% pre-strain and 72.6% for 30%

pre-strain) was reached even when the load was instantly removed (zero stress relaxation time). Similar to the pure SMPs [51], it was found that a longer stress relaxation time tends to increase the shape fixity ratio. However, an upper limit of the shape fixity ratio could be reached as the relaxation time continually increases. Further lengthening of the relaxation time could barely bring up any noticeable increase in the shape fixity ratio.

3.4.2.2 Free Shape Recovery Tests

Unconstrained strain recovery tests were performed on the programmed specimens. During the test, the programmed foam specimen was reheated to $T_{high} = 80\,°C$ at an average heating rate of $q = 0.4\,°C/min$. The displacement at the specimen surface was tracked by the same LVDT system.

Figure 3.36 also shows the unconstrained heating recovery. The programmed specimen initially shows slight thermal expansion. As the temperature further approaches T_g, the viscosity decreases and the molecule mobility increases, leading to entropy driven recovery of molecules to their original coiled morphology. Macroscopically, this leads to a rapid strain recovery. At temperatures well above T_g, the strain seems stabilized. A typical recovery path is shared by all the specimens with different relaxation times during programming, implying a universal strain release mechanism. It could be observed that the irrecoverable strain for all the specimens programmed by the same pre-strain appeared to be at nearly the same level (about 8% for 20% pre-strain and 10% for 30% pre-strain), indicating that a similar irrecoverable amount of damage occurred regardless of the relaxation time period. Therefore it is reasonable to assume that the damage occurred merely during the compression process (step 1 of programming). Since the damage in the SMP matrix under 30% pre-strain can be neglected [51], the damage should completely come from the crushing and implosion of the glass hollow microsphere.

The Hitachi S-3600N VP-Scanning Electron Microscope was used to examine the microstructure change due to programming (see Figure 3.37). From Figure 3.37 (b), some of the microballoons have been crushed after cold-compression programming by 30% pre-strain, which contributed to the irreversible strain after free shape recovery.

The extremely nonlinear behaviors for the entire thermomechanical cycle, including a three-step glassy temperature programming process and one-step heating recovery in both the stress–strain–time view and stress–strain–temperature view, are shown in Figure 3.38 (a) and (b),

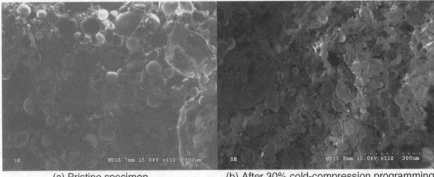

(a) Pristine specimen   (b) After 30% cold-compression programming

Figure 3.37 SEM observation of (a) a pristine specimen and (b) a specimen after 30% cold-compression programming. *Source:* [45] Reproduced with permission from ASME

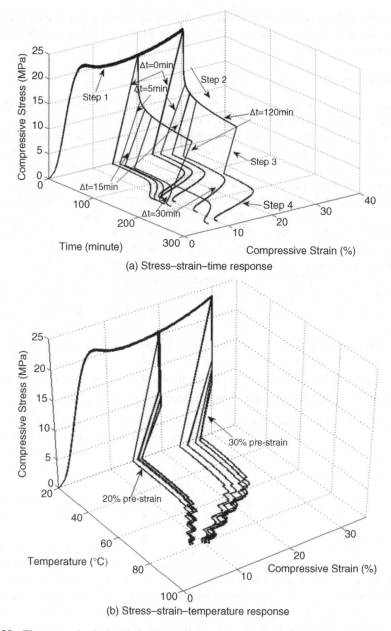

Figure 3.38 Thermomechanical cycle in terms of (a) stress–strain–time and (b) stress–strain–temperature responses for different stress relaxation times with a pre-strain level of 30% and 20%. *Source:* [45] Reproduced with permission from ASME

respectively. In-depth understanding of this complex thermomechanical behavior requires a comprehensive constitutive modeling effort. It is noted that, as instant unloading occurs at the end of the programming, straight lines were used to connect the final loading point of step 2 and the initial point of the free recovery path in step 4 in Figure 3.38. These straight lines are not actual and physical unloading curves because the sudden removal of the load could not be recorded by the MTS machine. Therefore, the slopes of these straight lines do not represent the unloading modulus.

3.4.2.3 Summary

In summary, a unique programming concept at glassy temperatures has been applied to a thermosetting SMP based syntactic foam. Considerable recoverability has been achieved, although some damage in glass hollow microsphere inclusions was inevitable. Generally speaking, cold-compression programming is a viable alternative to the classical heating, loading, cooling, and unloading programming method as it is fast and without the need for thermal management. However, it can increase the recovery stress because the energy input by cold programming is higher and the energy output during shape recovery is positively correlated to energy input during programming, as validated by Ping *et al.* [53]. However, it needs a higher loading capacity and may create some damage in the SMP or SMP based composites. Also, the shape fixity ratio is generally lower than that obtained by classical programming, although a longer stress relaxation time can increase the shape fixity ratio. Therefore, selection of cold-compression programming needs to be based on comprehensive analyses. Also, cold-tension programming of thermosetting SMP is not possible because of the lack of considerable pseudo-plastic deformation during tension. Sometimes, a hybrid programming, for instance tension using the classical programming method and compression using cold-programming, may be needed. This will be discussed in the next section.

3.5 Behavior of Thermoset Shape Memory Polymer Based Syntactic Foam Trained by Hybrid Two-Stage Programming

In order to conduct 2-D programming, a special truss fixture was fabricated and cruciform specimens were used [57]. It has been realized that this may not be realistic for large panels or structures. Together with the requirement for high temperature, and loading rate and pre-strain control, it turns out to be a challenging task. Therefore, an alternative programming approach is desired.

In the above sections, it has been demonstrated that thermosetting SMPs and their composites can be cold-compression programmed, although cold-tension programming is not possible. Based on the above physics for classical and cold programming methods, it is envisaged that one-stage 2-D stress condition programming may be replaced by two-stage 1-D programming, as long as the nonequilibrium configuration (temporary shape) can be created and fixed. Also, two-stage programming may not need to be conducted at the same temperature. In other words, tension programming can be conducted at temperatures above T_g and compression programming can be conducted at temperatures below T_g. We need to follow this pattern because thermosetting SMP has very small tensile deformability at temperatures below T_g and it cannot be programmed by cold-drawing.

3.5.1 Hybrid Two-Stage Programming

Our strategy for preparing test specimens is to first prepare tension specimens for programming at temperatures above T_g, and, after programming, machine the tensile specimens into compression specimens for second programming at temperatures below T_g. The reason is

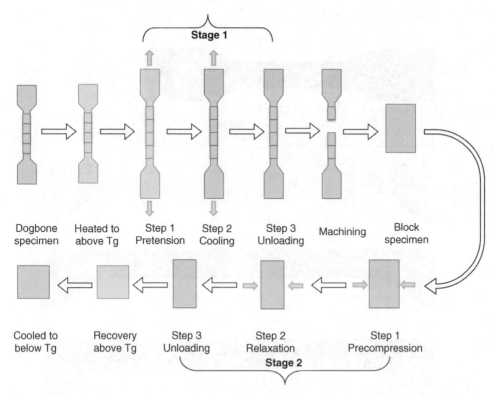

Figure 3.39 Schematic of the entire thermomechanical cycle (two-stage programming and one-step free shape recovery). *Source:* [59] Reproduced with permission from the American Society of Civil Engineers

that the foam has a maximum tensile strain of about 5% at temperatures below T_g and cannot be programmed because it fractures without yielding under tension [57]. Also, if the foam is programmed first at temperatures below T_g, the second programming at temperatures above T_g will lead to recovery of the first programming. The two-stage, hybrid, and biaxial programming process is schematically shown in Figure 3.39.

By referencing the ASTM D638M-89 standard, the dog-bone specimen was used for tension programming. The only revision was to increase the specimen thickness from 2 mm to 10 mm, in order to avoid buckling in the subsequent compression programming. A finite element modeling (FEM) was conducted that validated the uniform tensile stress distribution in the gage length of the dog-bone specimen (see Figure 3.40 (a) to (d)). During the FEM analysis, only half of the specimen was modeled due to symmetry. Because a section with a length of 25 mm at each end (Figure 3.40 (a)) was gripped during tension programming, the 25 mm long section at each end was not meshed; instead, boundary conditions were applied to simulate the gripped loading condition. In this study, the left-hand boundary was fixed and the right-hand boundary was allowed to move in the x direction (loading direction) only. At the centerline of the dog-bone specimen, the symmetry boundary condition was enforced. The analysis was conducted using the ANSYS10.0 software package and the element type was PLANE82. After convergence analysis, a total of 12,544 elements with 38,401 nodes was used to model the specimen.

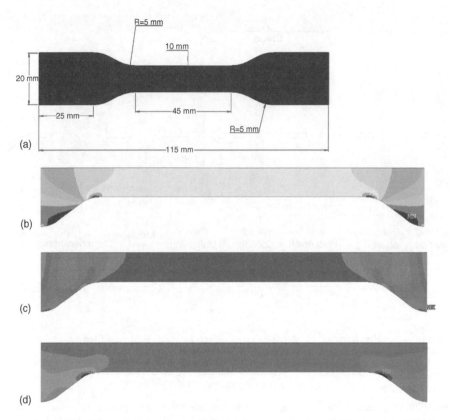

Figure 3.40   (a) Schematic of the dog-bone specimen, (b) normal stress distribution in the longitudinal direction (σ_x), (c) normal stress distribution in the transverse direction (σ_y), and (d) in-plane shear stress distribution (τ_{xy}). Note the uniform stress distribution in the gage length. Due to symmetry, only half of the specimen is analyzed. *Source:* [59] Reproduced with permission from the American Society of Civil Engineers

From Figure 3.40 (b) to (d), it is seen that both the normal and in-plane shear stress distributions within the gage length of the modified dog-bone specimens are uniform, which validates the use of the modified dog-bone specimens for tension programming. After tension programming, block specimens with a length of 10 mm in the tension direction were machined from the dog-bone specimen for compression programming, as schematically shown in Figure 3.39.

Four groups of specimens were programmed with various combinations of "nominal" pre-strain levels: T25C5, T5C25, T40C5, and T25C25, where T represents tension and C stands for compression. The numbers represent "nominal" pre-strain levels. For example, T25C5 stands for 25% "nominal" pre-tension at temperatures above T_g followed by 5% "nominal" pre-compression in the transverse direction below T_g. The word "nominal" is used because this is the strain applied without considering the Poisson's ratio effect. The "actual" strain in each direction is the sum of the "nominal" strain and the strain due to the Poisson's ratio effect. At least three effective specimens were tested for each group.

Stage 1 programming was similar to the traditional one-stage uniaxial programming [12,41]. The heating chamber was first brought to a temperature of 79 °C and held for 45 minutes. After

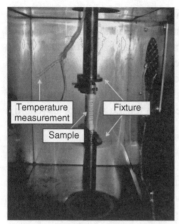

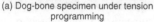

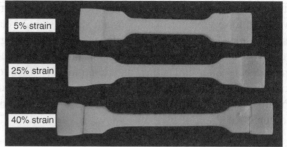

(b) Comparison of dog-bone specimens programmed
to various tensile strains

(a) Dog-bone specimen under tension
programming

(c) A specimen after tensile programming to 25% strain and compressive
programming to 5% strain. Letter "C" stands for compression direction
and "T" for tension direction

Figure 3.41 Specimens after two-stage programming. *Source:* [59] Reproduced with permission from
the American Society of Civil Engineers

that, the dog-bone specimen was placed in the chamber and mounted in the tensile fixture. The
heating chamber was then closed. The specimen was allowed to dwell for 30 more minutes to
achieve a uniform temperature. Then the MTS machine applied a tensile stress at a constant rate
of 1.3 mm/min until the desired tensile strain level was reached (step 1: pre-tension). The
displacement in the gage length was measured by an LVDT system (Cooper Instruments LDT
200 series) and the strain was calculated based on the definition of engineering strain (change
of the gage length over the original gage length). Once the desired strain level was achieved, the
strain was maintained and the specimen was allowed to slowly cool to room temperature, which
took about 10 hours (step 2: cooling). Once the specimen reached room temperature, the load
was removed by releasing the grips, causing a small springback of the specimen (step 3:
unloading). This completed the tensile programming. The temperature, stress, and strain were
recorded throughout this process using thermocouples and the MTS machine load cell, similar
to the process in Reference [41]. Once the specimen was tension programmed, it was machined
into four block specimens for compression programming. Figure 3.41 (a) shows the test
chamber and Figure 3.41 (b) shows the tension programmed specimens under various pre-
strain levels.

In Stage 2 programming, the programming was in the transverse direction (perpendicular to
the tension direction) at room temperature. The tension programmed block specimens were
first smoothed using sandpaper on two opposite sides to ensure parallel surfaces. The block
specimens were then mounted in a compression fixture. Teflon sheets were inserted between
the specimen and the steel platens of the fixture to reduce the transverse shear force during
uniaxial compression. A compressive load was applied at a constant rate of 1.3 mm/min until
the desired compressive strain level was reached (step 1: pre-compression). The compressive
load and axial displacement were directly recorded by the MTS machine and the data were used
to calculate the engineering stress and engineering strain curves. Once the specimen was

strained to 5% or 25% in the compressive direction, the strain was maintained and the specimen was held for 30 minutes with constant strain (step 2: relaxation). The load was then released (step 3: unloading). This completed the compression programming. Again, the stress and strain versus time were recorded by the machine. The specimen dimension was also recorded immediately after unloading and continuously monitored by linear variable differential transducers (LVDTs) until the dimension was stabilized after about 24 hours of unloading. Figure 3.41 (c) shows a specimen that has been programmed to a tensile strain of 25% and a compressive strain of 5%.

3.5.2 Free Shape Recovery Test

After the second programming, the free shape recovery test was conducted as schematically shown in Figure 3.39 by heating the specimen to above T_g without any constraint. During free shape recovery, the specimens were placed into the heating chamber and two LVDTs were attached to two faces of the programmed specimens. The opposite two faces were placed against an angle steel so that the LVDTs measured the total displacement in each direction (see Figure 3.42). The recovery step involved slowly ramping up the temperature in the heating chamber to ensure that the specimen had the same temperature as its environment. The temperature was increased incrementally at 2.8 °C every 5 minutes (i.e., the chamber temperature controller was controlled to increase the temperature step-wise every 5 minutes) until the programming temperature of 79.4 °C was reached. The reason was that this gave the specimen sufficient time to obtain temperature uniformity. The temperature of 79.4 °C was held for 20 minutes, which was sufficient to stabilize the recovery strain [41]. At the end of this holding time period, data recording was stopped and the recovery step was completed. The specimen was then taken out of the chamber and the specimen dimensions were recorded

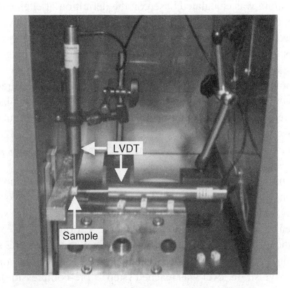

Figure 3.42 Free shape recovery test setup. *Source:* [59] Reproduced with permission from the American Society of Civil Engineers

immediately. It was noted that the heating rate affected the recovery rate, but not the final recovery strain. The combination of a higher heating rate and a longer socking time is equivalent to a lower heating rate and shorter socking time as long as the recovery strain was stabilized, which has been validated by Li and Xu [51].

3.5.3 Thermomechanical Behavior

A typical thermomechanical cycle for the T25C25 specimen in both the tension direction and compression direction is shown in Figures 3.43 and 3.44, respectively. Because of the hybrid two-stage biaxial programming, the thermomechancial behavior is very complex. The complexity in thermomechanical behavior first comes from the coupling between the pre-strain and the Poisson's ratio effect during programming. Based on Xu and Li [58], the same SMP based syntactic foam has a Poisson ratio of about 0.3; two-dimensional programming transformed the foam from isotropic to orthotropic; and the foam exhibits a negative Poisson ratio after 2-D programming when the strain was in a certain range. During the first programming in tension in the longitudinal direction, there is contraction in the two transverse directions (compression direction and free direction) due to the Poisson's ratio effect, which can be treated as a pre-strain in the transverse direction before the second room temperature programming in the compression direction. As a result, while the first programming behaves similar to traditional one-stage programming at temperatures above T_g (see Figure 3.43), the second programming looks considerably different from the first programming (see Figure 3.44). Another feature that

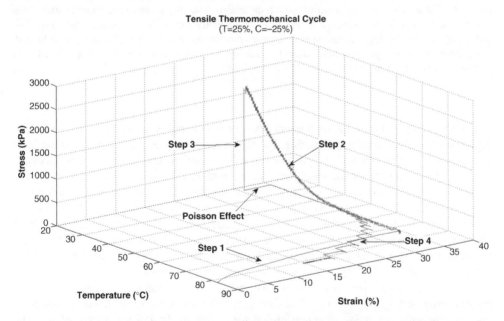

Figure 3.43 Thermomechanical cycle in the tension direction for a specimen of T25C25 (step 1 → pretension to 25% strain at temperatures above T_g, step 2 → cooling down to room temperature while holding the pre-strain constant, step 3 → unloading, which completes the first stage of programming. The Poisson effect is due to the second programming in the transverse direction by compression. Step 4 → free shape recovery). *Source:* [59] Reproduced with permission from the American Society of Civil Engineers

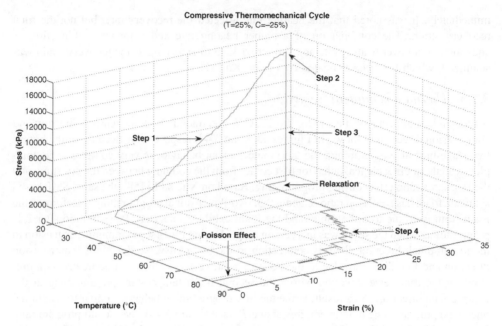

Figure 3.44 Thermomechanical cycle in the compression direction for a specimen of T25C25 (the deformation history starts with the Poisson effect due to the first tension programming at temperatures above T_g, followed by cooling down to room temperature. Then the second programming starts: step 1 → compression to 25% additional strain, step 2 → hold the strain for 30 minutes, and step 3 → unloading. After unloading, relaxation (viscoelastic rebound) occurs and after 24 hours it is stabilized, completing the second programming. Step 4 → free shape recovery). *Source:* [59] Reproduced with permission from the American Society of Civil Engineers

is not usually seen in classical high temperature programming is the viscoelastic rebound or relaxation after the second programming. Usually, this rebound reduces as the pre-strain level or relaxation time increases [51]. It is noted that the second programming also has Poisson's effect on the tension direction, which makes the length in tension direction increase further.

Because Figure 3.44 is a stress–strain–temperature plot, step 2 (30 minutes of holding or relaxation) during cold-compression programming cannot be visualized. To have a better understanding of the stress–strain evolution with time, a typical thermomechanical cycle in terms of stress–strain–time for the four groups of specimens is shown in Figure 3.45 (a) to (d), respectively. In order to have both tension and compression direction in the same quadrant, the compression stress and compression strain are also treated as positive.

From Figure 3.45, there are two unique behaviors that cannot be seen in uniaxial programming and shape recovery. The first is the Poisson's ratio effect. When loaded in one direction, the transverse direction also deforms due to the Poisson's ratio effect. The Poisson ratio is defined as the negative ratio of the transverse strain over the longitudinal strain. If the Poisson ratio is positive, it suggests that the transverse direction shrinks if the longitudinal direction is stretched, and vice versa. However, if the Poisson ratio is negative, which is true under certain conditions [48], the transverse direction expands when the longitudinal direction

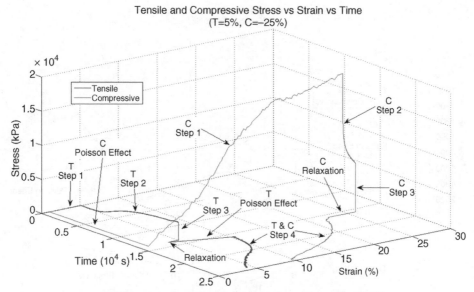

(a) Thermomechanical cycle for a specimen from group T5C25

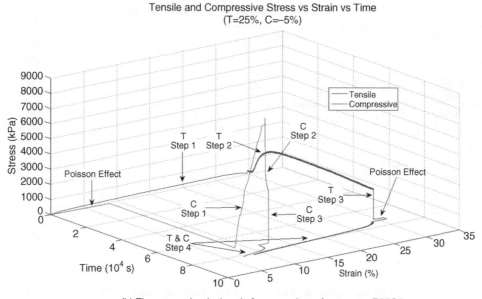

(b) Thermomechanical cycle for a specimen from group T25C5

Figure 3.45 Entire thermomechancial cycles in both the tension direction and compression direction for each group of specimens in terms of the stress–strain–time scale (in the figures, the letter "T" represents tension, the letter "C" stands for compression, and the letters "T & C" represent coupled tension and compression, respectively). *Source:* [59] Reproduced with permission from the American Society of Civil Engineers

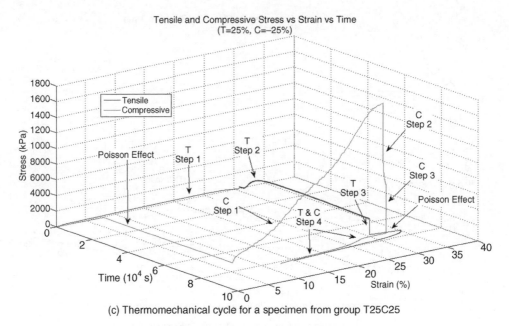

(c) Thermomechanical cycle for a specimen from group T25C25

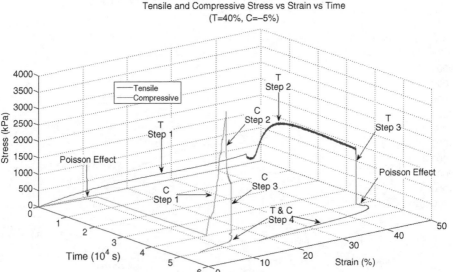

(d) Thermomechanical cycle for a specimen from group T40C5

Figure 3.45 (*Continued*)

is stretched. In other words, the material dilates. From Figure 3.45, it is clear that when the SMP based syntactic foam is first tension programmed, the transverse direction shrinks, suggesting a positive Poisson's ratio. During cold-compression programming, the Poisson's ratio effect leads to further extension in the tension direction, leading to higher pre-tension. The second feature is the stress relaxation effect. From Figure 3.45 (a), it is seen that after Step 3

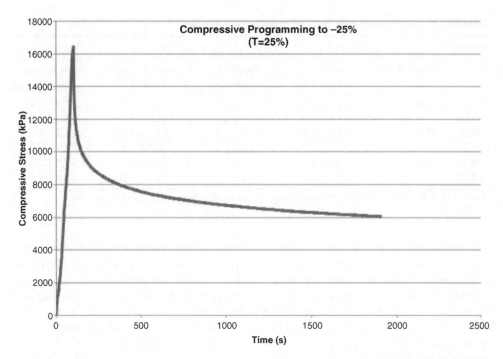

Figure 3.46 Compressive stress versus time in the compressive direction for a specimen programmed to a strain of 25% in the tensile direction followed by 25% in the compressive direction

(unloading) in the tension direction, the springback continues and is gradually stabilized. The same is seen after step 3 in the compression direction, which is even more pronounced because a larger stress is applied during cold-compression programming. The reason for this is due to the viscoelastic deformation of the SMP. The viscoelastic behavior of the SMP based syntactic foam can be better represented by the stress relaxation test (see Figure 3.46), which suggests that the foam is a viscoelastic material. Therefore, as the external load is removed, the springback consists of two components, the elastic component, which rebounds immediately, and the viscoelastic component, which rebounds gradually with time. Based on the stress relaxation test results in Figure 3.46, the springback time was taken as 30 minutes for all the specimens shown in Figure 3.45.

The effect of the pre-strain level on the thermomechanical behavior can be evaluated by the shape fixity and shape recovery ratios. In this study, the shape fixity and shape recovery ratios are determined using the following equations:

$$\text{Shape fixity ratio} = \frac{(u_0 - u_3)}{(u_0 - u_1)} \times 100\% \tag{3.5}$$

$$\text{Shape recovery ratio} = \frac{(u_3 - u_4)}{(u_3 - u_0)} \times 100\% \tag{3.6}$$

Table 3.4 Shape fixity and shape recovery ratios in each direction

Group	Tension direction		Compression direction		Free direction
	Shape fixity ratio (%)	Shape recovery ratio (%)	Shape fixity ratio (%)	Shape recovery ratio (%)	Shape recovery ratio (%)
T5C5	72.9 ± 2.1	77.1 ± 10.3	87.9 ± 6.8	31.0 ± 1.7	70.9 ± 7.8
T5C25		81.4 ± 4.0	86.9 ± 2.1	42.1 ± 5.7	85.9 ± 6.1
T25C5	81.2 ± 6.7	83.5 ± 2.2	88.4 ± 3.8	65.2 ± 4.2	68.2 ± 6.7
T25C25		74.9 ± 3.3	93.1 ± 1.6	48.5 ± 3.2	72.9 ± 6.8
T40C5	90.0 ± 2.5	83.0 ± 2.7	97.4 ± 1.3	66.0 ± 4.7	47.5 ± 5.8
T40C25		88.4 ± 1.0	94.4 ± 0.7	54.7 ± 3.3	26.9 ± 1.9

Source: [59] Reproduced with permission from the American Society of Civil Engineers.

Depending on the direction, the meaning of each term in the shape fixity and shape recovery ratios is as follows: in the tension direction, u_0, u_1, u_2, u_3, and u_4 are, respectively, the initial length, the length after pre-strain imposed in step 1, the length after step 2, the length after step 3, and the length after Poisson's ratio effect and step 4; in the compression direction, they are, respectively, the initial length, the length after Poisson's ratio effect and pre-strain in step 1, the length after step 2, the length after step 3, and the length after step 4. All the lengths were measured at the face center of the specimens. The shape fixity and shape recovery ratios are summarized in Table 3.4.

Upon recovery, the tensile direction is expected to contract while the compressive direction should expand. The free direction depends on the stored strain due to the two programmings. Figure 3.47 (a) shows a specimen after the two-stage programming and Figure 3.47 (b) and (c) show the specimen after free shape recovery. Generally, shape recovery is a reverse process of the programming. The recovered shape should be the same as the original permanent shape if the shape recovery ratio is 100%. However, the recovered shape is distorted for several reasons. One reason is that the shape recovery ratio is not perfect due to the damage of microballoons during cold-compression programming [45] (see Table 3.4). Therefore, the specimen does not restore its prismatic shape. Another reason is the coupling between the shape memory and Poisson's ratio effect. The shape memory (shortening) in the tension direction causes expansion in the compression direction due to Poisson's ratio effect. This expansion is

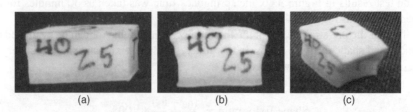

(a)   (b)   (c)

Figure 3.47 Comparison of (a) the programmed specimen and (b) the recovered specimen (side view) and (c) the recovered specimen (isometric view) (the face marked "T" is perpendicular to the tension direction, the face marked "C" is perpendicular to the compression direction, and the face marked "40/25" is the free direction subjected to Poisson's effects only). *Source:* [59] Reproduced with permission from the American Society of Civil Engineers

then coupled with the shape memory (expansion) in the compression direction. As a result, the compression direction expands more than the pure shape memory. Similarly, the tension direction shortens more than the pure shape memory, that is, the two effects enhance each other in each direction. However, based on Xu and Li [48], the foam after programming has different Poisson's ratios in the tension and compression directions, which suggests that the contribution by the Poisson's ratio to shape recovery is not the same in the two directions and thus leads to a distorted recovery shape. The third reason is the different constraints at the center and edge of each face. At the center of the face, the specimen is freer to move because it is away from the constraint of the boundary (the four edges on each programmed face). At the edge, the constraint is the strongest. One direction tends to expand and the other tends to contract. This physical discontinuity leads to distortion of the recovered specimen.

Comparing the three directions, it is clear from Table 3.4 that the tension direction sees the largest recovery ratio, followed by the free direction, and the least by the compression direction. The reason is that compression programming needs a very large stress (about 10 times that for tension programming), which may cause microballoon crushing [45] or molecular level damage, so that it cannot be fully recovered. Also, compression programming pushed some molecular segments, which have rotated towards the tension direction in the first programming, to align more towards the tension direction. Consequently, the recovery in the tension direction is more significant.

The effect of pre-strain levels on the shape fixity is clear. In the compression direction, the shape fixity increases as the pre-tension increases. It is noted that in a previous study, uniaxial cold-compression programming by a pre-compression of 5% did not fix a temporary shape because 5% was lower than the yielding strain of about 7% [51]. In the current study, 5% "nominal" pre-compression leads to considerable shape fixity. The reason is that pre-tension serves as a pre-compression in the compression direction, due to the Poisson's ratio effect. The total compression pre-strain is greater than the yielding strain, resulting in considerable shape fixity. In the tension direction, the shape fixity was calculated once the first-stage programming had completed. Therefore, the effect of the second-stage programming (Poisson's ratio effect) was not considered. As a result, the shape fixity is the same for the same pretension, regardless of the second programming. Similar to the shape fixity in the compression direction, the shape fixity increases as the pretension increases.

As given in Table 3.4, the shape recovery ratios in the tension direction are slightly smaller than those programmed by the one-stage 2-D stress condition, which is between 91% and 96% depending on the pre-stress levels [57]. The slight reduction is due to the damage during cold-compression programming.

3.5.4 Recovery Sequence and Weak Triple Shape

To have a better view of the free shape recovery in three directions (tension, compression, and free), Figure 3.48 (a) to (d) shows typical recovery strain versus temperature results for groups T25C5, T5C25, T40C5, and T25C25, respectively. It is noted that the stepped recovery curve is due to the stepped heating profile used during the shape recovery test.

From Figure 3.48 (a), the foam shows a sequential shape recovery. The tension direction starts to recover at temperatures around 50 °C. However, recovery for the compression direction and free direction does not occur until the temperature is about 70 °C. From Figure 3.48 (b), it is seen that the sequence of shape recovery is reversed. The compression

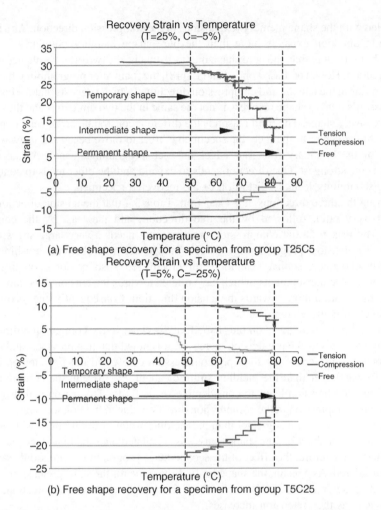

Figure 3.48 Free shape recovery with temperature for four groups of specimens in each direction. *Source:* [59] Reproduced with permission from the American Society of Civil Engineers

direction starts to recover at about 50 °C. The tension direction and free direction does not recover until about 60 °C. It is noted that the sudden recovery at about 50 °C for the free direction is most likely due to equipment error. Obviously, the difference in the recovery sequence between Figure 3.48 (a) and (b) is due to the reverse in the pre-strain levels.

Based on Sun and Huang [68], if SMPs are continuously programmed in the glass transition region, the SMPs will recover their original shapes, following the exact order of the previous pre-deformation but in an inverse fashion. Clearly, the result in Figure 3.48 (b) followed this pattern but Figure 3.48 (a) did not. From this result, it is inferred that the recovery sequence depends on the effort during programming, not necessarily the sequence of programming. In other words, it seems that the SMP based syntactic foam cannot remember the programming sequence. As is well known, the driving force for shape recovery is due to the conformational

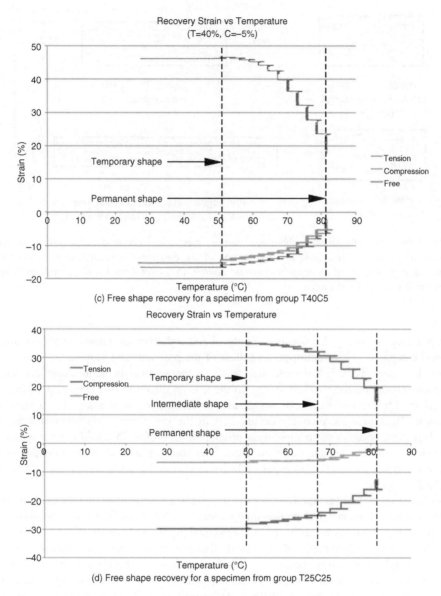

Figure 3.48 (*Continued*)

entropy. Programming leads to alignment and ordering of molecular segments towards the loading direction or reduction in entropy. During the two-stage 1-D programming, the first programming aligns some molecular segments along the tension direction. In the subsequent compression direction training, some molecular segments align along the compression direction. It also pushes some already aligned segments in the tension direction, further aligning along this direction (see the schematic in Figure 3.49). Depending on the degree of alignment (or degree of departure from the equilibrium configuration or degree of reduction in

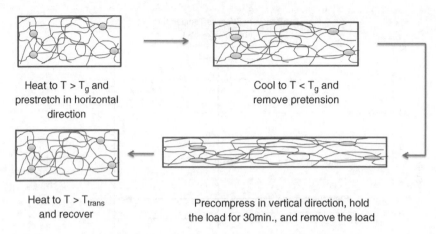

Heat to T > T$_g$ and
prestretch in horizontal
direction

Cool to T < T$_g$ and
remove pretension

Heat to T > T$_{trans}$
and recover

Precompress in vertical direction, hold
the load for 30min., and remove the load

Figure 3.49 Schematic showing the shape memory mechanism involved in the two-stage biaxial programming and recovery. Note the contribution of the compression programming to the further segmental alignment along the tension direction. *Source:* [59] Reproduced with permission from the American Society of Civil Engineers

entropy), shape recovery first occurs along the direction that has the largest number of aligned molecular segments (least entropy), regardless of the programming sequence.

For the group T25C5, the molecular segments predominately align along the tension direction. Therefore, the recovery occurs in the tension direction first (Figure 3.48 (a)). For the group T5C25, the molecular segments predominately align along the compression direction. Therefore, the recovery occurs first in the compression direction (Figure 3.48 (b)). However, because the compression programming in T5C25 also further aligns some segments towards the tension direction, the temperature range between the first and second recovery narrows to 50–60 °C in Figure 3.48(b), instead of 50–70 °C for the group T25C5 in Figure 3.48(a).

For both Figure 3.48 (a) and (b), it is clear that the foam recovers in two stages. Obviously, for these cases, two-stage 1-D programming leads to behavior different from one-stage 2-D programming, which shows recovery in both directions simultaneously [57].

With absolute predominance of segmental alignment in one direction, such as T40C5 in Figure 3.48 (c), the recovery seems to be simultaneously in all directions. In other words, the foam becomes a routine dual shape. The reason is that there are very few segments aligned along the compression direction for T40C5. An absolute majority of the segments align along the tension direction. Two-stage 1-D programming is thus similar to one-stage 2-D programming. The foam recovers all the way to its permanent shape, similar to dual shape SMPs.

This entropy driven shape recovery can be further validated by Figure 3.48 (d). The prestrain level in both the tension direction and compression direction is the same (25%). Therefore, it is expected that the two directions will recover simultaneously. This is exactly the case. From Figure 3.48 (d), both directions start to recover at a temperature of about 50 °C. For the free direction, it is mainly affected by the Poisson's effect up to a temperature of about 68 °C. After that, this direction also recovers to its permanent shape. Overall, this case suggests that one-stage 2-D programming can be replaced by two-stage 1-D programming.

It is noted that triple or multiple shape SMPs have been a popular topic recently. Currently, there are several ways to achieve triple shapes: one is through compositional design by involving two distinct switching phases in the polymer chain [56,69–75], the second is through a composite design by involving two components or a bi-layer with distinctively different transition temperatures [76,77], and the last is a polymer with a single yet sufficiently broad thermomechanical transition through two or more sequential programmings in the glass transition region [78,79]. Particularly, the study in References [78] and [79] provides a new way of creating multiple shapes for traditional dual-shape SMPs through mechanical means – programming. As discussed previously [78,79], for SMPs with a single glass transition, the triple shape can be seen only when the glass transition zone is broad (55–130 °C). Based on Li and Nettles [41], the glass transition zone for this SMP based syntactic foam is 63.7–76.2 °C. Obviously, this is a very narrow region. It is not clear whether the thermosetting SMP based syntactic foam with a narrow glass transition region has a triple shape or not.

From Figure 3.48 (a), it is seen that the foam shows a weak triple shape. The tension direction starts to recover at temperatures around 50 °C. However, the recovery for the compression direction and free direction does not occur until the temperature is about 70 °C. Therefore, before it recovers into the permanent shape, the foam specimen shows an intermediate shape between 50 °C and 70 °C. In other words, the T25C5 group has a weak triple shape. Therefore, the results in this study show that for SMPs with a single and yet narrow glass transition zone, a triple shape can be created using two-stage hybrid programming. It is envisaged that the quadruple shape may be created if programmed in three directions. It is noted that we use the word "weak" triple shape, which suggests that the intermediate shape was not fully distinctive (the tension direction further recovers above 70 °C), as against those in References [78] and [79], where the intermediate shape was fully developed.

From Figure 3.48 (b), it is seen that the triple shape is reversed. The compression direction starts to recover at about 50 °C. The tension direction and free direction does not recover until about 60 °C. It is noted that the sudden recovery at about 50 °C for the free direction is most likely due to equipment error. Therefore, the foam again has an intermediate shape before it recovers its permanent shape. The interesting part is the reverse in the recovery sequence. Obviously, the difference in the recovery sequence between Figure 3.48 (a) and (b) is due to the reverse in the pre-strain level, as discussed above.

With absolute predominance of segmental alignment in one direction, such as T40C5 in Figure 3.48 (c), which is similar to uniaxial programming, and with equal pre-strain in both directions during programming, such as T25C25 in Figure 3.48 (d), which is similar to one-stage 2-D programming, no weak triple shape is observed. Therefore, the weak triple shape is related to the recovery sequence, which is in turn determined by the pre-strain level or entropy level or the level of nonequilibrium configuration in each direction.

3.5.5 Summary

In summary, two-stage hybrid programming provides a way of creating a weak triple shape for SMPs with a single but narrow glass transition region. The intermediate shape depends on the degree of relative predominance of segmental alignments in each direction. However, when one direction absolutely predominates, the foam behaves as a dual shape material, similar to one-stage uniaxial programming. When both directions have a similar pre-strain, the two-stage

1-D programming is similar to the one-stage 2-D programming. The recovery sequence depends on the relative effort in the pre-strain levels, not necessarily the programming sequence. It starts in the direction with the least entropy or largest pre-strain level, and is independent of the programming sequence. The programming is coupled with the Poisson's ratio effect. The actual pre-strain in each direction is larger than the nominal pre-strain applied for this particular loading mode. The shape recovery is also coupled with the Poisson's ratio effect. The center of the programmed face recovers more than the edge shared by the two programmed faces, leading to a distorted recovery shape. Programming at temperatures below T_g may have a viscoelastic springback. This effect can be reduced with an increased pre-strain level and pre-strain holding time.

3.6 Functional Durability of SMP Based Syntactic Foam

The SMP based syntactic foam may be used in various lightweight composite structures. Most of the time, the structures may be used outdoors with uncontrolled environmental attacks. The most obvious environmental attacks include ultraviolet (UV) radiation, moisture, temperature, and combinations of these single factors, such as hydrothermal attacks. Polymers are extremely sensitive to these environmental attacks due to the photophysical and photochemical effects. Often, the combination of UV radiation with oxidative and hydrolytic factors leads to more severe degradation than that from a single factor. The damage may range from mere surface discoloration to extensive loss of mechanical properties.

UV radiation mainly comes from the Sun. Although the Earth's atmosphere filters out some of the UV portion of the solar radiation, the actinic range of solar UV radiation is about 6% of the total radiation of the sun that reaches the earth's surface [80]. The spectrum of sunlight penetrating the earth's atmosphere ranges from 290 to 3000 nm. UV is in the range of 295–380 nm. Saturated compounds possessing bonds such as $C-C$, $C-H$, $O-H$, and $C-Cl$ absorb light at wavelength <200 nm. Carbonyl groups and conjugated double bonds have absorption maxima between 200 and 300 nm. In an outdoor environment, the combination of oxygen and UV leads to a photooxidation reaction, and results in more degradation than the UV radiation alone without oxygen. Of course, this degradation can be further accelerated at elevated temperatures and in the presence of moisture. For the polystyrene SMP used in this chapter, it has generally been accepted that UV photons interact with polystyrene, resulting in carbon–carbon scissions. This generates a variety of oxygen-containing functionalities at the specimen surface, such as $C-O$, $C=O$, $COOH$, etc. [81], which causes both cross-linking and chain-scission to occur concurrently as a result of ageing exposure to UV in the presence of air [82].

Moisture attack on polymers includes physical degradation and chemical degradation. For physical degradation, water molecules diffuse into the polymer and sit in free volume holes or interstitial sites, which lead to a swelling or plasticizing effect on the polymers. One direct piece of evidence is the lowered glass transition temperature of polymers after immersion in water. Chemical degradation is due to the chemical attacks by various types of ions in water. For example, rainwater includes major inorganic ions (Cl^-, NO_3^-, SO_4^{2-}, F^-, NH_4^+, Ca^{2+}, Mg^{2+}, Na^+, K^+) and organic species (CH_3COO^-, $HCOO^-$, $CH_2(COO)_2^{2-}$, $C_2O_4^{2-}$) [83]. Rainwater may easily hydrate polymers. Also, larger organic species deposit in the open pores, causing a greater mass gain than that of smaller inorganic ions. It is noted that the polystyrene SMP includes some bonds such as $C-O$, $C=O$, $N-H$, etc., which are sensitive to water absorption.

In this section, we will discuss the environmental attacks of UV radiation in the presence of air and UV radiation under both rainwater and seawater immersion, again in the presence of air, on the SMP based syntactic foam.

3.6.1 Programming the SMP Based Syntactic Foam

The same procedure as that in Section 3.3.2 was used to conduct 2-D programming of the SMP based syntactic foam. A 300.7 kPa programming stress was applied to the cruciform specimen through the loading rod. After programming, the central square (about 25.4 mm × 25.4 mm) of the cruciform specimen was machined into four cube specimens with an edge length of 12.7 mm.

3.6.2 Environmental Conditioning

To investigate the effect of UV ageing on the programmed foam, the programmed cube specimens were exposed to UV for the shape recovery test. At least three effective specimens were used for the stress recovery test. A 300 W Mog Base UV lamp, which had a wavelength ranging from 280 to 340 nm (mixed UV-A and UV-B light), was placed about 33 cm above the top surface of the specimens for 90 days. Temperature rising from the light source was measured. The temperature on the exposed specimen surface was about 20–23 °C higher than that without UV exposure. Even with the increased temperature, the temperature of the foam was still lower than its T_g, which was about 70 °C. Therefore, shape recovery was impossible during the ageing process.

Additionally, in order to investigate the synergistic effects of UV and moisture on the degradation, the programmed cube specimens were immersed in saturated saltwater and rainwater at room temperature for 90 days, respectively. They were prepared for the stress recovery test. All these immersed specimens were synchronously exposed to UV. The original foams without environmental attacks were also prepared as control specimens for each group of programmed foams.

3.6.3 Stress Recovery Test

A fully constrained stress recovery test was conducted for specimens after various environmental attacks in order to evaluate the effect of environmental conditioning on the memory capability of the SMP based syntactic foam. The same test equipment and test fixture as others used in this chapter were used in the stress recovery teat. The test parameters were as follows: heating was performed at an average rate of 0.3 °C/min from room temperature until 79 °C and then the temperature was maintained constant for approximately 20 minutes. The load cell of the Q-TEST 150 machine was used to record the recovered force as a function of time and temperature.

Typical stress recovery test results are shown in Figure 3.50. In order to demonstrate the entire thermomechanical cycle, the three steps during programming for the dry specimen (control specimen without environmental attacks) are also shown in Figure 3.50, as well as the stress recovery results for four groups of specimens. They include the dry or control group, the UV radiation alone group, as well as the two groups with a combination of UV radiation that were simultaneously immersed in saltwater and rainwater.

From Figure 3.50, the stress comes from two parts during recovery: thermal expansion stress and entropically stored stress (memorized stress). When the temperature is below T_g, that is, the

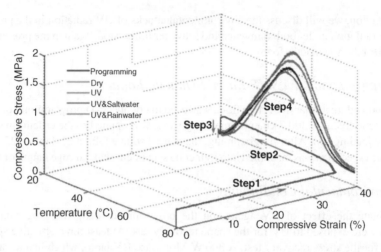

Figure 3.50 Programming (step 1: loading → step 2: cooling → step 3: unloading) for the foam and recovery stress–temperature response (step 4) under fully constrained conditions for the programmed foams exposed to UV alone and simultaneously immersed in saltwater and rainwater. *Source:* [84] Reproduced with permission from Elsevier

glassy state, the stiffness of the foam is high and the recovery stress increases gradually as the temperature rises, due to thermal expansion. When the temperature reaches the start of the glass transition region, the stress shows a peak. While the coefficient of thermal expansion still increases within the glass transition zone, the stiffness decreases sharply, leading to a continuous decrease in stress. Once the temperature approaches the programming temperature (79 °C), the recovery stress shows some relaxation initially due to large-scale segmental motion until stabilized.

As compared with the control group (dry group), a couple of observations can be made from Figure 3.50. First, environmental attacks lead to a decrease in the peak recovery stress. The programmed foams exposed to UV alone show a decrease in the peak recovery stress by 8.6%. However, a combination of UV and moisture further reduces the recovery stress. The moisture immersed foams exposed to UV show a larger reduction in the peak recovery stress by 15.7% and 19.5% for saltwater and rainwater, respectively. While the temperature corresponding to the peak recovery stress is basically unchanged for UV aged specimens, the temperature for specimens immersed in saltwater and rainwater is lowered by 3.1 °C and 4.8 °C, respectively, as compared to that of the dry specimen. This indicates that T_g becomes lower after immersion in saltwater and rainwater at room temperature for 90 days [85]. The water serves as a plasticizer or weakens the hydrogen bonding between N—H and C=O groups to lower the glass transition temperature in addition to other detrimental effects on the foam. Second, the stabilized recovery stress shows a similar tendency to the peak recovery stress. However, the difference between the control group and those of environmentally attacked groups becomes smaller. This suggests that, while environmental attacks have an adverse effect on the memory capability of the SMP based syntactic foam, the effect is not very significant in terms of the stabilized recovery stress. As will be seen later, the stabilized recovery stress is more important than the peak recovery stress in some applications, such as in the two-step bio-inspired close-then-heal (CTH) self-healing applications [41,42].

3.6.4 Summary

Based on the test results, it is seen that intensive environmental attacks up to 90 days negatively affect the memory capability of the SMP based syntactic foam. Generally, UV alone can reduce the peak recovery stress and stabilized recovery stress. Addition of moisture further reduces the memory of the SMP based foam. Comparing rainwater and seawater, it was found that rainwater causes larger degradation than saltwater because rainwater consists of more detrimental ion species. It is also seen that the reduction in recovery stress is generally not very significant, particularly for the more meaningful stabilized recovery stress, which is critical for the CTH self-healing applications. In other words, the SMP based syntactic foam still possesses considerable memory capability after intensive environmental attacks. Just like conventional polymers, an anti-ageing agent may be added to the SMP based syntactic foam to extend its service life and maintain its shape or stress memory capabilities.

References

1. Ratna, D. and Karger-Kocsis, J. (2008) Recent advances in shape memory polymers and composites: a review. *Journal of Materials Science*, **43**, 254–269.
2. Hickenboth, C.R., Moore, J.S., White, S.R., Sottos, N.R., Baudry, J., and Wilson, S.R. (2007) Biasing reaction pathways with mechanical force. *Nature*, **446**, 423–427.
3. Koerner, H., Price, G., Pearce, N.A., Alexander, M., Vaia, R.A. (2004) Remotely actuated polymer nano-composites – stress-recovery of carbon-nanotube-filled thermoplastic elastomers. *Nature Materials*, **3**, 115–120.
4. Lendlein, A., Jiang, H., Jünger, O., and Langer, R. (2005) Light-induced shape-memory polymers. *Nature*, **434**, 879–882.
5. Mohr, R., Kratz, K., Weigel, T., Lucka-Gabor, M., Moneke, M., and Lendlein, A. (2006) Initiation of shape-memory effect by inductive heating of magnetic nanoparticles in thermoplastic polymers. *Proceedings of the National Academy of Sciences of the United States of America*, **103**, 3540–3545.
6. Tobushi, H., Hara, H., Yamada, E., and Hayashi, S. (1996) Thermomechanical properties in a thin film of shape memory polymer of polyurethane series. *Smart Materials and Structures*, **5**, 483–491.
7. Irie, M. (1998) *Shape Memory Polymers*, Cambridge University Press, Cambridge, UK.
8. Monkman, G.J. (2000) Advances in shape memory polymer actuation. *Mechatronics*, **10**, 489–498.
9. Li, F., Qi, L., Yang, J., Xu, M., Luo, X., and Ma, D. (2000) Polyurethane/conducting carbon black composites: structure, electrical conductivity, strain recovery behaviour and their relationships. *Journal of Applied Polymer Science*, **75**, 68–77.
10. Jeong, H.M., Ahn, B.K., and Kim, B.K. (2001) Miscibility and shape memory effect of thermoplastic polyurethane blends with phenoxy resin. *European Polymer Journal*, **37**, 2245–2252.
11. Lendlein, A. and Langer, R. (2002) Biodegradable, elastic shape-memory polymers for potential biomedical applications. *Science*, **296**, 1673–1676.
12. Behl, M. and Lendlein, A. (2007) Shape-memory polymers. *Materials Today*, **20**, 20–28.
13. Xie, T. (2011) Recent advances in polymer shape memory. *Polymer*, **52**, 4985–5000.
14. Gunes, I.S. and Jana, S.C. (2008) Shape memory polymers and their nanocomposites: a review of science and technology of new multifunctional materials. *Journal of Nanoscience and Nanotechnology*, **8**, 1616–1637.
15. Leng, J.S., Lan, X., Liu, Y.J., and Du, S.Y. Shape-memory polymers and their composites: stimulus methods and applications. *Progress in Materials Science*, **56**, 1077–1135.
16. Meng, Q.H. and Hu, J.L. (2009) A review of shape memory polymer composites and blends. *Composites Part A: Applied Science and Manufacturing*, **40**, 1661–1672.
17. Liu, Y.J., Lv, H.B., Lan, X., Leng, J.S., and Du, S.Y. (2009) Review of electro-active shape memory polymer composite. *Composites Science and Technology*, **69**, 2064–2068.
18. Huang, W.M., Yang, B., Zhao, Y., and Ding, Z. (2010) Thermo-moisture responsive polyurethane shape-memory polymer and composites: a review. *Journal of Materials Chemistry*, **20**, 3367–3381.
19. Hu, J.L., Lu, J., and Zhu, Y. (2008) New developments in elastic fibers. *Polymer Reviews*, **48**, 275–301.
20. Hu, J.L., Meng, H., Li, G., and Ibekwe, S. (2012) An overview of stimuli-responsive polymers for smart textile applications. *Smart Materials and Structures*, **21**, paper 053001 (23 pages).

21. Hu, J.L., Zhu, Y., Huang, H., and Lu, J. (2012) Recent advances in shape–memory polymers: structure, mechanism, functionality, modeling and applications. *Progress in Polymer Science*, **37**, 1720–1763.
22. Meng, H. and Li, G. (2013) A review of stimuli-responsive shape memory polymer composites. *Polymer*, **54**, 2199–2221.
23. Vance, A.P. and Parks, R.M. (1996) Foam plastics in aircraft. *Journal of Cellular Plastics*, **2**, 345–347, (1966).
24. Anderson, T.F., Walters, H.A., and Glesner, C.W. (1970) Castable, sprayable, low density foams and composites for furniture, marble, marine. *Journal of Cellular Plastics*, **6**, 171–178.
25. Shutov, F.A. (1991) Syntactic polymer foams, in *Handbook of Polymer Foams and Foam Technology* (eds D. Klempner and K. C. Frisch), Hanser Publishers, pp. 355–374.
26. Hearon, K., Singhal, P., Horn, J., Small IV, W., Olsovsky, C., Maitland, K.C., Wilson, T.S., and Maitland, D.J. (2013) Porous shape-memory polymers. *Polymer Reviews*, **53**, 41–75.
27. Rizzi, E., Papa, E., and Corigliano, A. (2000) Mechanical behavior of a syntactic foam: experiments and modeling. *International Journal of Solids and Structures*, **37**, 5773–5794.
28. Song, B., Chen, W., and Frew, D.J. (2004) Dynamic compressive response and failure behavior of an epoxy syntactic foam. *Journal of Composite Materials*, **38**, 915–936.
29. Bardella, L. and Genna, F. (2001) On the elastic behavior of syntactic foams. *International Journal of Solids and Structures*, **38**, 7235–7260.
30. Liu, T., Deng, Z.C., and Lu, T.J. (2008) Analytical modeling and finite element simulation of the plastic collapse of sandwich beams with pin-reinforced foam cores. *International Journal of Solids and Structures*, **45**, 5127–5151.
31. Marur, P.R. (2005) Effective elastic moduli of syntactic foams. *Materials Letters*, **59**, 1954–1957.
32. Nji, J. and Li, G. (2008) A CaO enhanced rubberized syntactic foam. *Composites Part A – Applied Science and Manufacturing*, **39**, 1404–1411.
33. Li, G. and John, M. (2008) A crumb rubber modified syntactic foam. *Materials Science and Engineering A*, **474**, 390–399.
34. Li, G. and Nji, J. (2007) Development of rubberized syntactic foam. *Composites Part A – Applied Science and Manufacturing*, **38**, 1483–1492.
35. El-Hadek, M.A. and Tippur, H.V. (2003) Dynamic fracture parameters and constraint effects in functionally graded syntactic epoxy foams. *International Journal of Solids and Structures*, **40**, 1885–1906.
36. Gupta, N. (2007) A functionally graded syntactic foam material for high energy absorption under compression. *Materials Letters*, **61**, 979–982.
37. Nji, J. and Li, G. (2010) A biomimic shape memory polymer based self-healing particulate composite. *Polymer*, **51**, 6021–6029.
38. Li, G. and John, M. (2008) A self-healing smart syntactic foam under multiple impacts. *Composites Science and Technology*, **68**, 3337–3343.
39. Li, G. and Muthyala, V.D. (2008) Impact characterization of sandwich structures with an integrated orthogrid stiffened syntactic foam core. *Composites Science and Technology*, **68**, 2078–2084.
40. Li, G. and Chakka, V.S. (2010) Isogrid stiffened syntactic foam cored sandwich structure under low velocity impact. *Composites Part A: Applied Science and Manufacturing*, **41**, 177–184.
41. Li, G. and Nettles, D. (2010) Thermomechanical characterization of a shape memory polymer based self-repairing syntactic foam. *Polymer*, **51**, 755–762.
42. Li, G. and Uppu, N. (2010) Shape memory polymer based self-healing syntactic foam: 3-D confined thermo-mechanical characterization. *Composites Science and Technology*, **70**, 1419–1427.
43. Xu, W. and Li, G. (2010) Constitutive modeling of shape memory polymer based self-healing syntactic foam. *International Journal of Solids and Structures*, **47**, 1306–1316.
44. Caamaño, C., Grady, B., Resasco, D.E. (2012) Influence of nanotube characteristics on electrical and thermal properties of MWCNT/polyamide 6,6 composites prepared by melt mixing. *Carbon*, **50**, 3694–3707.
45. Xu, W. and Li, G. (2011) Thermoviscoplastic modeling and testing of shape memory polymer based self-healing syntactic foam programmed at glassy temperature. *ASME Journal of Applied Mechanics*, **78**, paper 061017 (14 pages).
46. Liu, Y., Gall, K., Dunn, M.L., Greenberg, A.R., and Diani, J. (2006) Thermomechanics of shape memory polymers: uniaxial experiments and constitutive modeling. *International Journal of Plasticity*, **22**, 279–313.
47. Miaudet, P., Derré, A., Maugey, M., Zakri, C., Piccione, P.M., Inoubli, R., and Poulin, P. (2007) Shape and temperature memory of nanocomposites with broadened glass transition. *Science*, **318**, 1294–1296.
48. Xu, T. and Li, G. (2011) A shape memory polymer based syntactic foam with negative Poisson's ratio. *Materials Science and Engineering A*, **528**, 6804–6811.

49. Strobl, G. (2007) *The Physics of Polymers: Concepts for Understanding Their Structures and Behavior*, Springer, Berlin.
50. Nguyen, T.D., Qi, H., Castro, F., and Long, K.N. (2008) A thermoviscoelastic model for amorphous shape memory polymers: incorporating structural and stress relaxation. *Journal of the Mechanics and Physics of Solids*, **56**, 2792–2814.
51. Li, G. and Xu, W. (2011) Thermomechanical behavior of thermoset shape memory polymer programmed by cold-compression: testing and constitutive modeling. *Journal of the Mechanics and Physics of Solids*, **59**, 1231–1250.
52. Lendlein, A. and Kelch, S. (2002) Shape memory polymers. *Angewandte Chemie International Edition*, **41**, 2034–2057.
53. Ping, P., Wang, W., Chen, X., and Jing, X. (2005) Poly(ε-caprolactone) polyurethane and its shape-memory property. *Biomacromolecules*, **6**, 587–592.
54. Rabani, G., Luftmann, H., and Kraft, A. (2006) Synthesis and characterization of two shape-memory polymers containing short aramid hard segments and poly(ε-caprolactone) soft segments. *Polymer*, **47**, 4251–4260.
55. Wang, W., Jin, Y., Ping, P., Chen, X., Jing, X., and Su, Z. (2010) Structure evolution in segmented poly(ester urethane) in shape-memory process. *Macromolecules*, **43**, 2942–2947.
56. Zotzmann, J., Behl, M., Feng, Y., and Lendlein, A. (2010) Copolymer networks based on poly(ω-pentadeca-lactone) and poly(ε-caprolactone) segments as a versatile triple-shape polymer system. *Advanced Functional Materials*, **20**, 3583–3594.
57. Li, G. and Xu, T. (2011) Thermomechanical characterization of shape memory polymer based self-healing syntactic foam sealant for expansion joint. *ASCE Journal of Transportation Engineering*, **137**, 805–814.
58. Xu, T. and Li, G. (2011) Cyclic stress–strain behavior of shape memory polymer based syntactic foam programmed by 2-D stress condition. *Polymer*, **52**, 4571–4580.
59. Li, G., King, A., Xu, T., and Huang, X. (2013) Behavior of thermoset shape memory polymer based syntactic foam sealant trained by hybrid two-stage programming. *ASCE Journal of Materials in Civil Engineering*, **25**, 393–402.
60. Berriot, J., Montes, H., Lequeux, F., Long, D., and Sotta, P. (2003) Gradient of glass transition temperature in filled elastomers. *Europhysics Letters*, **64**, 50–56.
61. Li, G., Torres, S., Alaywan, W., and Abadie, C. (2005) Experimental study of FRP tube-encased concrete columns. *Journal of Composite Materials*, **39**, 1131–1145.
62. Li, G. (2006) Experimental tudy of FRP confined concrete cylinders. *Engineering Structures*, **28**, 1001–1008.
63. Li, G., Maricherla, D., Singh, K., Pang, S.S., and John, M. (2006) Effect of fiber orientation on the structural behavior of FRP wrapped concrete cylinders. *Composite Structures*, **74**, 475–483.
64. Li, G. (2007) Experimental study of hybrid composite cylinders. *Composite Structures*, **78**, 170–181.
65. Li, G. and Maricherla, D. (2007) Advanced grid stiffened FRP tube encased concrete cylinders. *Journal of Composite Materials*, **41**, 1803–1824.
66. Li, G. and Velamarthy, R.C. (2008) Fabricating, testing, and modeling of advanced grid stiffened fiber reinforced polymer tube encased concrete cylinders. *Journal of Composite Materials*, **42**, 1103–1124.
67. Li, G., Pang, S.S., and Ibekwe, S.I. (2011) FRP tube encased rubberized concrete cylinders. *Materials and Structure*, **44**, 233–243.
68. Sun, L. and Huang, W.M. (2010) Mechanisms of the multi-shape memory effect and temperature memory effect in shape memory polymers. *Soft Matter*, **6**, 4403–4406.
69. Bellin, I., Kelch, S., Langer, R., and Lendlein, A. (2006) Polymeric triple shape materials. *Proceedings of the National Academy of Science of USA*, **103**, 18043–18047.
70. Bellin, I., Kelch, S., and Lendlein, A. (2007) Dual-shape properties of triple-shape polymer networks with crystallizable network segments and grafted side chains. *Journal of Materials Chemistry*, **17**, 2885–2891.
71. Behl, M., Bellin, I., Kelch, S., Wagermaier, W., and Lendlein, A. (2009) One-step process for creating triple-shape capability of AB polymer networks. *Advanced Functional Materials*, **19**, 102–108.
72. Pretsch, T. (2010) Durability of a polymer with triple-shape properties. *Polymer Degradation and Stability*, **95**, 2515–2524.
73. Wagermaier, W., Zander, T., Hofmann, D., Kratz, K., Kumar, U.N., and Lendlein, A. (2010) In situ X-ray scattering studies of poly(ε-caprolactone) net-works with grafted poly(ethylene glycol) chains to investigate structural changes during dual- and triple-shape effect. *Macromolecular Rapid Communications*, **31**, 1546–1553.
74. Behl, M., Razzaq, M.S., and Lendlein, A. (2010) Shape-memory polymers: multifunctional shape-memory polymers. *Advanced Materials*, **22**, 3388–3410.
75. Pretsch, T. (2010) Triple-shape properties of a thermoresponsive poly(ester urethane). *Smart Materials and Structures*, **19**, paper 015006 (7 pages).

76. Xie, T., Xiao, X., and Cheng, Y.T. (2009) Revealing triple-shape memory effect by polymer bilayers. *Macromolecular Rapid Communications*, **30**, 1823–1827.
77. Luo, X. and Mather, P.T. (2010) Triple-shape polymeric composites (TSPCs). *Advanced Functional Materials*, **20**, 2649–2656.
78. Xie, T. (2010) Tunable polymer multi-shape memory effect. *Nature*, **464**, 267–270.
79. Li, J. and Xie, T. (2011) Significant impact of thermo-mechanical conditions on polymer triple-shape memory effect. *Macromolecules*, **44**, 175–180.
80. Feldman, D. (2002) Polymer weathering: photo-oxidation. *Journal of Polymers and the Environment*, **10**, 163–173.
81. Zhang, D., Dougal, S.M., and Yeganeh, M.S. (2000) Effects of UV irradiation and plasma treatment on a polystyrene surface studied by IR–visible sum frequency generation spectroscopy. *Langmuir*, **16**, 4528–4532.
82. Nakatsuka, S. and Andrady, A. (1994) Studies on enhanced degradable plastics. III. The effect of weathering of polyethylene and (ethylene-carbon monoxide) copolymers on moisture and carbon dioxide permeability. *Journal of Environmental Polymer Degradation*, **2**, 161–167.
83. Song, F. and Gao, Y. (2009) Chemical characteristics of precipitation at metropolitan Newark in the US East Coast. *Atmospheric Environment*, **43**, 4903–4913.
84. Xu, T., Li, G., and Pang, S.S. (2011) Effects of ultraviolet radiation on morphology and thermo-mechanical properties of shape memory polymer based syntactic foam. *Composites Part A: Applied Science and Manufacturing*, **42**, 1525–1533.
85. Xu, T. and Li, G. (2011) Durability of shape memory polymer based syntactic foam under accelerated hydrolytic ageing. *Materials Science and Engineering A*, **528**, 7444–7450.

4

Constitutive Modeling of Amorphous Thermosetting Shape Memory Polymer and Shape Memory Polymer Based Syntactic Foam

From Chapter 3, it is seen that thermosetting SMP and thermosetting SMP based syntactic foam exhibit a complex thermomechanical behavior. Physics based mathematical modeling of the constitutive behavior will deepen understanding and design of these smart materials and broaden applications in various engineering structures. Physics based constitutive modeling, as the name implies, depends on the mechanisms that control the shape memory process. Currently, several mechanisms have been used to trigger shape recovery, such as glass transition for amorphous SMPs, melting transition for crystalline SMPs, photo-induced network rearrangements for photo triggered SMPs, etc. [1]. Therefore, the modeling approach changes as the mechanism controlling shape recovery changes. A comprehensive review of constitutive modeling of SMPs has been given in a few recent review papers [2–4]. In this chapter, focus will be on amorphous thermosetting SMPs and thermosetting SMP based syntactic foam programmed by cold compression.

Like other polymeric materials, rheological modeling was first attempted to predict the constitutive behavior of SMPs. Although earlier efforts [5–8] using rheological models were able to describe the characteristic thermomechanical behavior of SMPs, loss of the strain storage and release mechanisms usually led to limited prediction accuracy. Also, the models were 1-D and could only predict the behavior under a uniaxial stress condition, such as 1-D tension. Furthermore, these models can only work for a small strain. They cannot predict the thermomechanical behavior of SMPs with finite strain, which is the case for most SMPs. Later, meso-scale models [9,10] were developed to predict the constitutive behavior of SMPs. However, one limitation is that a meso-scale model cannot understand the shape memory mechanisms in detail because the mechanisms controlling the shape memory are in a molecular

Self-Healing Composites: Shape Memory Polymer-Based Structures, First Edition. Guoqiang Li.
© 2015 John Wiley & Sons, Ltd. Published 2015 by John Wiley & Sons, Ltd.

length scale such as macromolecule recoil driven by entropy or structural relaxation. Therefore, molecular dynamic (MD) simulation [11] was conducted. While the MD model can understand the shape memory mechanism at the molecular length scale, it is still limited to a small number of molecules and scale-up to macroscopic material behavior remains a significant challenge. This challenge can be easily justified because any load bearing structural component may need many moles of materials to fabricate and each mole of matter consists of 6.023×10^{23} particles, as specified by the Avogadro number. Recently, various phenomenological approaches [12–21] have been developed to interpret the thermomechanical behavior of SMPs from a macroscopic viewpoint, but with some links to shape memory mechanisms. With different underlying hypotheses, these approaches can be roughly divided into two categories. One is the phase transition modeling approach, that is, shape recovery is considered as a phase transition process from the frozen phase to the active phase, and the other is the thermo-viscoelastic/thermoviscoplastic modeling approach, that is, for amorphous shape memory polymers, which operate through the glass transition, thermoviscoelastic/thermoviscoplastic models describe the dramatic change in the molecular mobility of the polymer chains by developing physically motivated temperature dependent constitutive relations for the viscosity and/or moduli of amorphous SMPs [3].

For the phase transition modeling approach, Liu *et al.* [15] adopted a first-order phase transition concept and modeled the SMP as a continuum mixture of a frozen phase and an active phase. The frozen phase and the active phase represented two different configurations at temperatures below and above the glass transition temperature (θ_g). A term of stored strain (ε_s) was introduced to quantify the strain storage and release mechanisms. Shape recovery is a process of phase transition from the frozen phase to the active phase, and the process of programming or strain storage is a reverse process, that is, phase transition from the active phase to the frozen phase. The parameters of their model can be easily determined by simple macroscopic testing. It is found that the simulation reasonably captures the essential shape memory responses. However, the model simplified the SMP as a special elastic problem, and thus did not consider the time dependence of the material. Furthermore, a notable controversy is that it did not represent the physical processes of the glass transition and thus resulted in nonphysical parameters, such as the volume fraction of the frozen phase.

To address such issues, Nguyen *et al.* [20] investigated the shape memory behavior through a structural and stress relaxation mechanism. They proposed that the shape memory phenomena of SMPs were primarily motivated by the dramatic change of molecular chain mobility induced by the glass transition. The chain mobility underpins the ability for the chain segments to rearrange locally to bring the macromolecular structure and stress response to equilibrium, suggesting that the structure relaxes instantaneously to the equilibrium state at a temperature above the glass transition temperature but responds sluggishly during cooling. Such behavior macroscopically freezes the structure at a temperature below the glass transition temperature in a nonequilibrium configuration, and thus allows the material to fix a temporary shape. Reheating above the glass transition temperature restores the mobility and allows the macromolecule to relax to its originally coiled configuration, leading to macroscopic shape recovery. The important feature of this approach is that it reasonably and physically interprets the underlying mechanism for shape memory effects and the time-dependent behavior of the shape memory can be reproduced. However, utilization of the model requires incorporation of many other sophisticated models, which not only complicates the entire constitutive model but also requires some additional fundamental assumptions. It is noticed that some of these hypotheses, such as the free energy activation parameter in the Adams–Gibbs model, which was considered to be invariant to the temperature and pressure, are still debatable [22,23]. It deserves mentioning that the structural relaxation concept has deep roots in the rheological approach, although the constitutive relation of each element is more physics or mechanism based rather than purely phenomenologically based.

With increased understanding of various types of SMPs, physics based constitutive modeling is possible. With such models, fewer parameters will be needed and the predictability of the models will be more reliable and accurate. Also, a physics based model can be easily integrated into finite element modeling (FEM) so that SMP based engineering structures and devices can be analyzed and designed. This will be a direction for future studies.

It is noted that while the majority of constitutive modeling focuses on thermally induced dual-shape memory behavior, triple-shape and multishape SMPs have been developed recently and they call for constitutive modeling [1]. In addition, the effect of programming temperature and strain rate on the constitutive behavior also needs modeling [2]. Furthermore, some recent studies have found that while the shape recovery ratio can be 100%, other mechanical properties such as recovery stress or modulus become smaller and smaller as the thermo-mechanical cycles increase, which has been explained by the shape memory effect in the microscopic scale [24]. Obviously, these new findings also call for constitutive modeling.

As this is an introductory book, this chapter will be focused on the mechanism based phenomenological model and will be arranged in the following way. We will first present a brief overview of finite strain mechanics in order for readers to be able to understand the models to follow. We will then illustrate modeling of the thermosetting SMP programmed by cold compression. Finally, the thermosetting SMP based syntactic foam programmed by cold compression, with damage, will be modeled using the thermoviscoplastic approach.

4.1 Some Fundamental Relations in the Kinematics of Continuum Mechanics

For the SMP to be modeled in this book, the strain is usually very large, such as a uniaxial compression strain of 40% as discussed in Chapter 3. Therefore, classical mechanics of materials based on a small strain assumption will cause significant deviation. The reason is that a small strain assumption (such as less than 5%) can be simplified as a linear relationship between displacement and strain, while a large deformation shows a nonlinear relationship. Therefore, in a large deformation or finite strain analysis, the deformation gradient is usually used instead of the displacement gradient. For ease of presentation in the following sections, we will briefly discuss the continuum mechanics for a finite strain. Readers interested in more detailed information and a rigorous discussion can refer to continuum mechanics textbooks and papers, such as References [25] to [32].

4.1.1 Deformation Gradient

For a small strain, the following relationship holds between the tensor strain and the displacement gradient:

$$\varepsilon_{ij} = \frac{1}{2}\left(\frac{\partial u_i}{\partial x_j} + \frac{\partial u_j}{\partial x_i}\right) \tag{4.1}$$

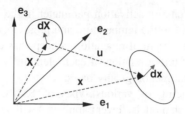

Figure 4.1 An undeformed body is mapped into a deformed body

where ε is the tensor strain, u is the displacement, x is the coordinate axis, and $i, j = 1, 2, 3$ are the running index. It is obvious that the strain–displacement has a linear relationship for a small strain assumption. Later, we will show that for a finite strain assumption, a second-order term exists in the parentheses on the right-hand side of Equation (4.1).

Consider a material's body in Figure 4.1 before deformation. After internal or external loading (force, thermal, electromagnetic field, gravity, etc.), the body is deformed, and may also have rigid body displacement, into the deformed spatial body shown in Figure 4.1. A material's point is defined in the Cartesian coordinate system ($\mathbf{e}_1$, $\mathbf{e}_2$, $\mathbf{e}_3$) by a position vector $\mathbf{X}$ in the undeformed body and position vector $\mathbf{x}$ in the deformed body; $\mathbf{u}$ is the displacement vector. Throughout this chapter, bold face letters will represent tensors. Obviously,

$$\mathbf{x} = \mathbf{X} + \mathbf{u} \tag{4.2}$$

Our interest is in finding the deformation in the material's body because deformation will cause stress, fatigue, cracking, failure, etc. In order to measure deformation in a material point, we need to observe the deformation of a differential element at that point. Let us consider an infinitesimal material vector in the reference configuration (undeformed body), $d\mathbf{X}$, in Figure 4.1 before deformation and $d\mathbf{x}$ after deformation (stretch and rotation). We can say that $d\mathbf{X}$ is deformed into $d\mathbf{x}$. Now we want to find the link between them. Since $\mathbf{x}$ is a function of $\mathbf{X}$, we can express $\mathbf{x}$ in terms of $\mathbf{X}$ using a Taylor series expansion. Obviously, if the higher-order terms are neglected ($(d\mathbf{X})^2$ and higher-order terms) in the series because $|d\mathbf{X}| \rightarrow 0$, we have

$$d\mathbf{x} = \frac{\partial \mathbf{x}}{\partial \mathbf{X}} d\mathbf{X} \tag{4.3}$$

In continuum mechanics, we say that $d\mathbf{X}$ is mapped into $d\mathbf{x}$ and the mapping is defined as the deformation gradient:

$$\mathbf{F} = \frac{\partial \mathbf{x}}{\partial \mathbf{X}} \tag{4.4}$$

where $\mathbf{F}$ is the deformation gradient tensor, which is a second-order tensor.

The deformation gradient, whose components are finite, characterizes the deformation in the neighborhood of a point $\mathbf{X}$, mapping infinitesimal line elements $d\mathbf{X}$ coming from $\mathbf{X}$ in the undeformed configuration to the infinitesimal line elements $d\mathbf{x}$ emanating from $\mathbf{x}$ in the deformed configuration. It is noted that no consideration is given to the particular sequence by which the deformed configuration is reached from the undeformed configuration. Therefore, the deformation gradient can be considered to be independent of time.

In matrix form, the second-rank deformation gradient tensor can be expressed by

$$
\mathbf{F} = \begin{bmatrix} F_{11} & F_{12} & F_{13} \\ F_{21} & F_{22} & F_{23} \\ F_{31} & F_{32} & F_{33} \end{bmatrix} = \begin{bmatrix} \dfrac{\partial x_1}{\partial X_1} & \dfrac{\partial x_1}{\partial X_2} & \dfrac{\partial x_1}{\partial X_3} \\ \dfrac{\partial x_2}{\partial X_1} & \dfrac{\partial x_2}{\partial X_2} & \dfrac{\partial x_2}{\partial X_3} \\ \dfrac{\partial x_3}{\partial X_1} & \dfrac{\partial x_3}{\partial X_2} & \dfrac{\partial x_3}{\partial X_3} \end{bmatrix} \tag{4.5}
$$

4.1.2 Relation Between Deformation Gradient and Displacement Gradient

Now we have two entities that link the undeformed body to the deformed body, that is, the displacement vector $\mathbf{u}$ and the deformation gradient tensor $\mathbf{F}$. Next, we would like to establish connections between these two entities, which will help us establish the link between the displacement and strain and the deformation gradient and strain for the case with finite strain. Let us differentiate both sides of Equation (4.2) with respect to $\mathbf{X}$:

$$
\frac{\partial \mathbf{x}}{\partial \mathbf{X}} = \frac{\partial \mathbf{X}}{\partial \mathbf{X}} + \frac{\partial \mathbf{u}}{\partial \mathbf{X}} \tag{4.6}
$$

Obviously, the left-hand side is the deformation gradient tensor $\mathbf{F}$. The first term on the right-hand side is the Kronecker delta second-order tensor and the second term on the right-hand side is the displacement gradient, which is also the strain tensor in the undeformed body. Therefore, Equation (4.6) can be rewritten as

$$
\mathbf{F} = \boldsymbol{\delta} + \frac{\partial \mathbf{u}}{\partial \mathbf{X}} \tag{4.7}
$$

where $\boldsymbol{\delta}$ is the Kronecker delta, that is,

$$
\boldsymbol{\delta} = \begin{bmatrix} 1 & 0 & 0 \\ 0 & 1 & 0 \\ 0 & 0 & 1 \end{bmatrix} \tag{4.8}
$$

4.1.3 Polar Decomposition of Deformation Gradient

In dynamics, it is well known that planar motion of a rigid body can be decomposed into translation with and rotation about a reference point. Similarly, in continuum mechanics, strain is decomposed into normal strain (stretch or translation) and shear strain (rotation). It is thus inferred that the deformation gradient could be decomposed into stretch and rotation. Before we proceed, let us discuss the rigid body motion induced deformation gradient.

First, we will consider translation only. From Equation (4.7), if **u** is translation only, the derivative (d**u**/d**X**) will be zero because the displacement **u** at any point on the material body is the same, or **u** (both magnitude and direction) does not change with the position vector **X**. Hence, the deformation gradient **F** = **δ**. It is interesting to note that if **u** = 0, that is, no displacement, we have the same **F** = **δ** from Equation (4.7). Therefore, the deformation gradient tensor does not distinguish between states of translation. In other words, rigid body translation does not cause change of the deformation gradient. However, if **u** is rotation only, we cannot obtain the same result. The reason is that rotation will cause a change of direction for each material point in the body. Therefore, the vector derivative of (d**u**/d**X**) in Equation (4.7) does not equal zero.

Based on the above discussion, the deformation gradient consists of two components: stretch and rotation. This may be confusing. Contrary to the common definition of deformation, which implies distortion or change in shape, the continuum mechanics definition includes rigid body motions where shape changes do not take place. In other words, the deformation gradient, as a measure of deformation, consists of both a stress inducing component (stretch) and a rigid body rotation (no stress is induced). Because we are more concerned with the component that induces stress, it is desired to separate the stress inducing component from the deformation gradient. This can be achieved by polar decomposition.

As shown in Figure 4.2, a material body can be first stretched and then rotated to the deformed shape; it can also be rotated first and then stretched to the deformed shape. This type of decomposition is similar to the decomposition of motion in dynamics. In Figure 4.2, we also highlight the change of a differential element with stretch and rotation. Let us define the rotation deformation gradient as **R** and the stretch deformation gradient as **U** (stretch first) or **V** (rotation first).

Using a similar argument to that of Equation (4.3), for stretch first followed by rotation, we have, from Figure 4.2:

$$
\begin{cases}
\mathbf{U} = \dfrac{\partial \mathbf{y}}{\partial \mathbf{X}} \\[2mm]
\mathbf{R} = \dfrac{\partial \mathbf{x}}{\partial \mathbf{y}}
\end{cases}
\tag{4.9}
$$

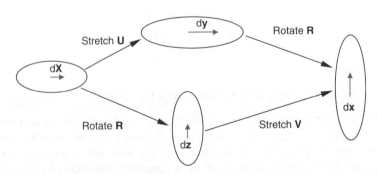

Figure 4.2 Polar decomposition of the deformation gradient

Based on the chain rule and considering Equation (4.9):

$$\mathbf{dx} = \frac{\partial \mathbf{x}}{\partial \mathbf{y}}\mathbf{dy} = \left(\frac{\partial \mathbf{x}}{\partial \mathbf{y}}\right)\left(\frac{\partial \mathbf{y}}{\partial \mathbf{X}}\mathbf{dX}\right) = \left(\frac{\partial \mathbf{x}}{\partial \mathbf{y}}\right)\left(\frac{\partial \mathbf{y}}{\partial \mathbf{X}}\right)\mathbf{dX} = (\mathbf{R})(\mathbf{U})\mathbf{dX} \qquad (4.10)$$

According to Equation (4.3), we have

$$\mathbf{F} = \mathbf{RU} \qquad (4.11)$$

Similarly, for the case of rotation first followed by stretch, we have

$$\mathbf{F} = \mathbf{VR} \qquad (4.12)$$

Equations (4.11) and (4.12) are called polar decomposition of the deformation gradient. Particularly, $\mathbf{U}$ is called the right stretch tensor and $\mathbf{V}$ is called the left stretch tensor.

It can be proved that $\mathbf{U}$ is a symmetric and positive definite tensor, which is a measure of the local stretching (or contraction) of material at $\mathbf{X}$. $\mathbf{V}$ is also is a symmetric and positive definite second-order tensor called the left stretch tensor, which is a measure of the local stretching (or contraction) of the material in the deformed configuration at $\mathbf{x}$. $\mathbf{R}$ is a proper orthogonal tensor, that is, $\mathbf{R}^\mathrm{T}\mathbf{R} = \mathbf{I}$ or $\det\mathbf{R} = 1$, where T means transpose, $\mathbf{I}$ is the identity tensor, and det is the determinant.

It can be proved that $\mathbf{U}$ and $\mathbf{V}$ can be evaluated from $\mathbf{F}$ as follows:

$$\mathbf{U} = \sqrt{\mathbf{F}^\mathrm{T}\mathbf{F}} \qquad (4.13)$$

$$\mathbf{V} = \sqrt{\mathbf{F}\mathbf{F}^\mathrm{T}} \qquad (4.14)$$

Corresponding to $\mathbf{U}$ and $\mathbf{V}$, two new tensors can be defined, which are used to calculate $\mathbf{U}$ and $\mathbf{V}$. We have the right Cauchy–Green deformation tensor $\mathbf{C}$ and the left Cauchy–Green deformation tensor $\mathbf{B}$:

$$\mathbf{C} = \mathbf{U}^2 = \mathbf{F}^\mathrm{T}\mathbf{F} \qquad (4.15)$$

$$\mathbf{B} = \mathbf{V}^2 = \mathbf{F}\mathbf{F}^\mathrm{T} \qquad (4.16)$$

4.1.4 Definition of Strain

In solid mechanics, various definitions have been used to measure strain. These measures are used depending on the particular types of problems to be solved. Here, we would like to define several commonly used strain measures.

The first one is the small strain definition, or Cauchy strain tensor. Equation (4.1) can be rewritten in tensor form as

$$\varepsilon = \frac{1}{2}\left[\frac{\partial \mathbf{u}}{\partial \mathbf{X}} + \left(\frac{\partial \mathbf{u}}{\partial \mathbf{X}}\right)^\mathrm{T}\right] \qquad (4.17)$$

Here we use $\mathbf{X}$ instead of $\mathbf{x}$ because they can be treated as equal for small deformations. Substituting Equation (4.7) in Equation (4.17) gives

$$\varepsilon = \frac{1}{2}(\mathbf{F} + \mathbf{F}^T) - \mathbf{I} \tag{4.18}$$

If there is no rotation, the rotation tensor $\mathbf{R} = \mathbf{I}$. Based on the polar decomposition Equation (4.11), $\mathbf{F} = \mathbf{U}$. Recall that $\mathbf{U}$ is symmetric $(\mathbf{U} = \mathbf{U}^T)$ and thus Equation (4.18) can be rewritten as:

$$\varepsilon = \mathbf{U} - \mathbf{I} \tag{4.19}$$

Therefore, for small or no rotation, the Cauchy strain can also be used to measure a large strain such as in uniaxial tension or compression testing of specimens where no rotation is involved. In other words, the Cauchy strain is not limited to small stretch; it is limited to small rotation.

The next strain measure is defined on the undeformed configuration, known as the Green–Lagrange strain $\mathbf{E}$:

$$\mathbf{E} = \frac{1}{2}(\mathbf{F}^T\mathbf{F} - \mathbf{I}) \tag{4.20}$$

Substituting Equation (4.11) in Equation (4.20) and paying attention to the fact that $\mathbf{R}^T\mathbf{R} = \mathbf{I}$, we have

$$\mathbf{E} = \frac{1}{2}(\mathbf{U}^T\mathbf{U} - \mathbf{I}) \tag{4.21}$$

It is seen that the Green–Lagrange strain is independent of the rotation. Therefore, it is indeed a measure of strain. Because we did not enforce any restriction on the level of deformation, it is applicable to both large and small deformation cases. It is noted that the Green–Lagrange strain can be understood from the point of view of geometry. Assuming that a differential element in the undeformed body has a length of dS and ds in the deformed body, it can be proved that in the 1-D case:

$$E = \frac{(ds)^2 - (dS)^2}{2(dS)^2} \tag{4.22}$$

Obviously, this is defined in the undeformed body because the denominator in Equation (4.22) is in terms of the squared length of the differential length in the undeformed body.

Another strain measure is defined in the deformed body, known as the Almansi–Euler strain $\mathbf{e}$:

$$\mathbf{e} = \frac{1}{2}(\mathbf{I} - (\mathbf{F}\mathbf{F}^T)^{-1}) \tag{4.23}$$

Similarly, it can also be rewritten in terms of the left stretch tensor $\mathbf{V}$ as

$$\mathbf{e} = \frac{1}{2}(\mathbf{I} - (\mathbf{V}\mathbf{V}^T)^{-1}) \tag{4.24}$$

Again, rotation has been eliminated from **e**. Hence, it is indeed a measure of strain. Also, just like the Green–Lagrange strain, the Almansi–Euler strain has a geometrical explanation:

$$e = \frac{(ds)^2 - (dS)^2}{2(ds)^2} \tag{4.25}$$

Obviously, it is defined in the deformed body because the denominator in Equation (4.25) is in terms of the squared length of the differential length in the deformed body.

It is noted that the above strain can also be expressed by the displacement gradient. For instance, the Green–Lagrange strain in Equation (4.20), after operation in terms of the tensor index, can be rewritten as

$$E_{ij} = \frac{1}{2}\left(\frac{\partial u_i}{\partial X_j} + \frac{\partial u_j}{\partial X_i} + \frac{\partial u_k}{\partial X_i}\frac{\partial u_k}{\partial X_j}\right) \tag{4.26}$$

Clearly, when compared to Equation (4.1), the quadratic terms appear in large deformation. Therefore, a large deformation strain in terms of the displacement gradient is nonlinear.

Finally, we would like to define volume strain, hydrostatic strain, and deviatoric strain. Consider that a body with an initial volume V_0 is deformed into a body with a final volume of V_f and assume that the principal strains are ε_1, ε_2, and ε_3 in the three principal directions. Then

$$\frac{V_f}{V_0} = (1 + \varepsilon_1)(1 + \varepsilon_2)(1 + \varepsilon_3) \tag{4.27}$$

For small strains, Equation (4.27) can be simplified by neglecting higher order terms as

$$\frac{V_f}{V_0} = 1 + \varepsilon_1 + \varepsilon_2 + \varepsilon_3 = 1 + 3\left(\frac{\varepsilon_1 + \varepsilon_2 + \varepsilon_3}{3}\right) = 1 + 3\varepsilon_h \tag{4.28}$$

where ε_h is called the hydrostatic strain. It can also be expressed as

$$\varepsilon_h = \frac{1}{3}tr(\boldsymbol{\varepsilon}) = \frac{1}{3}I_1 \tag{4.29}$$

in which tr means trace and I_1 is the first invariant of the strain.

The volume strain is defined as

$$\varepsilon_v = \frac{V_f - V_0}{V_0} = \frac{V_f}{V_0} - 1 = 3\varepsilon_h \tag{4.30}$$

and it can be proved that for a large strain

$$\frac{V_f}{V_0} = \det(\mathbf{U}) = \det(\mathbf{F}) \tag{4.31}$$

where det($\mathbf{F}$) is given a special symbol, J, and a special name, the *Jacobian*. Therefore, for a large strain, the volume strain is given by

$$\varepsilon_v = \frac{V_f - V_0}{V_0} = \frac{V_f}{V_0} - 1 = \det(\mathbf{F}) - 1 = J - 1 \qquad (4.32)$$

With the definition of hydrostatic strain, which causes volume change, we can also define the deviatoric strain, which causes shape change, as

$$\varepsilon' = \varepsilon - \varepsilon_h \qquad (4.33)$$

where ε' is the deviatoric strain and ε is the total strain.

4.1.5 Velocity Gradient

The velocity gradient is essential in analyzing path dependent materials such as plastic deformation of cold-compression programmed SMP. It is also useful in determining the energetically conjugate stress and strain, as will be demonstrated in the section on stress definition.

The velocity gradient, $\mathbf{L}$, an Eulerian quantity rather than a Lagrangian quantity, is defined in the deformed configuration as

$$\mathbf{L} = \frac{\partial \mathbf{v}}{\partial \mathbf{x}} \qquad (4.34)$$

where $\mathbf{v}$ is velocity.

We have known that the deformation gradient, $\mathbf{F}$, is defined in the undeformed body, that is, the Lagrangian quantity. We now want to know the relation between $\mathbf{L}$ and $\mathbf{F}$ so that conversion between the Lagrangian and Eulerian quantities becomes possible. The rate of deformation gradient is

$$\dot{\mathbf{F}} = \frac{d\mathbf{F}}{dt} = \frac{d}{dt}\left(\frac{\partial \mathbf{x}}{\partial \mathbf{X}}\right) = \frac{\partial}{\partial \mathbf{X}}\left(\frac{d\mathbf{x}}{dt}\right) = \frac{\partial \mathbf{v}}{\partial \mathbf{X}} = \frac{\partial \mathbf{v}}{\partial \mathbf{x}}\frac{\partial \mathbf{x}}{\partial \mathbf{X}} = \mathbf{LF} \qquad (4.35)$$

This can be rewritten by mutliplying both sides by $\mathbf{F}^{-1}$ as

$$\mathbf{L} = \dot{\mathbf{F}}\mathbf{F}^{-1} \qquad (4.36)$$

Therefore, it is ensured that the Eulerian quantity can be caluclated by using the Lagrangien quantity. From the velocity gradient tensor, two new tensors, rate of deformation tensor, $\mathbf{D}$, and spin tensor, $\mathbf{W}$, can be defined:

$$\mathbf{L} = \frac{1}{2}(\mathbf{L} + \mathbf{L}^T) + \frac{1}{2}(\mathbf{L} - \mathbf{L}^T) = \mathbf{D} + \mathbf{W} \qquad (4.37)$$

Obviously, **D** is symmetric and represents a strain rate tensor in an Eulerian context. **W**, on the other hand, is an antisymmetric tensor and is related to the rate of rotation of a body, also in an Eulerian context. To prove this point, **L** can be rewritten after some routine tensor operations as

$$\mathbf{L} = \mathbf{R}(\dot{\mathbf{U}}\,\mathbf{U}^{-1})\mathbf{R}^{\mathrm{T}} + \dot{\mathbf{R}}\,\mathbf{R}^{\mathrm{T}} \tag{4.38}$$

Clearly, the first term represents the deformation through stretch **U**. The contribution by the rotation **R** in the first term is to rotate the stretched body to its current orientation. On the other hand, the second term represents spin. This is true because if only rigid body rotation occurs, the first term is equal to zero because $\dot{\mathbf{U}} = 0$, and only the second term contributes to **L**, which is spin.

Finally, the deformation gradient, the Green–Lagrange strain, and rate of deformation tensor can be linked by the following relation:

$$\mathbf{D} = \mathbf{F}^{\mathrm{T}}\dot{\mathbf{E}}\,\mathbf{F}^{-1} \tag{4.39}$$

This relation will be used in the next section to identify the energetically conjugate stress and strain.

4.2 Stress Definition in Solid Mechanics

In strength of materials text, it is well known that Cauchy stress, $\boldsymbol{\sigma}$, is defined as the force per unit deformed area. This is an appropriate definition for a small strain because the area in the undeformed body and deformed body is almost the same. For a large strain, however, we generally do not know the area of the deformed configuration. Thus we need to define a stress measure that we can use in the reference configuration. However, it is noted that Cauchy stress is still the most used stress definition or true stress because the equilibrium is about the deformed body but not the undeformed body. Therefore, other definitions of stress are for convenience of mathematical operation. Stress is generally reported as the Cauchy stress. This shows a distinct departure from the strain definitions.

First, let us define conjugate stress and strain. Stresses and strains are energetically conjugate if their double-dot product represents the strain energy stored in the body. In this sense, the Cauchy stress and Cauchy strain used in strength of materials text are energetically conjugate because this is exactly how we define strain energy density in strength of materials text. One reason is that for a small strain, the Cauchy strain can be considered to be defined either in the undeformed or deformed body. Because Cauchy stress is defined in the deformed body, they are energetically conjugate.

Now we will define two new stress measures. One is the first Piola–Kirchoff stress tensor, **T**, and the other is the second Piola–Kirchoff stress tensor, **S**. Consider that a differential force d**P** is applied to the body. Obviously, the force does not change no matter it is from the undeformed body or the deformed body. This argument can be used to define the two new stress measures. Using the tensor note, we have

$$\mathrm{d}\mathbf{P} = \boldsymbol{\sigma}\mathbf{n}\,\mathrm{d}s \tag{4.40}$$

where **n** is the normal vector to the surface of the deformed configuration and ds is the surface area in the deformed configuration. In the undeformed body

$$d\mathbf{P} = \mathbf{T N}\, d\mathbf{S} \tag{4.41}$$

where **N** is the normal vector to the surface of the undeformed configuration and dS is the surface area in the undeformed configuration.

Using the mapping between the undeformed and deformed configuration surface areas, we can write the definition of the force in the deformed configuration as

$$d\mathbf{P} = \sigma\mathbf{n}\, ds = \sigma J\mathbf{N F}^{-1}\, d\mathbf{S} \tag{4.42}$$

The differential forces in Equations (4.41) and (4.42) are the same, so

$$\mathbf{T} = \sigma J\mathbf{F}^{-T} \tag{4.43}$$

This is the first Piola–Kirchoff stress tensor. It means the force in the deformed configuration but in the undeformed area. It is noted that **T** is not symmetric so it is not easy to use. Similarly, we can obtain the second Piola–Kirchoff stress tensor, which is

$$\mathbf{S} = J\mathbf{F}^{-1}\sigma\mathbf{F}^{-T} \tag{4.44}$$

It is noted that **S** is symmetric and energetically conjugate with the Green–Lagrange strain **E**. From Equation (4.44), it is seen that **S** is produced by mapping the stress in the deformed body to the undeformed body. In other words, **S** represents the force mapped to the undeformed body on the undeformed area.

To help visualize the three stress measures, they are illustrated in Figure 4.3.

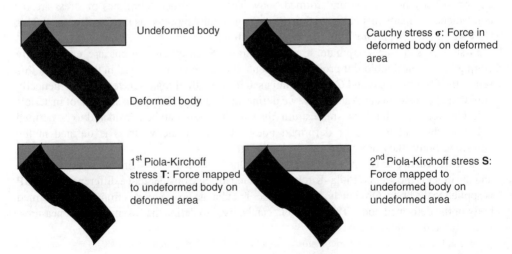

Figure 4.3 Visualization of stress measures

4.3 Multiplicative Decomposition of Deformation Gradient

Multiplicative decomposition of the deformation gradient has become an effective tool to solve finite strain problems. Multiplicative decomposition of the deformation gradient is based on an intermediate material configuration, which is obtained by a conceptual unloading of the deformed material configuration (deformed body) to zero stress. It has been widely used in thermoelasticity, elastoplasticity, and biomechanics. An excellent review paper on multiplicative decomposition of the deformation gradient was given by Lubarda [31]. In elastoplasticity, which is the focus of this chapter on cold-compression programmed SMP, the intermediate configuration is obtained from the deformed material configuration by elastic unloading to zero stress. It differs from the undeformed configuration by residual or plastic deformation, and from the deformed configuration by reversible or elastic deformation [31]. The corresponding decomposition of the elastoplastic deformation gradient into its elastic and plastic parts was introduced by Lee [33]. Related early contributions also include Backman [34], Lee and Liu [35], Fox [36], and Willis [37]. The multiplicative decomposition technique has been successfully employed in the constitutive modeling of various polymers [38–42].

Figure 4.4 shows a schematic of mapping from an undeformed body to a deformed body, as well as the decomposition of the deformation gradient to the plastic deformation gradient, which leads to an intermediate relaxed body with pure plastic deformation and is obtained by conceptually unloading the deformed body elastically to zero stress, and the elastic deformation gradient, which leads the intermediate body to the deformed body. If the total deformation gradient is $\mathbf{F}$, the plastic component is $\mathbf{F}_p$, and the elastic component is $\mathbf{F}_e$, similar to polar decomposition, the multiplicative decomposition leads to

$$\mathbf{F} = \mathbf{F}_e \mathbf{F}_p \tag{4.45}$$

As indicated in Reference [31], in the case when elastic unloading to zero stress is not physically achievable due to the possible onset of reverse plastic deformation before the state of zero stress is reached, the intermediate configuration can be conceptually introduced by a virtual unloading to zero stress, locking all inelastic structural changes that would occur during the actual unloading.

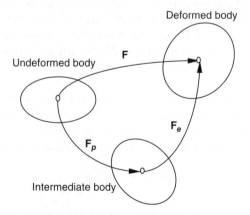

Figure 4.4 Multiplicative decomposition of an undeformed body to an intermediate body by elastically unloading the deformed body to zero stress

It is noted that multiplicative decomposition of the deformation gradient is different from polar decomposition. Based on the displacement sequence in polar decomposition, Equation (4.45) can be interpreted as plastically deforming into the intermediate configuration and then elastically deforming into the current configuration. In other words, it seems that the plastic and elastic deformation is decoupled. This is not always correct. In some materials, such as elastic–perfectly plastic materials, this is true. However, for other materials, such as those with strain hardening, plastic and elastic deformation is coupled, and sometimes large plastic deformation changes the elastic behavior of the materials. Therefore, Equation (4.45) cannot be interpreted as plastic deformation followed by elastic deformation; rather, it should be understood from the point of view that the intermediate configuration is obtained by elastically unloading the deformed body to zero stress. It is a conceptual process, not a real physical process. However, for small and moderate plastic deformation, the effect of plastic deformation on elastic behavior can be neglected and Equation (4.45) can be understood as plastically deformed first followed by elastic deformation. It can also be treated as elastically deformed first followed by plastic deformation, which reads as

$$\mathbf{F} = \mathbf{F}_p \mathbf{F}_e \tag{4.46}$$

Another point that deserves attention is that the plastic and elastic deformation gradients are not unique. Rigid body rotation will change the deformation gradient, although it does not change the stretch and thus stress of the materials. More discussions on this issue can be found in References [43] to [45].

Multiplicative decomposition can be extended to more than one intermediate configuration, as long as each intermediate configuration has a physical meaning. For example, in thermo-mechanical constitutive modeling of SMP, the material experiences thermal, plastic, and elastic deformation. Therefore, the deformed configuration can be achieved by the following intermediate configuration. First, the undeformed body is deformed into an intermediate heated configuration with the deformation gradient $\mathbf{F}_\mathrm{T}$, which is obtained from the deformed material configuration by mechanically unloading to zero stress; then the heated configuration is isothermally deformed into the deformed configuration with the deformation gradient $\mathbf{F}_\mathrm{M}$. Because $\mathbf{F}_\mathrm{M}$ consists of both plastic and elastic deformation, it can be further decomposed into the plastic deformation gradient $\mathbf{F}_p$ and the elastic deformation gradient $\mathbf{F}_e$, as shown in Figure 4.5. Of course, we have assumed that the thermal deformation does not affect the mechanical behavior and plastic deformation does not affect the elastic behavior of the material.

Similarly, assuming the reference or original or material or undeformed configuration is deformed into the current or final or spatial or deformed configuration through a series of intermediate configurations (for instance thermal, electrical, magnetic, mechanical, etc.) (Figure 4.6), multiplicative decomposition of the deformation gradient yields

$$\mathbf{F} = \mathbf{F}_n \mathbf{F}_{n-1} \cdots \mathbf{F}_1 \tag{4.47}$$

here $\mathbf{F}_i$ ($i = 1, 2, \ldots, n$) is the deformation gradient of the ith intermediate configuration.

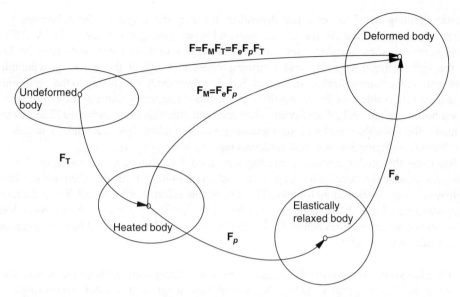

Figure 4.5 Multiplicative decomposition of the thermomechanical deformation gradient

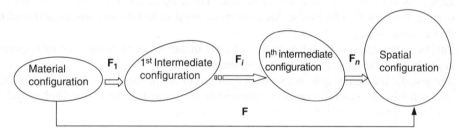

Figure 4.6 Schematic of multiplicative decomposition of the deformation gradient with n intermediate configurations

4.4 Constitutive Modeling of Cold-Compression Programmed Thermosetting SMP

4.4.1 General Considerations

As previously documented, the molecular resistance to inelastic deformation for SMPs below the glass transition temperature (T_g) mainly originates from two sources: the intermolecular resistance to segmental rotation and the entropic resistance to molecular alignment [38,39,46]. The four-step thermomechanical cycle shown in Figure 3.13 can be analyzed as follows. It is assumed that the plastic flow does not commence until the stressed material completely overcomes the free energy barrier to the segmental rotation, a restriction imposed on molecular segment motion from neighboring chains. Following the initial yield, molecular alignment occurs and subsequently alters the conformational entropy of the material (step 1). Since the plastic straining develops in a rate-dependent manner, the length of the relaxation time physically indicates the degree of the nonequilibrium configuration (step 2). A relaxed configuration is then obtained after elastically unloading to a stress-free state (step 3). Due to the high material viscosity and vanishing chain mobility at the glassy programming temperature, the nonequilibrium structure is largely prevented from relaxing to the equilibrium state in an observable time frame, resulting in a retained temporary shape at the end of step 3. Upon heating to above T_g the viscosity decreases and chain mobility increases. The thermo-dynamically favorable tendency of increasing entropy allows the material to restore its equilibrium configuration and thus achieve shape recovery (step 4).

Based on this understanding, a mechanism based constitutive model incorporating the nonlinear structural relaxation model into the continuum finite-deformation thermoviscoelastic framework was developed as follows. The aim of this effort was to establish a quantitative understanding of the shape memory behavior of the thermally responsive thermoset SMP programmed at temperatures below T_g. To simplify the formulation, several basic assumptions were made in this study:

1. The SMP system is considered to be macroscopically isotropic and homogeneous. A uniform stress field assumption is held. This assumption suggests that when comparing with experimental results, uniaxial compression test results are preferred because specimens under uniaxial loading create a uniform stress distribution within the gauge length of the specimens.
2. Heat transfer in the material is not considered. The temperature is then treated as uniform over the entire body. Obviously, this assumption requires that the test specimen should be thin in order to neglect the heat transfer effect.
3. The structural relaxation and inelastic behavior of the material is assumed to be solely dependent on the temperature, time, and stress.
4. The material is assumed to be without any damage during the thermomechanical cycle. This assumption is acceptable for small pre-strain and a low number of thermomechanical cycles. For large pre-strain or cyclic loading, damage will occur.

In continuum mechanics, constitutive modeling of materials follows certain steps, including deformation response, stress response, as well as other particular steps based on materials studied, such as structural relaxation for polymers and a plastic flow rule, and a hardening rule for materials with plastic deformation. In the following, we will present the deformation response, structural relaxation, stress response, and flow rule for the thermosetting SMP programmed by cold-compression programming.

4.4.2 Deformation Response

As discussed in Chapter 3, pseudo-plastic deformation is the key for cold-programmed thermosetting SMP to display shape memory functionality. Therefore, the deformation includes both plastic/viscoplastic and elastic/viscoelastic deformation. The thermomechanical cycle also includes thermal deformation. Based on Figure 4.5, the deformation gradient $\mathbf{F}$ can be multiplicatively decomposed into thermal $\mathbf{F}_T$ and mechanical $\mathbf{F}_M$, which are further decomposed into plastic $\mathbf{F}_p$ and elastic $\mathbf{F}_e$, as follows:

$$\mathbf{F} = \mathbf{F}_M \mathbf{F}_T = \mathbf{F}_e \mathbf{F}_p \mathbf{F}_T \tag{4.48}$$

Because the material is assumed to be isotropic, the thermal deformation gradient can be expressed as

$$F_T = J_T^{1/3} I \tag{4.49}$$

where J_T is the *Jacobian* and I is the identity tensor. The viscous part of the velocity gradient is then defined as

$$L_v = \dot{F}_v F_v^{-1} = D_v + W_v \tag{4.50}$$

where D_v is the symmetric part of L_v, representing the plastic stretch of the velocity gradient and W_v is the asymmetric component, representing the plastic spin. By applying the polar decomposition, we can also split F_e into a stretch (V_e) and a rotation (R_e) as

$$F_e = V_e R_e \tag{4.51}$$

4.4.3 Structural Relaxation Response

A fictive temperature T_f based approach firstly introduced by Tool [47] has been proved to be extremely successful in supplying the information about the free volume or the structure in the formulation of the free energy density. The fictive temperature T_f is an internal variable to characterize the actual thermodynamic state during the glass transition, defined as the temperature at which the temporary nonequilibrium structure at T is in equilibrium [20]. It was assumed that the rate change of the fictive temperature is proportional to its deviation from the actual temperature and the proportionality factor depends on both T and T_f [48], as indicated in the evolution equation [47]:

$$\frac{dT_f}{dt} = K(T, T_f)(T - T_f) \tag{4.52}$$

The Narayanaswamy–Moynihan model (NMM), discussed in detail by Donth and Hempel [49], is an improvement on this approach. Instead of postulating a simple exponential relaxation mechanism governed by a single relaxation time [47], the nonexponential structural relaxation behavior and the spectrum effect were studied. It was assumed that the whole thermal history $T(t)$ started from a thermodynamic equilibrium state where $T(t_0) = T_f(t_0)$. Tool's fictive temperature is defined by

$$T_f(t) = T(t) - \int_{t_0}^{t} \varphi(\Delta\zeta) dT(t) \tag{4.53}$$

The response function is chosen, according to Moynihan *et al.* [50], in a manner of a Kohlrausch function [51], which is also called the stretched exponential function and is often used as a phenomenological description of relaxation in disordered systems. The value of β, which is called the stretching exponent and describes the nonexponential characteristic of the relaxation process, is defined as

$$\varphi = \exp\left[-(\Delta\zeta)^\beta\right], \quad 0 < \beta \leq 1 \tag{4.54}$$

The dimensionless material time difference $\Delta\zeta$ is introduced to linearize the relaxation process:

$$\Delta\zeta = \zeta(t) - \zeta(t') = \int_{t'}^{t} \frac{dt}{\tau_s} \tag{4.55}$$

where the structural relaxation time τ_s, a macroscopic measurement of the molecular mobility of the polymer, accounts for the characteristic retardation time of the volume creep [20,52]. As presented earlier, the structural relaxation in terms of τ_s is controlled by both the actual temperature T and the fictive temperature T_f, and a Narayanaswamy mixing parameter x was introduced to weigh their individual influence [48]:

$$\tau_s = \tau_0 \exp\left[B\left(T_g - T_\infty\right)^2 \left(\frac{x}{T - T_\infty} + \frac{1-x}{T_f - T_\infty}\right)\right], \quad 0 < x \leq 1 \tag{4.56}$$

It can be observed that the term of $(1-x)$ describes the contribution of T_f. Here, T_g is the glass transition temperature. T_∞ denotes the Vogel temperature, defined as $(T_g - 50)$ (°C), τ_0 corresponds to the reference relaxation time at T_g, and B is the local slope at T_g of the trace of the time–temperature superposition shift factor in the global William–Landel–Ferry (WLF) equation [53].

After obtaining the evolution profile of T_f, we can then evaluate the isobaric volumetric thermal deformation corresponding to a temperature change from T_0 to T [20,48,54]:

$$J_{\mathrm{T}}\left(T, T_f\right) = 1 + \alpha_r\left(T_f - T_0\right) + \alpha_g\left(T - T_f\right) \tag{4.57}$$

where α_r and α_g represent the long-time volumetric thermal expansion coefficients of the material in the rubbery state and the short-time response in the glassy state, respectively.

4.4.4 Stress Response

The mechanical behavior of amorphous glassy polymers under various temperature conditions has been extensively studied by numerous researchers [38–40,55–60]. Although other approaches can still accommodate the present constitutive framework, the method by Boyce and co-workers [38–40,60] is adopted in this study to model the general stress–strain behavior of the thermosetting SMP.

As demonstrated previously, the overall mechanical resistance to the straining of a polymer mainly comes from two distinct sources: the temperature rate dependent intermolecular segmental rotation resistance and the entropy driven molecular network orientation resistance. It is therefore possible to capture this nonlinear behavior by decomposing the stress response into a nonequilibrium time dependent component σ_{ve} representing the viscoplastic behavior and an equilibrium time independent component σ_n representing the rubber-like behavior. The two stress components can be represented by a three-element conceptual model schematically illustrated in Figure 4.7 for a one-dimensional analog [61]. An elastic–viscoplastic component consists of an Eyring dashpot monitoring an isotropic resistance to chain segment rotation and a linear spring used to characterize the initial elastic response, while a paralleled nonlinear hyperelastic element accounts for the orientation network strain hardening behavior. Although the elements in Figure 4.7 are more mechanism or physics based as compared to those in

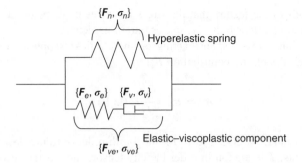

Figure 4.7 A linear rheological illustration for the stress response. *Source:* [61] Reproduced with permission from Elsevier

classical rheological models, the deep roots of the mechanism based model in rheological models must be admitted.

If we further denote the deformation gradient acting on the elastic–viscoplastic component by F_{ve} and the deformation gradient acting on the network orientation spring by F_n, the following constitutive relations are established [61]:

$$\sigma = \sigma_{ve} + \sigma_n \tag{4.58}$$

$$\sigma_{ve} = \sigma_e = \sigma_v \tag{4.59}$$

$$F_{ve} = F_n = F_M \tag{4.60}$$

$$F_{ve} = F_e F_v \tag{4.61}$$

Then the equilibrium response acting on the network orientation element can be defined following the Arruda–Boyce eight-chain model [40] as

$$\sigma_n = \frac{1}{J_n} \mu_r \frac{\lambda_L}{\lambda_{chain}} \mathcal{L}^{-1}\left(\frac{\lambda_{chain}}{\lambda_L}\right) \overline{B}' + k_b(J-1)I \tag{4.62}$$

where μ_r is the initial hardening modulus and k_b denotes the bulk modulus to account for the incompressibility of the rubbery behavior. Because most amorphous polymers exhibit vastly different volumetric and deviatoric behavior, the volumetric and deviatoric contributions are considered separately by taking out the volumetric strain through the split formulation [62,63]:

$$\overline{F}_n = J_n^{-1/3} F_n \tag{4.63}$$

where $J_n = \det(F_n)$. $\overline{B} = \overline{F}_n \overline{F}_n^{\mathrm{T}}$ is the isochoric left Cauchy–Green tensor and $\overline{B}' = \overline{B} - \frac{1}{3}\overline{I}_{n1}I$ represents the deviatoric component of $\overline{B}$. $\overline{I}_{n1} = tr(\overline{B})$ is the first invariant of $\overline{B}$, $\lambda_{chain} = \sqrt{\overline{I}_{n1}/3}$ is the effective stretch on each chain in the eight-chain network, and λ_L is the locking stretch representing the rigidity between entanglements. The Langevin function $\mathcal{L}$ is defined by

$$\mathcal{L}(\beta) = \coth(\beta) - \frac{1}{\beta} \tag{4.64}$$

whose inverse leads to the feature that the stress increases dramatically as the chain stretch approaches its limiting extensibility λ_L, that is, strain hardening.

The nonequilibrium stress response acting on the elastic–viscoplastic component can be determined through the elastic contribution F_e:

$$\sigma_{ve} = \sigma_e = \frac{1}{J_e} L^e (\ln V_e) \tag{4.65}$$

where $J_e = \det(F_e)$ and $L^e = 2G\mathcal{I} + \lambda I \otimes I$ is the fourth-order isotropic elasticity tensor. G and λ are Lamé constants, $\mathcal{I}$ is the fourth-order identity tensor, and I is the second-order identity tensor.

4.4.5 Viscous Flow

As proposed earlier, the molecular process of a viscous flow is to overcome the shear resistance of the material for local rearrangement, that is, overcome segmental rotation resistance. Therefore, a plastic shear strain rate $\dot{\gamma}_v$ is given to help constitutively prescribe the rate of viscous stretch D_v as

$$D_v = \dot{\gamma}_v n \tag{4.66}$$

where

$$n = \frac{\sigma'_{ve}}{\sqrt{\sigma'_{ve} \cdot \sigma'_{ve}}} = \frac{\sigma'_{ve}}{\|\sigma'_{ve}\|}$$

is the normalized deviatoric portion of the nonequilibrium stress. It states that the viscous rate of stretch scales with the plastic shear strain rate and evolves in the direction of the flow stress.

Taking into account the fact that the non-Newtonian fluid relationship must be valid for the dashpot of the mechanical model, the shear strain rate $\dot{\gamma}_v$ can be formulated in an Eyring model [64] with the temperature dependence in a WLF kinetics manner:

$$\dot{\gamma}_v = \frac{s}{\eta_g} \frac{T}{Q} \exp\left(\frac{c_1(T - T_g)}{c_2 + T - T_g}\right) \sinh\left(\frac{Q\bar{\tau}}{T s}\right) \tag{4.67}$$

where $\bar{\tau} = \|\sigma'_{ve}\|/\sqrt{2}$ is defined as the equivalent shear stress, c_1 and c_2 are the two WLF constants; Q is the activation parameter, s represents the athermal shear strength, and η_g denotes the reference shear viscosity at T_g. The evolution in Equation (4.67) reveals the nature of the viscoplastic flow to be temperature dependent and stress activated.

More recently, Nguyen et al. [20] further extended the viscous flow rule to a structure dependent glass transition region by introducing the fictive temperature T_f into the temperature dependence:

$$\dot{\gamma}_v = \frac{s}{\eta_g} \frac{T}{Q} \exp\left(c_1 \left(\frac{c_2(T - T_f) + T(T_f - T_g)}{T(c_2 + T - T_g)}\right)\right) \sinh\left(\frac{Q\bar{\tau}}{T s}\right) \tag{4.68}$$

It can be observed that once the material reaches equilibrium where $T_f = T$, Equation (4.68) will be reduced to Equation (4.67) for a structure independent time–temperature shift factor.

As extensively documented, upon yielding, the initial rearrangement of the chain segments alters the local structure configuration, resulting in a decrease in the shear resistance. To further feature the macroscopic post-yield strain softening behavior, the phenomenological evolution rule for the athermal shear strength s proposed by Boyce, Weber and Parks [39] is used:

$$\dot{s} = h\left(1 - \frac{s}{s_s}\right)\dot{\gamma}_v \qquad (4.69)$$

in which the initial condition $s = s_0$ applies and s_0 denotes the initial shear strength, while s_s denotes the saturation value and h is the slope of the yield drop with respect to the plastic strain. It should be noted that a softening characteristic can only be captured when $s_0 > s_s$ holds.

For completeness, the constitutive relations for the sophisticated temperature and time dependent thermomechanical behavior of the thermally activated thermoset SMP are summarized in Table 4.1 [61].

4.4.6 Determination of Materials Constants

In the above modeling process, 20 parameters of materials are involved. These materials constants need to be determined first before the model can be used to evaluate the effect of various parameters.

Table 4.1 Summary of the thermoviscoelastic model

Deformation response	$\boldsymbol{F} = \boldsymbol{F}_e\boldsymbol{F}_v\boldsymbol{F}_T$
	$\boldsymbol{F}_T = J_T^{-1/3}\boldsymbol{I}$
Structure relaxation	$T_f(t) = T(t) - \int_{t_0}^{t}\varphi(\Delta\varsigma)\mathrm{d}T(t)$
	$\varphi = \exp\left[-(\Delta\varsigma)^{\beta}\right]$
	$\Delta\varsigma = \varsigma(t) - \varsigma(t') = \int_{t'}^{t}\dfrac{\mathrm{d}t}{\tau_s}$
	$\tau_s = \tau_0\exp\left[B(T_g - T_\infty)^2\left(\dfrac{x}{T - T_\infty} + \dfrac{1-x}{T_f - T_\infty}\right)\right]$
Stress response	$\boldsymbol{\sigma} = \boldsymbol{\sigma}_{ve} + \boldsymbol{\sigma}_n$
	$\boldsymbol{\sigma}_n = \dfrac{1}{J_n}\mu_r\dfrac{\lambda_L}{\lambda_{\text{chain}}}\mathcal{L}^{-1}\left(\dfrac{\lambda_{\text{chain}}}{\lambda_L}\right)\overline{\boldsymbol{B}}' + k_b(J-1)\boldsymbol{I}$
	$\boldsymbol{\sigma}_{ve} = \dfrac{1}{J_e}\boldsymbol{L}^e(\ln \boldsymbol{V}_e)$
Viscous flow rule	$\boldsymbol{D}_v = \dot{\gamma}_v\boldsymbol{n}$
	$\dot{\gamma}_v = \dfrac{s}{\eta_g}\dfrac{T}{Q}\exp\left(c_1\left(\dfrac{c_2(T - T_f) + T(T_f - T_g)}{T(c_2 + T - T_g)}\right)\right)\sinh\left(\dfrac{Q\bar{\tau}}{T\,s}\right)$

Source: [61] Reproduced with permission from Elsevier.

Some of the materials constants can be directly obtained through test results, while some others need curve fitting. Curve fitting is similar to optimization. During this process, we fix all the other parameters and only change one parameter until the predicted results match the test results. This is usually an iterative process and involves a compromise among different parameters. For example, if one parameter perfectly matches the test results while another shows a significant deviation from the test results, a compromise is needed between these two parameters so that both of them make the prediction close to the test results, instead of one perfect match and the other with considerable deviation. Also, a sensitivity analysis is helpful so that we know which parameters are more sensitive to the prediction results, and thus these parameters can only be fine-tuned instead of by using a coarse adjustment. The following tests were conducted and used to determine the materials constants.

First, a cooling profile of the thermal deformation is plotted versus the temperature in Figure 4.8. The reference height L_0 denotes the initial sample height. It can be observed that the thermal response is not linear as the temperature passes through the glass transition region. The linear coefficients of thermal expansion α_r and α_g were computed from the slopes both above and below the T_g. Volumetric CTEs are three times the values of the linear CTEs.

Second, μ_r and λ_L are the parameters characterizing the rubbery behavior of the material and can be determined from the stress–strain response at temperatures above T_g (Figure 4.9). Lamé constants G and λ can be related to the initial slope of the isothermal uniaxial compression stress–strain curve in the glassy state by assuming a typical polymer Poisson ratio of 0.4 [17]. Although it has been suggested that different sets of parameters μ_r and λ_L are preferable to capture the fundamentally different response of the rubbery state and the glassy state [17,65], they are treated to be temperature independent for the sake of convenience in parameter identification and computational simplicity.

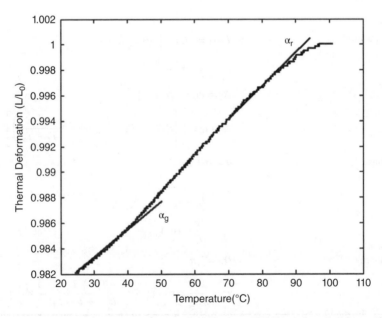

Figure 4.8 Thermal response to a stress-free cooling. *Source:* [61] Reproduced with permission from Elsevier

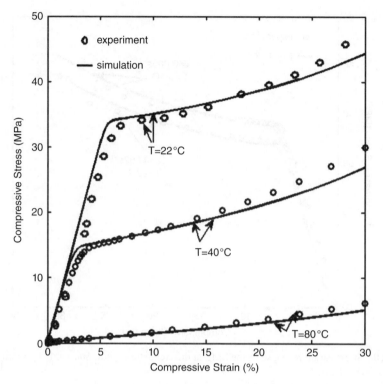

Figure 4.9 Stress–strain response of the SMP at different temperatures. *Source:* [61] Reproduced with permission from Elsevier

As suggested in previous efforts [17,20,39], the viscoplastic parameters such as Q, s, s_s, and h can be roughly determined from curve fitting of the compression tests at different strain rates (Figure 4.10). The ratio Q/s determines the strain rate dependence of the yield strength and s/s_s indicates the drop of the shear strength, while h characterizes the strain-softening rate after yielding. Finally, the structure relaxation parameters x and β are fitted to a stress-free, constant heating profile of the thermal deformation (Figure 4.11).

Other parameters such as the glass transition temperature, programming temperature, relaxation time during cold-compression programming, and glassy shear modulus can be found from the test results presented in Chapter 3. The determined model parameters for the polystyrene based thermosetting SMP are summarized in Table 4.2.

4.4.7 Model Validation

The constitutive relations were then coded and implemented into the MATLAB® program. The simulation flowchart is illustrated in Figure 4.12 to simulate the corresponding experimental data reported earlier.

Based on the parameters identified in Table 4.2, the numerical simulation result, which covers the entire thermomechanical profile of the SMP programmed at 30% pre-strain and is for different relaxation histories in a strain–time scope, is shown in Figure 4.13(a). The material was initially stressed to the pre-strain level after overcoming the yielding point and experiencing a slight strain

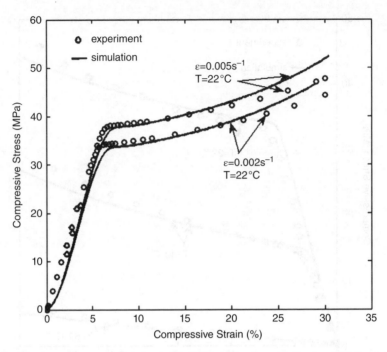

Figure 4.10 Stress–strain response of the SMP at different strain rates. *Source:* [61] Reproduced with permission from Elsevier

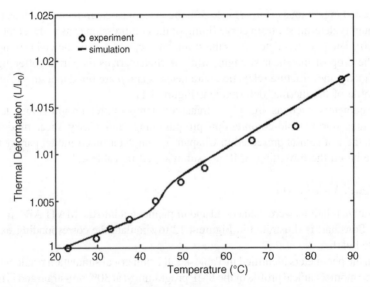

Figure 4.11 Thermal response for a stress-free, constant heating rate ($q = 0.56\,°C/min$) test. *Source:* [61] Reproduced with permission from Elsevier

Table 4.2 Material parameters of the preliminary constitutive model

Model parameters	Values
T_g (°C) glass transition temperature	62
T_0 (°C) programming temperature	20
Δt (minute) relaxation times	0/5/15/30/120
α_g (10^{-4} °C^{-1}) volumetric CTE of glassy state	5.462
α_r (10^{-4} °C^{-1}) volumetric CTE of rubbery state	8.441
G (MPa) glassy shear modulus	196.4
λ (MPa) Lamé constant for glassy state	785.7
μ_r (MPa) rubbery modulus	1.2
k_b (MPa) bulk modulus	1000
λ_L locking stretch	0.95
μ_g (MPa/s) reference shear viscosity at T_g	1550
s_0 (MPa) initial shear strength.	35
s_s (MPa) steady-state shear strength	33
Q/s_0 (K/MPa) flow activation ratio	380
h (MPa) flow softening constant	250
c_1 first WLF constant	25.8
c_2 (°C) second WLF constant	90
τ (s) structure relaxation characteristic time	200
x NMM constant	0.95
β Kohlrausch index	0.95

Source: [61] Reproduced with permission from Elsevier.

softening followed by a significant strain hardening (step 1) phenomenon. After that it was held with different time periods of relaxation for pseudo-plastic strain development (step 2). Finally, the remaining stress was instantly removed, leading to a stress-free state (step 3). Lengthy relaxation seemed to enhance the level of strain fixity. The stored deformation was then released and the original shape recovered during a subsequent heating process (step 4).

From Figure 4.13(a), the model simulation generally has a reasonable agreement with the test results. It proves that the model is capable of capturing the basic nonlinear material behavior of the SMP during such a thermomechanical cycle, although it can be observed that the real SMP samples failed to achieve the full recovery indicated by the prediction. The reason for this discrepancy may come from a couple of sources. Considering the large peak compressive stress applied during programming (about 40 MPa), some irreversible damage may have been induced in the SMP specimen. The deficiency of the single relaxation assumption also appears evident in the discrepancies between the simulation and experiments when the relaxation time is insufficient. This can be validated by Figure 4.13(a), which shows that when the relaxation time is short, the discrepancy is large; when the relaxation time is long enough (120 min), the discrepancy becomes comparatively small. Therefore, a spectrum of multiple nonequilibrium processes would be required to describe the actual stress relaxation process of a real thermosetting SMP.

In this study, the same parameters calibrated in modeling the constitutive behavior of the SMP programmed by 30% pre-strain level were also used to predict the thermomechanical behavior of the same SMP programmed by a 10% pre-strain level (see Figure 4.13(b)). It is clear that, with the same set of parameters, the model predicted the constitutive behavior well for the SMP programmed by the 10% pre-strain. This further validated the developed model.

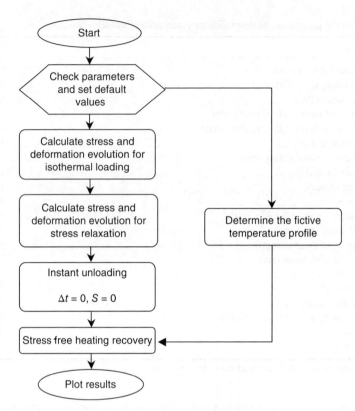

Figure 4.12 Flowchart of MATLAB® implementation. *Source:* [61] Reproduced with permission from Elsevier

4.4.8 Prediction and Discussion

Once the model was validated through the above process, it was used to evaluate the effect of some typical design parameters on the thermomechanical behavior of the cold-compression programmed SMP. First, the shape recovery depends on the heating rate. Figure 4.14 shows the free shape recovery prediction results for two different heating rates, $q = 0.6°C/min$ and $q = 3°C/min$. It is observed that a faster heating rate shifts the initiation of the recovery process to a higher temperature and leads to a more gradual temperature dependence at the start of the strain release, but hardly affects the final recovery ratio. This is understandable because the strain release mechanism is based on structural relaxation, which depends on both time and temperature. A faster heating rate allows less time for structural relaxation and thus needs a higher temperature to recover the same amount of strain as the lower heating rate. In other words, it seems that the time–temperature equivalent principle holds for the shape recovery.

To further validate this point, let us investigate the dependence of shape recovery on the heating history. Two types of heating profiles are considered. Heating profile #1 represents a heating profile from 22 to 79 °C with a constant heating rate of $q = 1°C/min$, while heating profile #2 represents a heating profile from 22 to 68 °C with a constant heating rate of $q = 1°C/min$, followed by a 50 minute soaking period. The calculation results for the two types of heating profiles are shown in Figure 4.15, which shows that, although heating profile #2 does not reach

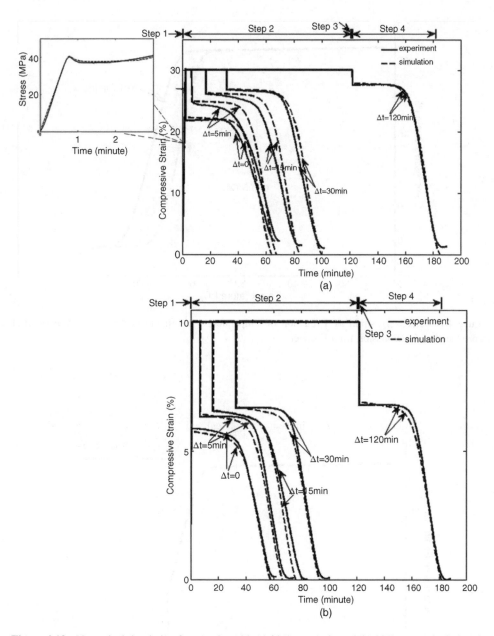

Figure 4.13 Numerical simulation for samples with (a) 30% pre-strain and (b) 10% pre-strain during the entire thermomechanical cycle (the four steps for the entire thermomechanical cycle for the specimen with 120 min of stress relaxation time during programming are also shown). *Source:* [61] Reproduced with permission from Elsevier

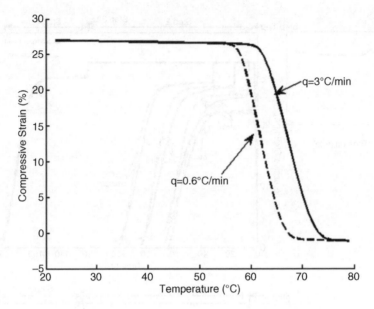

Figure 4.14 Recovery strain as a function of temperature for different heating rates. *Source:* [61] Reproduced with permission from Elsevier

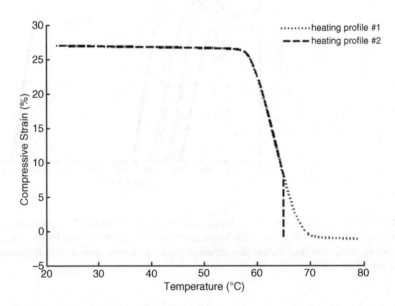

Figure 4.15 Recovery strain as a function of temperature for different heating profiles. *Source:* [61] Reproduced with permission from Elsevier

the same high temperature of 79 °C as that of heating profile #1, it still reaches the same recovery strain level after adequate soaking. Again, this is an indication of time–temperature equivalence.

Finally, we want to investigate the dependence of shape fixity and shape recovery on the programming temperature T_0. Although the programming is conducted at temperatures below T_g, how much below may make a difference. The effect of the programming temperature T_0 on shape recovery is shown in Figure 4.16. The SMP samples are programmed at 20 °C and 40 °C, respectively, for the same relaxation time period of 20 minutes. Two cases are considered. For case (a), shape recovery immediately follows the programming at a heating rate of 3 °C/min, which means that the starting temperatures for recovery are different (20 °C and 40 °C, respectively). It can be seen that a higher T_0 significantly increases the shape fixity ratio due to the decrease in molecular segmental resistance during the plastic flow and shortens the recovery time period. As shown by the temperature-recovery strain inset in Figure 4.16(a), the

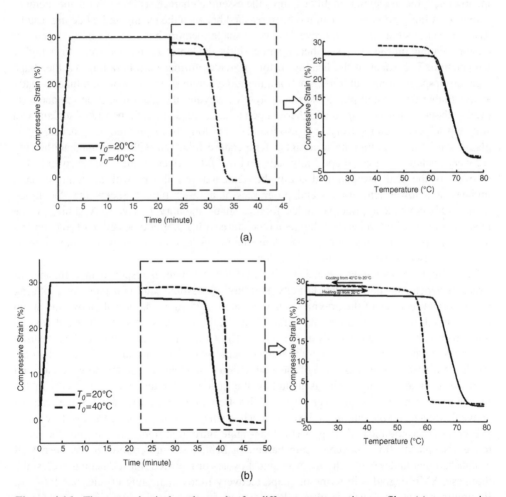

Figure 4.16 Thermomechanical cycle results for different programming profiles: (a) programming followed with immediate heating recovery and (b) programming followed with cooling and then heating recovery. *Source:* [61] Reproduced with permission from Elsevier

two programmed SMPs generally follow a similar recovery path except for the small deviation caused by structure relaxation and thermal expansion. For case (b), the sample programmed at 40 °C is first cooled to 20 °C before being heated to recovery, which means that the starting temperature for recovery is the same (20 °C). It can be seen from Figure 4.16(b) that for the sample programmed at 40 °C, it takes a longer time for completion of step 4 but with the major recovery completed at a lower temperature, again showing a time–temperature equivalency.

4.4.9 Summary

A finite deformation theory and mechanism based thermoviscoelastic–viscoplastic constitutive model has been developed to study the thermomechanical behavior of the thermoset SMP programmed by cold compression. Because the pseudo-plasticity and structure evolution are incorporated, the model reasonably captures the essential characteristics of the shape memory response. A fairly good agreement has been reached between the testing and modeling results. The parametric simulation study reveals that the shape memory behavior is highly dependent on the heating profile. A faster heating rate shifts the onset of the recovery to a higher temperature. The effect of the heating history provides further corroboration that the shape recovery response turns out to be more of a thermodynamic structure evolution than a steady state variable determined phase transition process. Beyond the glass transition temperature, even without further heating to a higher temperature, an adequate time period of soaking can still help achieve a full recovery. It is also found that as long as the programming occurs in the glassy state, the programming at a higher temperature followed with an immediate heating recovery leads to a higher shape fixity ratio and has little effect on the strain recovery. The recovery of the SMP programmed at a higher temperature followed with a cooling process initiates at a lower temperature and progresses at a faster rate. It seems that the time–temperature equivalent principle holds for the shape memory behavior. A similar shape recovery ratio can be achieved at a higher temperature with a shorter time period of soaking or a longer time period of soaking at a lower temperature, as long as the recovery temperature is above the onset of the glass transition region.

The comprehensive model considers the material mechanical response in a manner of structure dependent thermoviscoelasticity–thermoviscoplasticity and is capable of capturing the important features of the polymer behavior, such as yielding, strain softening, and strain hardening. However, it is noted that with 20 materials constants, this model is not easy to implement in finite element analysis (FEA), which is a popular design tool for load bearing structures. A deeper understanding of the shape memory mechanism and physics based model are desired. Also, removal of some of the assumptions can make the model more realistic and increase its predictability, such as including the heat transfer and pressure effects in the modeling. A single nonequilibrium stress relaxation process is assumed for the sake of convenience and yet the multiple relaxation mechanism (i.e., more separate Maxwell elements in Figure 4.7) is required to distinguish the long-range entropic stiffening process and the short-range viscoplastic flow induced strain hardening behavior. Furthermore, it has been well established that under cyclic thermomechanical cycles or repeated impact/healing cycles, the thermoset SMP degrades in terms of shape recovery ratios or healing efficiencies [66–75]. Environmental attacks such as ultraviolet radiation, hydrothermal immersion, saltwater, or other chemicals may damage the SMP molecules and lead to degradation in the shape memory capability [76,77]. All of these need to be considered in future modeling efforts.

4.5 Thermoviscoplastic Modeling of Cold-Compression Programmed Thermosetting Shape Memory Polymer Syntactic Foam

Based on Section 4.4, we are now able to model the cold-compression programmed thermosetting SMP based syntactic foam. Because the same thermosetting SMP is used, we only need to focus on separating the effect of glass microballoons on the thermomechanical behavior and also consider its damage because it has been found that glass microballoons experience wall bending, crushing, and a consolidating process during cold compression [78,79].

4.5.1 General Considerations

As shown by the material characterization test results (DMA and XPS results) in Chapter 3, incorporation of glass microballoons altered the chemical bonds at the interface between the SMP matrix and glass hollow microsphere inclusions. Earlier studies [80,81] discovered that there even exists a long-range gradient (over 100 K difference) of the polymer matrix glass transition temperature in the vicinity of the particles. Therefore, it was believed that an interfacial transition zone (ITZ) layer similar to the phenomenon in cement based materials [82–84] also occurs in the SMP based syntactic foam. To consider the influence of such a layer on the performance of the foam, a unit cell of the SMP based syntactic foam should be perceived as a three-phase composite with ITZ coated glass hollow microspheres embedded in the pure SMP matrix, as illustrated in Figure 4.17. However, since current techniques have difficulties in characterizing the ITZ layer in detail, a convenient approach of integrating the ITZ and pure SMP as a new equivalent SMP medium [21] was adopted. The equivalent scheme is also shown in Figure 4.17.

To simplify the modeling work, the same fundamental assumptions as in Section 4.4 were made for the equivalent SMP matrix. For the microballoons, we further assume that:

1. All the damage originates from the crushing and implosion of the glass hollow microspheres. The SMP matrix does not have any damage.
2. The equivalent SMP matrix has a perfect interfacial bonding with the microballoons.

4.5.2 Kinematics

As discussed above, our focus will be on the portion of contribution by the microballoons. The contributions will be simply considered by the rule-of-mixture principle, although more accurate micromechanics models can be used, as discussed by Shojaei and Li [85].

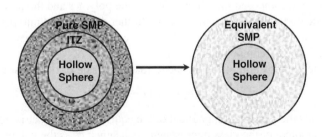

Figure 4.17 Equivalent two-layer unit cell model

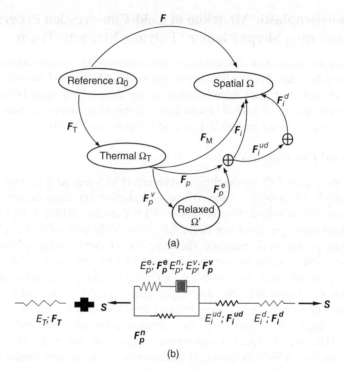

Figure 4.18   (a) Multiplicative decomposition of the deformation gradient of each component. *Source:* [86] Reproduced with permission from ASME and (b) 1-D rheological analogy

The multiplicative decomposition and one-dimensional rheological analogy are shown in Figure 4.18(a) and (b), respectively. In Figure 4.18, the scalar E represents the modulus and tensor F represents the deformation gradient. The subscripts T, M, p, and i represent thermal, mechanical, equivalent polymer matrix, and microballoon inclusions, respectively. The superscripts e, v, and n represent elastic, viscose, and hyperelastic component of the equivalent polymer matrix. The superscript ud and d represent undamaged and damaged glass microballoons. Finally, S represents the shear stress applied.

Figure 4.18(a) shows that an arbitrary thermomechanical deformation mapping from an initial undeformed and unheated configuration Ω_0 to a spatial configuration (deformed) Ω can be multiplicatively decomposed into thermal and mechanical deformations. The mechanical deformation can be decomposed into the portion by the polymer and the portion by the glass microballoon. The portion by the polymer can be further decomposed into plastic and elastic. Finally, the portion of the microballoon, while it is elastic, can be further decomposed into damaged and undamaged components.

To consider the composition of the syntactic foam, the rule-of-mixtures method applies:

$$F_M = \varphi_p F_p + (1 - \varphi_p)F_i \tag{4.70}$$

where F_p represents the deformation gradient of the equivalent SMP matrix and F_i represents the deformation gradient of the glass microsphere inclusions; ϕ_p is the volume fraction of the polymer matrix. Since F_p has been formulated in Section 4.4, our focus here is on F_i.

Usually glass microspheres crush during the loading step; therefore, a damage allowable constitutive model of the microsphere inclusions is required. If an internal stress and time dependent evolution parameter $\phi_d(\sigma, t)$ is introduced to represent the volume fraction of the damaged microspheres out of the total microsphere volume, the deformation gradient of the inclusions could be expressed as

$$F_i = (1 - \phi_d)F_i^{ud} + \phi_d F_i^d \tag{4.71}$$

where F_i^{ud} refers to undamaged microspheres while F_i^d refers to damaged microspheres.

4.5.3 Constitutive Behavior of Glass Microsphere Inclusions

Since the glass hollow microspheres are brittle and have a high Young's modulus, the constitutive behavior of the undamaged portion can be considered to be purely elastic:

$$\sigma = \sigma_i = L_i^e(\ln F_i^{ud}) \tag{4.72}$$

where $L_i^e = 2G_i\iota + \lambda_i I \otimes I$ is the fourth-order isotropic elasticity tensor of the glass microspheres. G_i and λ_i are Lamé constants, ι is the fourth-order identity tensor, and I is the second-order identity tensor.

Physically, evolution of the crushing and implosion of the hollow microspheres can be extremely complex. Since our focus is just on establishing a thermomechanical framework for the SMP based syntactic foam, to keep matters simple we assume that an instant and complete damage mechanism occurs to the hollow microspheres partly because the glass hollow microspheres are brittle and thus the crack propagation speed is high. Hence, $\phi_d(\sigma,t) = \phi_d(\sigma)$, that is, independent of time.

Normally $\phi_d(\sigma)$ evolves nonlinearly. If the statistic normal distribution applies, an arbitrary nonlinear curve of the volume fraction of the damaged microballoons should start slowly when the applied load initially overcomes the bearing strength σ_b and then accelerates as the load further increases; it finally slows down gradually as the damage proceeds and reaches a complete failure of all the microsphere inclusions. Since it is difficult to capture the real nonlinear damage profile, a linear equivalent damage model was considered. As the irrecoverable strain is assumed to fully come from the damage and volume reduction of the hollow microspheres, the total damage volume fraction (ϕ_d^{total}) of the microspheres can be calculated based on its relation with the final irrecoverable strain (ε_{ir}) as

$$1 + \varepsilon_{ir} = \phi_p + (1 - \phi_p)((1 - \phi_d^{total}) + \phi_d^{total}(1 + (1 - w)^3)^{1/3}) \tag{4.73}$$

where w is the wall thickness ratio for the glass hollow microspheres. The proportional factor k for the linear equivalent damage model is given by

$$k = (\phi_d^{total})/(\sigma_m - \sigma_b) \tag{4.74}$$

where σ_m is the maximum stress during the programming process.

Because the maximum stress is achieved at the end of loading in step 1 of the programming process, the peak stress at the corresponding pre-strain (30% or 20%) is used; σ_b corresponds to the initial damage stress, which is the crushing pressure of the glass microspheres provided by the manufacturer (1.72 MPa). It is noted that the microballoons are not completely crushed (damaged) in the first programming cycle. The damage should accumulate as the programming recovery cycles increase and stabilize after several cycles, which may lead to a decrease in the shape recovery ratio in the first several cycles and stabilize in the shape recovery ratio thereafter. For simplicity, however, this dependence of damage on programming recovery cycles was not considered in this study, which could be a potential source for the discrepancy between the model prediction and the test result.

If we additionally consider the glass microspheres to be isotropic, the damage gradient can be given by

$$\mathbf{F}_d = J_d^{1/3} \mathbf{I} \tag{4.75}$$

where J_d represents the ratio of the volume reduction during the damage, which can be determined as

$$J_d = \frac{V_{ad}}{V_{bd}} = \frac{\frac{4}{3}\pi \left(r^3 - (r-t)^3 \right)}{\frac{4}{3}\pi r^3} = 1 - (1 - w)^3 \tag{4.76}$$

where V_{bd} and V_{ad} represent the volume of the hollow microsphere before and after damage, respectively, r is the outer radius of the microsphere, t is the wall thickness, and $w = t/r$ is the wall thickness ratio.

It should be noted that even after being completely crushed, the fractured pieces of the glass microspheres still behave elastically. Hence, the deformation gradient of the damaged portion of the microspheres $\mathbf{F}_i^d$ should be expressed as

$$\mathbf{F}_i^d = \mathbf{F}_i^{ud} \mathbf{F}_d \tag{4.77}$$

4.5.4 Model Summary

For completeness, the temperature and time dependent, damage allowable thermomechanical constitutive relations for the SMP based syntactic foam are summarized in Table 4.3 [86].

4.5.5 Determination of Materials Constants

Similar to Section 4.4.6, the following tests were conducted to determine the materials constants in the model. These tests are the thermal response for a stress-free constant-rate cooling (Figure 4.19), the stress–strain response at various temperatures (Figure 4.20), the stress–strain response at different strain rates (Figure 4.21), and the thermal response for a stress-free constant-rate heating (Figure 4.22). Based on the test results, the materials constants are obtained and are summarized in Table 4.4.

4.5.6 Model Validation

The structure evolving, damage-allowable thermoviscoplastic constitutive model was computed in MATLAB®. The corresponding model parameters were mainly obtained by curve-fitting various thermal and mechanical testing results, as summarized in Table 4.4.

Table 4.3 Summary of the constitutive model

Deformation response	$\mathbf{F}=\mathbf{F}_M\mathbf{F}_T;\ \mathbf{F}_M=\varphi_p\mathbf{F}_p+(1-\varphi_p)\mathbf{F}_i$
	$\mathbf{F}_i=(1-\varphi_d)\mathbf{F}_i^{ud}+\varphi_d\mathbf{F}_i^{d};\ \mathbf{F}_i^{d}=\mathbf{F}_i^{ud}\mathbf{F}_d;$
	$\mathbf{F}_d=J_d^{1/3}\mathbf{I};\ J_d=1-(1-w)^3$
	$\mathbf{F}_p=\mathbf{F}_p^e\mathbf{F}_p^v;\ \mathbf{F}_T=J_T^{1/3}\mathbf{I}$
	$J_T=1+\alpha_r(T_f-T_0)+\alpha_g(T-T_f)$
Structure relaxation	$T_f(t)=T(t)-\int_{t_0}^{t}\phi(\Delta\varsigma)\mathrm{d}T(t)$
	$\phi=\exp(-(\Delta\varsigma)^\beta);\ \Delta\varsigma=\varsigma(t)-\varsigma(t')=\int_{t'}^{t}\frac{\mathrm{d}t}{\tau_s}$
	$\tau_s=\tau_0\exp\left[B(T_g-T_\infty)^2\left(\frac{x}{T-T_\infty}+\frac{1-x}{T_f-T_\infty}\right)\right]$
Stress response	$\boldsymbol{\sigma}=\boldsymbol{\sigma}_i=\boldsymbol{\sigma}_p=\boldsymbol{\sigma}_p^{ve}+\boldsymbol{\sigma}_p^{n}$
	$\boldsymbol{\sigma}_i=\mathbf{L}_i^e(\ln\mathbf{F}_i^{ud})$
	$\boldsymbol{\sigma}_p=\frac{1}{J_p^e}\mathbf{L}_p^e(\ln\mathbf{V}_p^e)+\left[\frac{1}{J_n}\mu_r\frac{\lambda_L}{\lambda_{\text{chain}}}\mathbf{L}^{-1}\left(\frac{\lambda_{\text{chain}}}{\lambda_L}\right)\overline{\mathbf{B}}'+k_b(J_n-1)\mathbf{I}\right]$
Viscous flow	$\mathbf{D}_p^v=\dot{\gamma}_v\frac{\boldsymbol{\sigma}_p^{ve'}}{\lVert\boldsymbol{\sigma}_p^{ve'}\rVert}$
	$\dot{\gamma}_v=\frac{s}{\eta_g}\frac{T}{Q}\exp\left(c_1\left(\frac{c_2(T-T_f)+T(T_f-T_g)}{T(c_2+T-T_g)}\right)\right)\sinh\left(\frac{Q}{T}\frac{\overline{\tau}}{s}\right)$

Source: [86] Reproduced by permission from ASME.

The numerical simulation results shown in Figure 4.23(a), (b), and (c) cover the full thermomechanical cycle of the SMP based syntactic foam programmed at room temperature with pre-strains of both 20% and 30% in both the strain–time scale and the stress–strain–time scale. All of the five different relaxation histories (0 min, 5 min, 15 min, 30 min, and 120 min) for both pre-strains are included. The simulation generally shows reasonable agreement with the experimental results and captures most of the essential nonlinear material behavior, although less agreement on the final recovery strain is found for samples programmed to 20% pre-strain than those programmed to 30% pre-strain. This may be because under 20% pre-strain, damage in the microballoons was considerably less than that under 30% pre-strain and was below the linear interpolation prediction. In other words, the linear damage evaluation assumption is more appropriate for heavily damaged microballoons than for slightly damaged counterparts. It is also noted that the deficiency of the single relaxation assumption appears evident. When the relaxation time is insufficient, such as 0 minute, the discrepancy is obvious, while as the relaxation time further increases, the discrepancy becomes comparatively less significant. Multiple nonequilibrium relaxation processes would be required to more closely describe an actual stress relaxation.

In order to demonstrate the predictability of the model further, the thermomechanical cycle test results of the same SMP based syntactic foam programmed by the classical programming

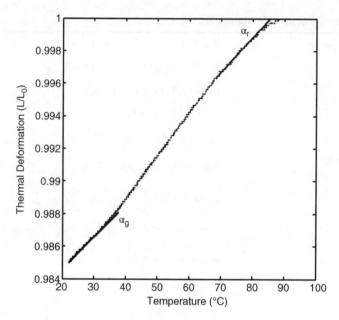

Figure 4.19   Thermal response to a stress-free natural cooling. *Source:* [86] Reproduced with permission from ASME

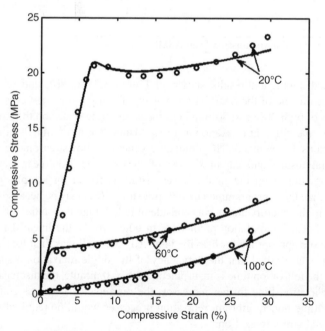

Figure 4.20   Stress–strain response of the SMP based syntactic foam at various temperatures. *Source:* [86] Reproduced with permission from ASME

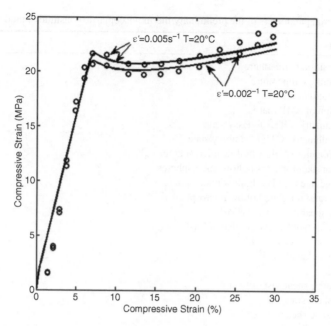

Figure 4.21 Stress–strain response of the SMP based syntactic foam at different strain rates. *Source:* [86] Reproduced with permission from ASME

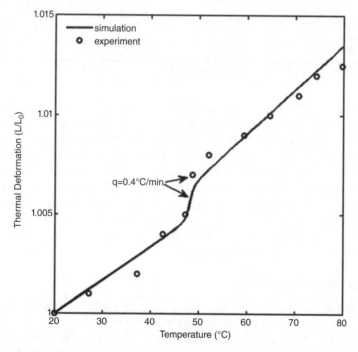

Figure 4.22 Thermal response for a stress-free constant-rate heating. *Source:* [86] Reproduced with permission from ASME

Table 4.4 Material parameters of the preliminary thermoviscoplastic constitutive model

Model parameters	Values
T_g (°C) glass transition temperature	64.3
T_0 (°C) programming temperature	20
Δt (minute) relaxation times	0/5/15/30/120
Φ_p volume fraction of SMP matrix	0.6
α_g (10^{-4}/°C) volumetric CTE of glassy state	5.062
α_r (10^{-4}/°C^{-1}) volumetric CTE of rubbery state	6.841
G_i (GPa) shear modulus of glass hollow microspheres	27.7
λ_i (GPa) Lamé constant for glass hollow microspheres	41.5
k (MPa^{-1}) damage rate for glass hollow microspheres	0.02
w wall thickness ratio for glass hollow microspheres	0.019
G_P (MPa) glassy shear modulus of SMP	96.4
λ_P (MPa) Lamé constant for glassy state of SMP	385.7
μ_r (MPa) rubbery modulus of SMP	0.3
k_b (MPa) bulk modulus of SMP	1000
λ_L locking stretch	1.4
η_g (MPa/s) reference shear viscosity at T_g	4050
s_0 (MPa) initial shear strength	20
s_s (MPa) steady state shear strength	18
Q/s_0 (K/MPa) flow activation ratio	800
h (MPa) flow softening constant	200
c_1 first WLF constant	17.3
c_2 (°C) second WLF constant	70
τ (s) structure relaxation characteristic time	20
x NMM constant	0.95
β Kohlrausch index	0.95

Source: [86] Reproduced with permission from ASME.

process but with the 2-D stress condition (tension in one direction and compression in the transverse direction) are compared to the modeling results using the materials constants in Table 4.4. The cruciform specimen was initially subjected to a constant load of 54.3 N (168.3 kPa) vertically in compression and horizontally in tension at 79 °C and then followed the conventional training method to achieve shape fixity (cooling to room temperature in about 10 hours while holding the load, and then removing the load completely and instantly). After that it was reheated to 79 °C at a heating rate of 0.3 °C/min and equilibrated for 30 minutes for free recovery [70]. The simulation results in Figure 4.24 show the strain evolution in the horizontal and vertical directions during the entire thermomechanical cycle. Again, a good agreement is found between the testing and modeling results.

4.5.7 Prediction by the Model

Once the model is validated by test results, it is used for predictions. First, we consider the effect of the volume fraction of the SMP matrix Φ_p on the thermomechanical cycle. As shown in Figure 4.25, two SMP volume fractions ($\Phi_p = 0.5$ and $\Phi_p = 0.6$), which experienced a 40 minute relaxing period during cold-compression programming and had a recovery heating rate of 0.4 °C/min, were conducted. It was found that less SMP appears to slightly increase the

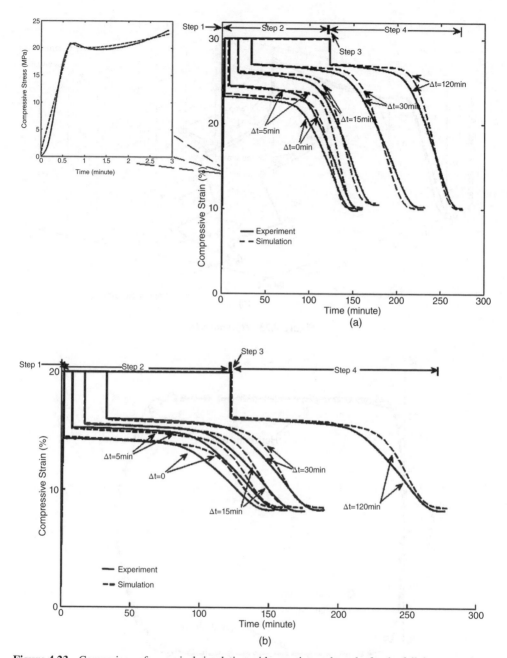

Figure 4.23 Comparison of numerical simulation with experimental results for the full thermomechanical cycle: (a) strain evolution with 30% pre-strain, (b) strain evolution with 20% pre-strain, and (c) thermomechanial cycle in terms of the stress–strain–time response. *Source:* [86] Reproduced with permission from ASME

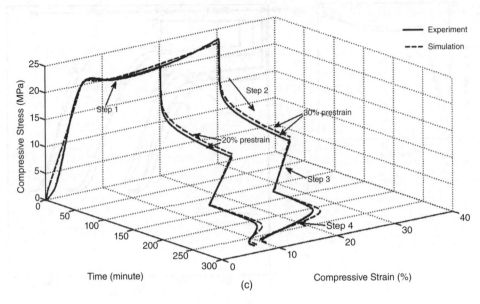

Figure 4.23 (*Continued*)

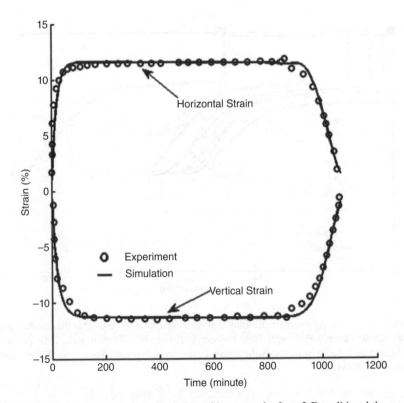

Figure 4.24 Comparison of numerical simulation with test results for a 2-D traditional thermomechanical cycle. *Source:* [86] Reproduced with permission from ASME

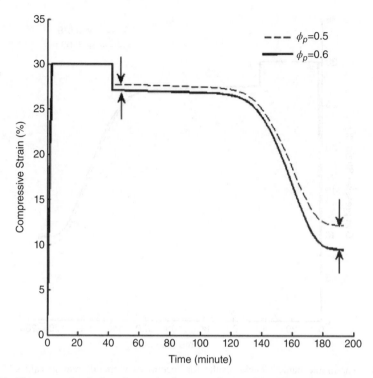

Figure 4.25 Thermomechanical cycle results for a specimen with different Φ_p values. *Source:* [86] Reproduced with permission from ASME

shape fixity ratio, which seems abnormal. Further observation on the heating recovery reveals that the seeming enhancement in shape fixity originates from an increase in glass hollow microsphere damage. This is because the specimen with less SMP resulted in more irreversible strain and the loss of recoverability is noticeably larger than the gain in the shape fixity. Therefore, it is believed that lower Φ_p would lead to more damage and less recovery ratio.

Second, we considered the effect of the microballoon wall thickness ratio w on the thermomechanical cycle, as shown in Figure 4.26. The specimen with a higher w was found to be able to achieve a larger recovery ratio (lower permanent strain), as it contains fewer voids and hence suffers less damage during the programming. It is also interesting to notice that the shape fixity seems to be hardly affected by the variation in w because the same crushing strength was assumed. Although the irreversible deformation of microballoons with lower w may tend to increase the shape fixity ratio, the reduction in the reversible viscous deformation in SMP manages to counterbalance it.

4.5.8 Summary

A structure evolving, damage-allowable thermoviscoplastic model has been developed and reasonably captured the most essential shape memory response of the SMP based syntactic foam during this process. A finite deformation, continuum constitutive model has been developed to study the thermomechanical behavior of the SMP based self-healing syntactic

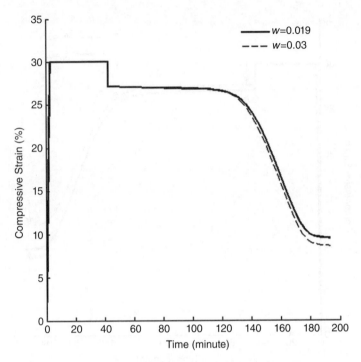

Figure 4.26 Thermomechanical cycle results for specimens with different *w* values. *Source:* [86] Reproduced with permission from ASME

foam programmed at the glassy temperature. As thermoviscoplasticity, structural relaxation and inclusion damage mechanism are considered, the model plausibly captures the essential elements of the shape memory response. A fairly good agreement has been reached between the modeling results and the experimental results. The parametric simulation study reveals the possible optimization schemes for future generations of the SMP based syntactic foam: a high volume fraction of microsphere inclusions leads to a low recovery ratio and a high wall thickness ratio of the glass microballoons leads to a larger recovery strain. An optimal configuration can be achieved by adjusting and balancing these parameters.

References

1. Meng, H. and Li, G. (2013) A review of stimuli-responsive shape memory polymer composites. *Polymer*, **54**, 2199–2221.
2. Heuchel, M., Sauter, T., Kratz, K., and Lendlein, A. (2013) Thermally induced shape-memory effects in polymers: quantification and related modeling approaches. *Journal of Polymer Science Part B: Polymer Physics*, **51**, 621–637.
3. Nguyen, T.D. (2013) Modeling shape-memory behavior of polymers. *Polymer Reviews*, **53**, 130–152.
4. Zhang, Q. and Yang, Q.S. (2012) Recent advance on constitutive models of thermal-sensitive shape memory polymers. *Journal of Applied Polymer Science*, **123**, 1502–1508.
5. Tobushi, H., Hara, H., Yamada, E., and Hayashi, S. (1996) Thermomechanical properties in a thin film of shape memory polymer of polyurethane series. *Smart Materials and Structures*, **5**, 483–491.
6. Tobushi, H., Hashimoto, T., Hayashi, S., and Yamada, E. (1997) Thermomechanical constitutive modeling in shape memory polymer of polyurethane series. *Journal of Intelligent Material Systems and Structures*, **8**, 711–718.

7. Tobushi, H., Hashimoto, T., Ito, N., Hayashi, S., and Yamada, E. (1998) Shape fixity and shape recovery in a film of shape memory polymer of polyurethane series. *Journal of Intelligent Material Systems and Structures*, **9**, 127–136.

8. Bhattacharyya, A. and Tobushi, H. (2000) Analysis of the isothermal mechanical response of a shape memory polymer rheological model. *Polymer Engineering and Science*, **40**, 2498–2510.

9. Kafka, V. (2001) *Mesomechanical Constitutive Modeling*, World Scientific, Singapore.

10. Kafka, V. (2008) Shape memory polymers: a mesoscale model of the internal mechanism leading to the SM phenomena. *International Journal of Plasticity*, **24**, 1533–1548.

11. Diani, J. and Gall, K. (2007) Molecular dynamics simulations of the shape-memory behaviour of polyisoprene. *Smart Materials and Structures*, **16**, 1575–1583.

12. Morshedian, J., Khonakdar, H.A., and Rasouli, S. (2005) Modeling of shape memory induction and recovery in heat shrinkable polymers. *Macromolecular Theory and Simulations*, **14**, 428–434.

13. Gall, K., Yakacki, C.M., Liu, Y., Shandas, R., Willett, N., and Anseth, K.S. (2005) Thermomechanics of the shape memory effect in polymers for biomedical applications. *Journal of Biomedical Materials Research Part A*, **73**, 339–348.

14. Liu, Y., Gall, K., Dunn, M.L., and McCluskey, P. (2004) Thermomechanics of shape memory polymer nanocomposites. *Mechanics of Materials*, **36**, 929–940.

15. Liu, Y., Gall, K., Dunn, M.L., Greenberg, A.R., and Diani, J. (2006) Thermomechanics of shape memory polymers: uniaxial experiments and constitutive modeling. *International Journal of Plasticity*, **22**, 279–313.

16. Yakacki, C.M., Shandas, R., Lanning, C., Rech, B., Eckstein, A., and Gall, K. (2007) Unconstrained recovery characterization of shape-memory polymer networks for cardiovascular applications. *Biomaterials*, **28**, 2255–2263.

17. Qi, H.J., Nguyen, T.D., Castro, F., Yakacki, C.M., and Shandas, R. (2008) Finite deformation thermo-mechanical behavior of thermally induced shape memory polymers. *Journal of the Mechanics and Physics of Solids*, **56**, 1730–1751.

18. Chen, Y.H. and Lagoudas, D.C. (2008) A constitutive theory for shape memory polymers. Part I – large deformations. *Journal of the Mechanics and Physics of Solids*, **56**, 1752–1765.

19. Chen, Y.H. and Lagoudas, D.C. (2008) A constitutive theory for shape memory polymers. Part II – A linearized model for small deformations. *Journal of the Mechanics and Physics of Solids*, **56**, 1766–1778.

20. Nguyen, T.D., Qi, H.J., Castro, F., and Long, K.N. (2008) A thermoviscoelastic model for amorphous shape memory polymers: incorporating structural and stress relaxation. *Journal of the Mechanics and Physics of Solids*, **56**, 2792–2814.

21. Xu, W. and Li, G. (2010) Constitutive modeling of shape memory polymer based self-healing syntactic foam. *International Journal of Solids and Structures*, **47**, 1306–1316.

22. Tüdős, F. and David, P.K. (1996) Comments on the interpretation of temperature dependence of processes in polymers. *Journal of Thermal Analysis and Calorimetry*, **47**, 589–593.

23. Andreozzi, L., Faetti, M., Zulli, F., and Giordano, M. (2004) Enthalpy relaxation of polymers: comparing the predictive power of two configurational entropy models extending the AGV approach. *European Physics Journal B*, **41**, 383–393.

24. Farzaneh, S., Fitoussi, J., Lucas, A., Bocquet, M., and Tcharkhtchi, A. (2013) Shape memory effect and properties memory effect of polyurethane. *Journal of Applied Polymer Science*, **128**, 3240–3249.

25. Sedov, L.I. (1966) *Foundations of the Non-linear Mechanics of Continua*, Pergamon Press, Oxford.

26. Lu, S.C.H. and Pister, K.S. (1975) Decomposition of deformation and representation of the free energy function for isotropic thermoelastic solids. *International Journal of Solids and Structures*, **11**, 927–934.

27. Nemat-Nasser, S. (1982) On finite deformation elasto-plasticity. *International Journal of Solids and Structures*, **18**, 857–872.

28. Havner, K.S. (1992) *Finite Plastic Deformation of Crystalline Solids*, Cambridge University Press, Cambridge.

29. Khan, A.S. and Huang, S. (1995) *Continuum Theory of Plasticity*, John Wiley and Sons, Inc., New York.

30. Lubarda, V.A. (2002) *Elastoplasticity Theory*, CRC Press, Boca Raton, Florida, USA.

31. Lubarda, V.A. (2004) Constitutive theories based on the multiplicative decomposition of deformation gradient: thermoelasticity, elastoplasticity, and biomechanics. *ASME Applied Mechanics Review*, **57**, 95–108.

32. Bower, A.F. (2009) *Applied Mechanics of Solids*, CRC Press, Boca Raton, Florida, USA.

33. Lee, E.H. (1969) Elastic–plastic deformation at finite strains. *ASME Journal of Applied Mechanics*, **36**, 1–6.

34. Backman, M.E. (1964) From the relation between stress and finite elastic and plastic strains under impulsive loading. *Journal of Applied Physics*, **35**, 2524–2533.

35. Lee, E.H. and Liu, D.T. (1967) Finite-strain elastic–plastic theory particularly for plane wave analysis. *Journal of Applied Physics*, **38**, 19–27.
36. Fox, N. (1968) On the continuum theories of dislocations and plasticity. *Quarterly Journal of Mechanics and Applied Mathematics*, **21**, 67–75.
37. Willis, J.R. (1969) Some constitutive equations applicable to problems of large dynamic plastic deformation. *Journal of the Mechanics and Physics of Solids*, **17**, 359–369.
38. Boyce, M.C., Parks, D.M., and Argon, A.S. (1988) Large inelastic deformation of glassy polymers, Part I: Rate-dependent constitutive model. *Mechanics of Materials*, **7**, 15–33.
39. Boyce, M.C., Weber, G.G., and Parks, D.M. (1989) On the kinematics of finite strain plasticity. *Journal of the Mechanics and Physics of Solids*, **37**, 647–665.
40. Arruda, E.M. and Boyce, M.C. (1993) A three-dimensional constitutive model for the large stretch behavior of rubber elastic materials. *Journal of the Mechanics and Physics of Solids*, **41**, 389–412.
41. Wu, P.D. and Van der Giessen, E. (1993) On improved network models for rubber elasticity and their applications to orientation hardening in glassy polymers. *Journal of the Mechanics and Physics of Solids*, **41**, 427–456.
42. Lion, A. (1997) A physically based method to represent the thermomechanical behavior of elastomers. *Acta Mechanica*, **123**, 1–25.
43. Green, A.E. and Naghdi, P.M. (1971) Some remarks on elastic–plastic deformation at finite strain. *International Journal of Engineering Science*, **9**, 1219–1229.
44. Casey, J. and Naghdi, P.M. (1980) Remarks on the use of the decomposition $F = F_e F_p$ in plasticity. *ASME Journal of Applied Mechanics*, **47**, 672–675.
45. Lubarda, V.A. (1991) Constitutive analysis of large elasto-plastic deformation based on the multiplicative decomposition of deformation gradient. *International Journal of Solids and Structures*, **27**, 885–895.
46. Boyce, M.C., Kear, K., Socrate, S., and Shaw, K. (2001) Deformation of thermoplastic vulcanizates. *Journal of the Mechanics and Physics of Solids*, **49**, 1073–1098.
47. Tool, A.Q. (1946) Relation between inelastic deformability and thermal expansion of glass in its annealing range. *Journal of American Ceramic Society*, **29**, 240–253.
48. Narayanaswamy, O.S. (1971) A model of structural relaxation in glass. *Journal of American Ceramic Society*, **54**, 491–498.
49. Donth, E. and Hempel, E. (2002) Structural relaxation above the glass temperature: pulse response simulation with the Narayanaswamy Moynihan model for glass transition. *Journal of Non-Crystalline Solids*, **306**, 76–89.
50. Moyniham, C.T., Easteal, A.E., Debolt, M.A., and Tucker, J. (1976) Dependence of the fictive temperature of glass on cooling rate. *Journal of the American Ceramic Society*, **59**, 12–16.
51. Kohlrausch, R. (1854) Theorie des elektrischen Rückstandes in der Leidner Flasche. *Annalen der Physik und Chemie (Poggendorff)*, **91**, 56–82.
52. Hempel, E., Kahle, S., Unger, R., and Donth, E. (1999) Systematic calorimetric study of glass transition in the homologous series of poly(*n*-alkyl methacrylate)s: Narayanaswamy parameters in the crossover region. *Thermochimica Acta*, **329**, 97–108.
53. William, M.L., Landel, R.F., and Ferry J.D. (1955) The temperature dependence of relaxation mechanisms in amorphous polymers and other glass-forming liquids. *Journal of American Chemistry Society*, **77**, 3701–3707.
54. Scherer, G.W. (1990) Theories of relaxation. *Journal of Non-Crystalline Solids*, **123**, 75–89.
55. Treloar, L.R.G. (1958) *The Physics of Rubber Elasticity*, Clarendon Press, Oxford.
56. Govindjee, S. and Simo, J. (1991) A micro-mechanically based continuum damage model for carbon black-filled rubbers incorporating Mullins effect. *Journal of the Mechanics and Physics of Solids*, **39**, 87–112.
57. Bergstrom, J.S. and Boyce, M.C. (1998) Constitutive modeling of the large strain time-dependence behavior of elastomers. *Journal of the Mechanics and Physics of Solids*, **46**, 931–954.
58. Miehe, C. and Keck, J. (2000) Superimposed finite elastic–viscoelastic–plastoelastic stress response with damage in filled rubbery polymers. Experiments, modeling and algorithmic implementation. *Journal of the Mechanics and Physics of Solids*, **48**, 323–365.
59. Boyce, M.C., Kear, K., Socrate, S., and Shaw, K. (2001) Deformation of thermoplastic vulcanizates. *Journal of the Mechanics and Physics of Solids*, **49**, 1073–1098.
60. Qi, H.J. and Boyce, M.C. (2005) Stress–strain behavior of thermoplastic polyurethanes. *Mechanics of Materials*, **37**, 817–839.
61. Li, G. and Xu, W. (2011) Thermomechanical behavior of thermoset shape memory polymer programmed by cold-compression: testing and constitutive modeling. *Journal of the Mechanics and Physics of Solids*, **59**, 1231–1250.

62. Flory, P.J. (1961) Thermodynamic relations for highly elastic materials. *Transactions of the Faraday Society*, **57**, 829–838.
63. Simo, J.C., Taylor, R., and Pister, K.S. (1985) Variational and projection methods for the volume constraint in finite deformation elasto-plasticity. *Computer Methods in Applied Mechanics and Engineering*, **51**, 177–208.
64. Eyring, H. (1936) Viscosity, plasticity, and diffusion as examples of absolute reaction rates. *Journal of Computational Physics*, **28**, 373–383.
65. Anand, L. and Ames, N.M. (2006) On modeling the micro-indentation response of amorphous polymer. *International Journal of Plasticity*, **22**, 1123–1170.
66. Li, G., Ajisafe, O., and Meng, H. (2013) Effect of strain hardening of shape memory polymer fibers on healing efficiency of thermosetting polymer composites. *Polymer*, **54**, 920–928.
67. Shojaei, A., Li, G., and Voyiadjis, G.Z. (2013) Cyclic viscoplastic–viscodamage analysis of shape memory polymer fibers with application to self-healing smart materials. *ASME Journal of Applied Mechanics*, **80**, paper 011014 (15 pages).
68. Li, G., Meng, H., and Hu, J. (2012) Healable thermoset polymer composite embedded with stimuli-responsive fibers. *Journal of the Royal Society Interface*, **9**, 3279–3287.
69. Nji, J. and Li, G. (2012) Damage healing ability of a shape memory polymer based particulate composite with small thermoplastic contents. *Smart Materials and Structures*, **21**, paper 025011 (10 pages).
70. Li, G. and Xu, T. (2011) Thermomechanical characterization of shape memory polymer based self-healing syntactic foam sealant for expansion joint. *ASCE Journal of Transportation Engineering*, **137**, 805–814.
71. Xu, T. and Li, G. (2011) Cyclic stress–strain behavior of shape memory polymer based syntactic foam programmed by 2-D stress condition. *Polymer*, **52**, 4571–4580.
72. Nji, J. and Li, G. (2010) A biomimic shape memory polymer based self-healing particulate composite. *Polymer*, **51**, (25), 6021–6029 (November).
73. John, M. and Li, G. (2010) Self-healing of sandwich structures with grid stiffened shape memory polymer syntactic foam core. *Smart Materials and Structures*, **19**, paper 075013 (12 pages).
74. Nji, J. and Li, G. (2010) A self-healing 3D woven fabric reinforced shape memory polymer composite for impact mitigation. *Smart Materials and Structures*, **19**, paper 035007 (9 pages).
75. Li, G. and John, M. (2008) A self-healing smart syntactic foam under multiple impacts. *Composites Science and Technology*, **68**, 3337–3343.
76. Xu, T., Li, G., and Pang, S.S. (2011) Effects of ultraviolet radiation on morphology and thermo-mechanical properties of shape memory polymer based syntactic foam. *Composites Part A: Applied Science and Manufacturing*, **42**, 1525–1533.
77. Xu, T. and Li, G. (2011) Durability of shape memory polymer based syntactic foam under accelerated hydrolytic ageing. *Materials Science and Engineering A*, **528**, 7444–7450.
78. Li, G. and Uppu, N. (2010) Shape memory polymer based self-healing syntactic foam: 3-D confined thermo-mechanical characterization. *Composites Science and Technology*, **70**, 1419–1427.
79. Li, G. and Nettles, D. (2010) Thermomechanical characterization of a shape memory polymer based self-repairing syntactic foam. *Polymer*, **51**, 755–762.
80. Berriot, J., Montes, H., Lequeux, F., Long, D., and Sotta, P. (2003) Gradient of glass transition temperature in filled elastomers. *Europhysics Letters*, **64**, 50–56.
81. Berriot, J., Montes, H., Lequeux, F., Long, D., and Sotta, P. (2002) Evidence for the shift of the glass transition near the particles in silica-filled elastomers. *Macromolecules*, **35**, 9756–9762.
82. Li, G., Zhao, Y., and Pang, S.S. (1998) A three-layer built-in analytical modeling of concrete. *Cement and Concrete Research*, **28**, 1057–1070.
83. Li, G., Zhao, Y., and Pang, S.S. (1999) Four-phase sphere modeling of effective bulk modulus of concrete. *Cement and Concrete Research*, **29**, 839–845.
84. Li, G., Zhao, Y., Pang, S.S., and Li, Y. (1999) Effective Young's modulus estimation of concrete. *Cement and Concrete Research*, **29**, 1455–1462.
85. Shojaei, A. and Li, G. (2013) Viscoplasticity analysis of semicrystalline polymers: a multiscale approach within micromechanics framework. *International Journal of Plasticity*, **42**, 31–49.
86. Xu, W. and Li, G. (2011) Thermoviscoplastic modeling and testing of shape memory polymer based self-healing syntactic foam programmed at glassy temperature. *ASME Journal of Applied Mechanics*, **78**, paper 061017 (14 pages).

5

Shape Memory Polyurethane Fiber

Since the discovery by Bayer and co-workers in 1937, polyurethane (PU) has been widely used in various applications, such as biomaterials [1], textiles and sensors [2,3], coating materials [4], etc. The linear chain segmented block PU copolymer is composed of soft segment and hard segment domains, similar to spider silk [5], which defines its stimuli-responsive functionality and mechanical property. The soft segment domain contributes high ductility and elongation at break; on the other hand, the hydrogen-bonded hard segment domain plays a role in providing mechanical strength [3,5–7]. In recent years, polyurethane fibers have emerged as a new class of smart materials. A comprehensive review on polyurethane fibers can be found in Reference [8]. Most recently, tension programmed shape memory polyurethane fibers (SMPFs), which are pre-embedded in a thermoset polymer matrix, have been used to close wide-opened (millimeter-scale) cracks in the matrix by constrained shape recovery (shrinkage) of the SMPFs [9–12]. The required recovery force is enhanced by cold-tension programming of the SMPFs [9]. The change in microstructures due to cold-drawing programming has been characterized [9–11] and viscoplasticity theory and multiscale modeling have been developed to predict the thermomechanical cycle of SMPFs [9,13,14]. Long-term shape memory functionality has been studied [15]. In this chapter, we will first discuss strengthening of SMPFs by cold-tension programming. We will then focus on the characterization of SMPFs, including thermomechanical behavior and vibration damping. After that, we will discuss some modeling work on SMPFs. Finally, we will discuss stress memory versus strain memory, which is a topic of interest not only for SMPFs but also for all shape memory polymers.

5.1 Strengthening of SMPFs Through Strain Hardening by Cold-Drawing Programming

5.1.1 SMPFs with a Phase Segregated Microstructure

Among the various thermoplastic SMPs, thermoplastic polyurethane, a microphase segregated block copolymer, is widely used for fabricating SMP fibers. The reason is that polyurethane has a good spin-ability, and its transition temperature and mechanical properties can be tailored in a wide range by controlling the composition, morphology, and ratio of the soft segment phase and the hard segment phase. Usually, SMPFs can be manufactured by various spinning

Self-Healing Composites: Shape Memory Polymer-Based Structures, First Edition. Guoqiang Li.
© 2015 John Wiley & Sons, Ltd. Published 2015 by John Wiley & Sons, Ltd.

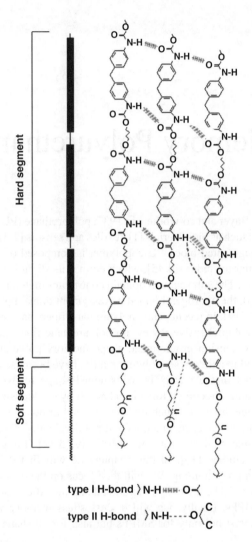

Figure 5.1 Schematic molecular structure of an SMPF. The soft segment and hard segment form the amorphous soft phase and crystalline hard segment domain, respectively. Abundant hydrogen bonds form between the amino group and carbonyl group in the crystalline hard domain. The amorphous phase and crystalline phase are partially oriented. *Source:* [10] Reproduced with permission, from the Royal Society

methods, such as wet, dry, chemical, and melt spinning approaches [16–18]. Melt spinning of polyurethane does not involve the use of harmful dimethylformamide as a solvent, and no coagulation bath. Therefore, melt spinning is the preferred method [16–18].

The shape memory mechanism of polyurethane is due to its microphase segregated heterogeneous structure, which consists of the hard segment domain and the soft segment domain (see the schematic in Figure 5.1). Also, the hard segment, or a portion of the hard segment with a smaller molecular weight, may mix with the soft segment to form a mixture. Both phases can be either a crystalline phase or an amorphous phase or even semicrystalline (consisting of both amorphous and crystalline subphases). Sometimes, a less perfect crystalline phase, also called a mesomorphic form [19,20], can be found, which has a morphology in between the amorphous phase and crystalline phase with a certain long-range order. Therefore, the transition temperature of each phase can be either a glass transition or a melt transition, that is, an amorphous phase with a glass transition and a crystalline phase with a melt transition, or a mesomorphic phase with a broad melt transition. Usually, the transition temperature of the hard segment domain is higher than that of the soft segment domain. Therefore, we usually use the soft segment domain to control the shape memory, that is, molecular switches, and the hard segment domain to control the mechanical strength and permanent shape, that is, netpoints [8]. However, we can also use the melting temperature of the soft segment domain or even the glass transition temperature of the hard segment domain as the transition temperature as long as they are programmed properly and sufficiently. For example, Zhang and Li [15] observed two stress recovery rates for a cold-drawing programmed SMPF recovered at 150 °C, signifying that both shape memory mechanisms in the soft and hard amorphous phases have been triggered. However, using recovery stress as a criterion, it is a matter of balance between the shape memory and stiffness at the recovery temperature. At a higher recovery temperature, more shape memory mechanisms can be triggered, which tends to result in a higher recovery stress; at the same time, the stiffness is reduced at a higher temperature, which tends to result in a lower recovery stress. Therefore, a balance needs to be sought. The formation of the netpoints in the hard segment domain of polyurethane is due to physical cross-linking, that is, intermolecular interactions. Generally, the crystalline domain is bonded by a hydrogen bond between the folded molecules. The hydrogen bond, although it can form a crystalline phase, is not as strong as other chemical bonds such as the covalent bond, metallic bond, or ionic bond, and therefore the crystalline domain can be changed comparatively easily by external stimuli such as heat or force. This provides an opportunity for using mechanical means to change the microstructure and thus the properties of polyurethane.

The phase segregated microstructure of SMPFs can also be validated by the atomic force microscope (AFM) test. The microphase separation morphology of an SMPF was investigated using an AFM (SPA-300HV, Seiko Instruments Inc.) in the tapping mode at ambient environment (22 ± 2 °C, $45 \pm 5\%$ relative humidity). The fiber composition is the same as that given in Section 5.1.2 with a diameter of 0.05 mm. From Figure 5.2, a clear phase segregated microstructure is seen. In the phase image, a higher modulus material induces a higher phase offset and it appears lighter as opposed to a softer phase, which appears darker. Therefore, the mixed lighter color and darker color suggest the existence of both a hard segment domain and a soft phase in SMPFs. It can also be estimated that, for this particular SMPF, the hard domain is probably the continuous phase and predominates.

It is noted that in this chapter all the SMPFs have the same composition. The only difference is their fiber diameter, pulling force, and cooling rate during manufacturing. Because fiber diameter has a significant effect on the mechanical and shape memory behavior, we will report the fiber diameter for each case. Also, because the molecular alignment and pulling force during manufacturing are different for fibers with different diameters, it is expected that, even with the same composition, the fiber may have different microstructures. Actually, even the cooling rate affects the morphology significantly. It is known that for molten polyurethane, the morphology is amorphous or a mixture of hard domain and soft phase. If the cooling rate is super high, supercooled liquid forms, that is, the amorphous morphology is frozen and the hardened fiber is amorphous; on the other hand, if the cooling rate is not super high but high

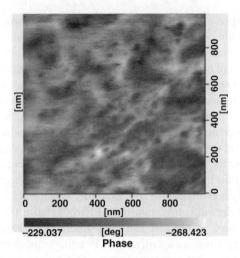

Figure 5.2 AFM image of the microphase segregated SMPF. The phase image indicates the microphase separation of the polymer with lighter hard segment phase regions and darker soft segment phase regions. *Source:* [10] Reproduced with permission from the Royal Society

enough, molecules will have sufficient time and mobility to form crystallite centered around the nucleus. Because of the high cooling rate, the nucleus formed can be frozen or stabilized and crystals can be produced. If the cooling rate is very slow, while it gives sufficient time for the formation of nuclei, it also allows disassociation of molecules from the nuclei, which may lead to nonstabilized nuclei and the polymer may be amorphous. It is noted that the effect of the cooling rate on polymers is different from that on a small-molecule liquid, which needs a lower cooling rate to form crystals. It is also noted that cooling induced crystallization depends on the temperature and pressure. In other words, the effect of the cooling rate is coupled with the temperature (close to the melting temperature or close to the glass transition temperature) and pressure. Besides, the cooling rate affects the type of crystals. At a higher cooling rate, more stable α-crystal may form whereas at a lower cooling rate, meta-stable β-crystal may form. Even with the same cooling rate, fibers with different diameters have different morphologies. The reason is that for thinner fibers, the temperature difference at the fiber surface and fiber center is smaller, while for thicker fibers the difference is larger. The temperature gradient in the radial direction may lead to a gradient in morphology along the radial direction. In other words, fibers with different diameters may have different morphologies, even though they have the same composition, same pulling rate, and same cooling rate during melt spinning. Figure 5.3 shows a schematic of the melt spinning process.

As discussed above, both the soft segment domain and hard segment domain may be semicrystalline, that is, consisting of both amorphous and crystalline subphases, and these two domains can form a mixture. Figure 5.4 shows DSC results of such an as-received polyurethane fiber. Again, the fiber is made of the same raw materials as those that will be discussed in Section 5.1.2 but with a fiber diameter of 0.06 mm. The PerkinElmer DSC4000 power compensated dynamic scanning calorimeter (DSC) was used to measure the heat capacity changes of the material as a function of temperature. The sample was first cooled from room temperature to −60 °C at a rate of 20 °C/min. After this, the sample was scanned from −60 °C to

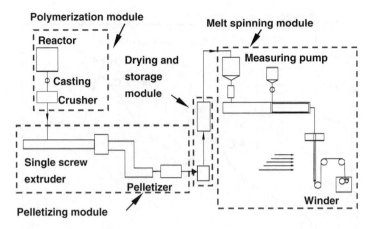

Figure 5.3   Schematic of the melt spinning process for fabricating SMPFs

250 °C at a heating rate of 10 °C/min. Then this thermal cycle was repeated one more time. In the test, the first thermal cycle included the manufacturing history while the manufacturing history was erased in the second cycle. The DSC results in Figure 5.4(a) and (b) illustrate the heat flow of the sample as a function of temperature for the first cycle and second cycle, respectively.

From Figure 5.4(a), the following observations can be made. (1) We believe that the SMPF has a multilength scale morphology. The first level sees a two-phase composite with a portion of the hard segment domain, which has a higher molecular weight, dispersed in a matrix. When we zoom into the matrix, the second-level structure is revealed. It is a mixture of the soft segments and the remaining hard segments (which have a lower molecular weight). We also believe that the hard segment domain with a lower or higher molecular weight may have a certain short-range and long-range order, respectively, due to annealing or physical ageing at room temperature. This is because the SMPF has been stored at room temperature for about six months before the first DSC cycle test. (2) The soft–hard segment mixture has a glass transition temperature at about 23.4 °C. This glass transition temperature stays in between the pure soft segment glass transition temperature (−55 °C) and the hard segment glass transition temperature (110 °C), indicating that the soft segments and hard segments are well mixed together and follows the type of rule-of-mixture behavior. With the increase in temperature, the hard segment with the lower molecular weight, which may have a certain short-range order, leads to an endothermic peak at about 69.5 °C, similar to melting of crystallites. As the temperature rises, the mobility of the soft segment and the hard segment with the lower molecular weight becomes higher. At about 82.0 °C, we see an exothermic peak. This peak represents the microphase separation, that is, separation of the soft segment from the hard segment with the lower molecular weight. In order words, the soft–hard segment mixture becomes a clearer two-phase morphology. Because the phase separation suggests that the system becomes more ordered, it releases heat (similar to crystallization). With a further rise in temperature, the hard segment domain with the higher molecular weight, which may also have a certain long-range order, has a broad melting transition temperature at about 141.9 °C. It is noted that this peak is not very sharp because the crystallite is not perfect. With the increase in annealing time, it is believed that this very broad melting transition will become sharper, as proved by Saiani

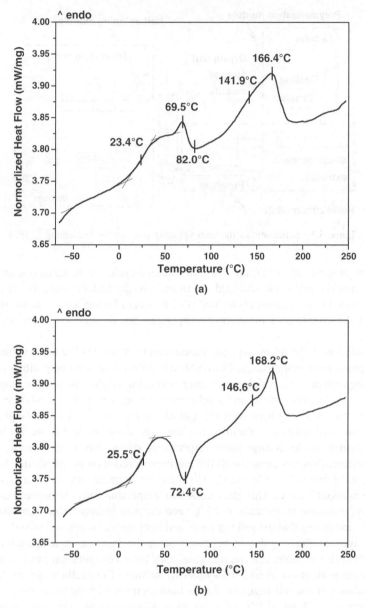

Figure 5.4 Microstructure change of as-spun SMPF during (a) the first DSC cycle and (b) the second DSC cycle

et al. [21]. At about 166.4 °C, an endothermic peak appears. This peak does not suggest a melting peak for crystals. Instead, it suggests microphase mixing of the soft segment and hard segment. Because the mixing increases randomness of the system (similar to melting of a crystal), it absorbs energy, leading to an endothermic peak. In other words, due to the significantly enhanced mobility of both the soft and hard segments at this higher temperature, regardless of their molecular weight, mixing of the segregated soft segments and hard segments becomes possible because the thermodynamic driving force (entropy) may be sufficient to overcome the thermodynamic barriers (incompatibility).

From Figure 5.4(b), we can make the following observations. (1) The SMPF has a similar morphology to the SMPF before the first DSC cycle, except for the disappearance of the short-range order in the hard segments with the lower molecular weight. This is because there is no sufficient annealing time between the first and second DSC cycles. However, we believe that the long-range order in the hard segment with the higher molecular weight may still exist. This is because the possible long-range order formed in the hard segment with the higher molecular weight at the higher temperature may have been frozen during the fast cooling-down process in the first DCS cycle. (2) Because of the morphology, the second DSC cycle sees a glass transition temperature of the soft–hard segment mixture at about 25.5 °C. As compared to the first DSC cycle, the soft segment and hard segment mixture is more complete and thus the mobility of the soft phase is more restricted by the hard domain, leading to an increase in the glass transition temperature. Also, because of the disappearance of the short-range order in the hard segment with the lower molecular weight, the endothermic peak in the first DSC cycle is not seen in the second cycle. Further, the microphase separation (demixing) temperature is 72.4 °C, which is about 10 °C lower than that in the first DSC cycle. We believe the reason is that, in the first DSC cycle, the hard segment has a certain short-range order, which needs a higher temperature to melt and to separate from the soft segment mixture. The hard segment broad melting transition temperature, 146.6 °C, is higher than that of the first cycle. This suggests that by removing the manufacturing history and better separation of the soft phase from the hard segment domain, the hard domain is more ordered, thereby leading to a higher transition temperature. Finally, the microphase mixing temperature, 168.2 °C, is slightly higher than that in the first cycle, suggesting that a similar thermodynamic driving force controls the formation of the mixture. We would like to emphasize that we do not call the microphase mixing temperature the melting temperature of crystals for a couple of reasons. First, the wide-angle X-ray diffraction test does not reveal significant presence of a crystal phase in this fiber. Second, this can better explain the microphase separation event because the enthalpy for the microphase separation (exothermic peak) and the enthalpy for microphase mixing (endother-mic peak) is similar. Therefore, the mixture formed at the higher temperature is separated at a lower temperature and the energy input for the formation of the mixture is similar to the energy output for the separation of the mixture (see Figure 5.4(b)).

The microphase separation is not uncommon in polymers such as the SMPF. The reason for this is determined by thermodynamic quantities and molecular mobility [22]. In a very simple, purely hypothetical case, which assumes a crystal of infinite size and no surface energy contribution, the change of free energy of the polymer is

$$\Delta G = \Delta H - T\Delta S \tag{5.1}$$

where G is Gibb's free energy, H is enthalpy, T is temperature, and S is entropy. For the isothermal and isobaric process, $\Delta G < 0$ suggests an autonomous process. Here ΔH is the thermodynamic driving force for the formation of an ordered structure (in this case the microphase separation) and ΔS is the entropic barrier against the separation. An autonomous process means that, in the sense of absolute value, the thermodynamic driving force is greater than the entropic barrier. In other words, in order to make $\Delta G < 0$, we need $\Delta H - T\Delta S < 0$, or $\Delta H < T\Delta S$. For our case, the microphase separation gives off heat and thus $\Delta H < 0$; at the same time, the formation of an ordered structure reduces entropy, that is, $\Delta S < 0$. Because T is positive, the inequality $\Delta H < T\Delta S$ is possible, that is, microphase separation is possible. Because both ΔH and $T\Delta S$ are negative when the microphase separates, the inequality ($\Delta H < T\Delta S$) suggests that the absolute value of the thermodynamic driving force (ΔH) is greater than the absolute value of the entropic barrier ($T\Delta S$). Of course, further study showed that the ease of microphase separation is most closely related to the conformational entropy and the molecular mobility [22].

5.1.2 Raw Materials and Fiber Fabrication

Generally, SMPFs consist of a hard domain and a soft domain, which can be manufactured from a spectrum of raw materials. A soft segment may consist of amorphous (e.g., polyester and polyether) or semicrystalline (e.g., poly(ε-caprolactone) (PCL)) structures while the hard segments (e.g., toluene diisocyanate (TDI), aromatic urethane, or aramid) may form thermally stable chemical or physical cross-links. In this chapter, the SMPF was synthesized using poly (butylene adipate) (PBA) (Sigma-Aldrich, USA) as a soft segment, and $4'4$-diphenylmethane diisocyanate (MDI) (Sigma-Aldrich, USA), and 1,4-butanediol (BDO) (Sigma-aldrich, USA) as a hard segment. Dibutyltin dilaurate was used as catalyst with a content of 0.02% by weight. The average formula–weight ratio is (MDI + BDO):PBA = 1021:300, indicating a hard segment rich composition. BDO was dehydrated with $4\,\text{Å}$ molecular sieves for one day in advance. All the chemicals were demoisturized prior to use in a vacuum oven. Molten MDI was filtered to remove the precipitate dimmers and any impurities before use. The reaction was conducted in a high speed mixer at room temperature. The obtained polyurethane was further cured in a vacuum oven at $110\,°C$ for 12 hours for further reaction. Before spinning, the polymer was dried in a vacuum oven at $80\,°C$ for 6 hours. A modified One-Shot Extrusion Machine with a spinning speed of $40\,\text{m/min}$ was used to prepare the filaments. The spinning temperature was $180\,°C$. The spinneret filter was $9/50/18/9\,\text{mazes/cm}^2$. The extruder head pressure was $5.0\,\text{MPa}$. The filament passed three pairs of rollers before being wound up at a speed of $40\,\text{m/min}$. Figure 5.3 shows a schematic of the melt spinning process while Figure 5.5 shows an SEM image of one individual fiber by a JEOL JSM – 6390 scanning electron microscope as well as the fiber roll.

Figure 5.5   SEM image of a single SMPF (the diameter of the fiber is $51.9\,\mu m$)

5.1.3 Cold-Drawing Programming

It is well known that shape memory polyurethane has a very small recovery stress, which limits its application to non-load-bearing structures such as textiles [23]. Various ways have been investigated to increase its recovery stress such as by adding carbon nanotubes [24], controlling the spinning method (molecular alignment) [16], and cold drawing [25]. It is noted that cold-drawing programming of thermoplastic SMPs has been conducted by several researchers. Lendlein and Langer [26] indicated that shape memory polymer can be programmed by cold drawing but did not give sufficient details. Ping et al. [25] investigated a thermoplastic poly (ε-caprolactone) (PCL) polyurethane for medical applications. In this polymer, PCL was the soft segment and can be stretched (tensioned) to several hundred percent at room temperature (15–20 °C below the melting temperature of the PCL segment). They found that the cold-drawing programmed polyurethane had a good shape memory capability with an increase in recovery stress. Rabani, Luftmann, and Kraft [27] also investigated the shape memory functionality of two shape memory polymers containing short aramid hard segments and poly(ε-caprolactone) (PCL) soft segments with cold-drawing programming. As compared to the study by Ping et al. [25], the hard segment was different but with the same soft segment PCL. Meng and Hu [17] prepared a polyurethane fiber with the soft segment domain having a semicrystalline structure (an amorphous phase and a crystalline phase) and used the melt temperature of the soft segment domain as the switch temperature for the shape memory effect. They compared the recovery stress of the cold-drawing programmed polyurethane fiber (with a programming temperature below the melt temperature of the soft phase) with its high temperature programmed counterpart (with a programming temperature above the melt temperature of the soft phase) and found that the recovery stress of the cold-drawing programmed fiber is about 3.5 times that of the fiber programmed by the classical method. Wang et al. [28] further studied the same SMP as Ping et al. [25]. They used FTIR to characterize the microstructure change during the cold-drawing programming and shape recovery. They found that in cold-drawing programming, the amorphous PCL chains orient first at small extensions, whereas the hard segments and the crystalline PCL largely maintain their original state. When stretched further, the hard segments and the crystalline PCL chains start to align along the stretching direction and quickly reach a high degree of orientation; the hydrogen bonds between the urethane units along the stretching direction are weakened and the PCL undergoes stress induced disaggregation and recrystallization while maintaining its overall crystallinity. When the SMP recovers, the microstructure evolves by reversing the sequence of the microstructure change during programming. Zotzmann et al. [29] emphasized that a key requirement for materials suitable for cold-drawing programming is their ability to be deformed by cold drawing. Based on their discussion, it seems that an SMP with an elongation at break as high as 20% is not suitable for cold-drawing programming.

To conduct cold-drawing programming, the SMPFs were loaded at room temperature by using MTS (RT/5, MTS Inc., USA) equipped with a 250 N load cell. Fibers with a diameter of 0.038 mm were gripped by aluminum fixtures (Manual Bollard Grips, MTS Inc.) with a gauge length of 20 mm and stretched to the designed strain level with a displacement rate of 400 mm/min. Similar to cold-compression programming in Chapter 3, once the strain level is reached, the strain is held for a period of time to increase its shape fixity (30 minutes [15]). Sometimes, cyclic loading/unloading cycles are used to further increase the shape fixity of the fibers [15] (see Figure 5.6). Each cycle leads to a certain pseudoplastic deformation and an accumulation of

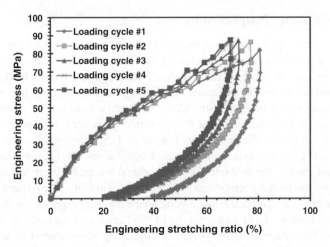

Figure 5.6 Engineering stress–engineering strain behavior of cyclic cold-drawing programming of SMPF. *Source:* [15] Reproduced with permission from John Wiley & Sons, Ltd

pseudoplastic deformation from each cycle leads to an increase in shape fixity. Most of the time, a fiber bundle with 10 filaments was programmed instead of one individual filament to reduce the adverse effect of the noise of signals on the test results because the maximum tension load for each filament is much smaller than the capacity of the load cell.

5.1.4 Microstructure Change by Cold-Drawing Programming

A demonstration on microstructure changes during cold-drawing programming is schematically shown in Figure 5.7. In this demonstration, the hard segment phase is semicrystalline with lamellae linked by amorphous molecules to form spherulite, which is mixed with the soft segment phase (amorphous with some molecules having long-range order). They undergo microphase segregation and contribute to the thermodynamic incompatibility. With the increase in the external tensile load, this first causes alignment of soft segment molecules, probably by overcoming segmental rotation resistance of the soft segment molecules, and then leading to network stretching and intermolecular sliding. With a further increase in the tensile load, stress induced crystallization (SIC) may occur in the soft phase and may lead to alignment of the hard domain along the loading direction and ultimately fracture of the hard segment spherulite. Similar to the thermosetting amorphous SMP studied in Chapter 3, this process mainly causes deformation in the soft segment domain, which is partially fixed by the hard segment domain (as physical cross-linking) once the external load is removed. Owing to the formation of the aligned and stretched network in the soft segment phase and orientation alignment of the hard segment phase, cold-drawing programming leads to an increase in the tensile strength, stiffness, and stored back stress. In other words, cold-drawing programming results in enhancement of mechanical properties and recovery stress. Of course, further stretching may lead to broken hydrogen bonds in the hard segment domain or a broken network in the soft segment domain, which signifies fracture of the SMPF.

It is noted that the cold-stretching induced microstructure change depends on the temperature and stretching rate due to the famous time–temperature equivalent principle. During the stretching process, the molecule chains experience two competing processes. One is the extension, that is, the

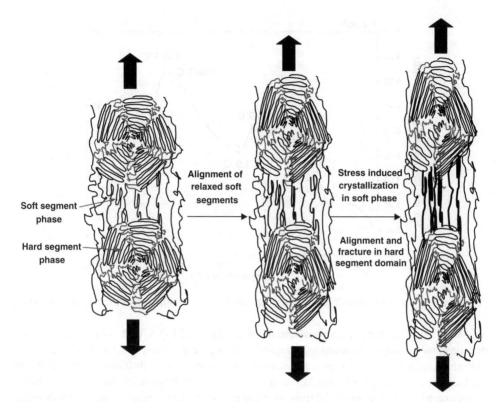

Figure 5.7 Schematic representations of segments alignment, reorientation, and broken SMPFs under monotonic tension

molecule chains extend and align along the loading direction, and the other is molecule chain relaxation, that is, relaxation to their recoiled configuration. If the loading rate is very high, the tendency for chain extension exceeds that for chain relaxation, leading to strain induced ordering or alignment, such as a certain long-range order (mesomorphic form) or even strain induced crystallization. On the other hand, if the loading rate is low, the tendency for molecule chain relaxation exceeds that for molecule chain extension, leading to an amorphous configuration after cold-drawing programming. Of course, if taking both temperature and loading rate into account, a combination of a high loading rate and a low drawing temperature tends to result in strain (or stress) induced crystallization; on the contrary, a combination of both a high drawing temperature and a low stretching rate may lead to amorphous configuration. Therefore, it is not certain that cold-drawing will lead to strain induced crystallization. Both drawing temperature and drawing speed need to be taken into account.

The microstructure change during cold-drawing programming can be validated by DSC analysis and infrared (IR) dichroism analysis. Figure 5.8 shows the first cycle of the DSC test result after SMPFs were cold-drawing programmed to 100%. Compared to the same SMPFs in Figure 5.4(a), it is seen that the corresponding characteristic temperatures have become: a glass transition temperature of the mixture of 28.6 °C, a melting temperature of the hard segments with a lower molecular weight of 68.9 °C, a microphase separation temperature of 81.9 °C, a broad melting transition temperature of the hard segment domain with a higher molecular

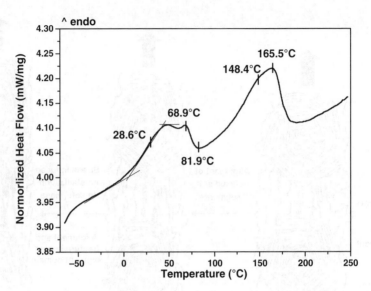

Figure 5.8 First cycle of DSC test results of the SMPF after cold-drawing programming to 100%

weight of 148.4 °C, and a microphase mixing temperature of 165.5 °C. It is interesting to note that the glass transition temperature of the soft–hard segment mixture, the melting temperature of the hard segment with a lower molecular weight, and the broad transition temperature of the hard segment with a higher molecular weight have been slightly shifted to higher temperatures, as compared to the as-spun SMPFs in Figure 5.4(a). This suggests that cold-drawing programming leads to further molecule alignment in both soft and hard segment domains, which is the physical basis for the increased mechanical properties and recovery stress for cold-drawing programmed SMPFs. It is noted that, as compared to Figure 5.4(a), both the microphase separation temperature and microphase mixing temperature of the cold-drawn SMPFs are very close to those of the as-spun SMPFs. This implies that cold-drawing programming could not cause microphase separation of the soft segment from the hard segment, which only causes alignment of the molecules along the loading direction. Of course, this observation is based on the pre-strain level used. If the cold-drawing parameters are changed such as temperature, loading rate, pre-strain level, etc., this observation may change because SMPFs are very sensitive to the loading history.

The soft segment and hard segment orientation of the SMPF can also be demonstrated by IR dichroism. The orientation number of the amorphous phase and crystalline phase in the SMPF with a diameter of 0.05 mm was determined by IR dichroism using Perkin–Elmer microscopy (Perkin-Elmer Inc., USA) equipped with a diamond cryogenic Micro-ATR unit and an IR polarizer. The fiber was cold-drawn to 100% strain before testing. As shown in Figure 5.9, the IR dichroism was conducted both along the axial (parallel with the fiber) direction and in the transverse (perpendicular to the fiber) direction after cold-drawing programming. The significant differences between the two directions $A_{\parallel}$ and $A_{\perp}$ suggest the orientation change of the macromolecules in the SMPF after cold stretching. Based on References [30] and [31], the orientation number of the soft segment was calculated to be 0.19 and the orientation number of the hard segment was calculated to be 0.56.

Because the stress applied during cold-drawing programming is usually very large as compared to the classical programming above the transition temperature, it is interesting to find

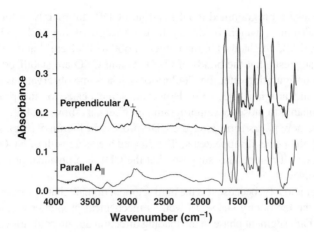

Figure 5.9 Infrared spectra of the cold-stretched SMPF with the polarizer parallel with and perpendicular to the fiber direction. *Source:* [10] Reproduced with permission from the Royal Society

whether the high stress causes any changes in chemical bonds. This can be determined by Fourier transform infrared spectroscopy (FTIR) analysis. We have conducted FTIR analysis of the cold-drawn SMPF using FTIR (Nicholet 6700, Thermo Scientific, USA). The fiber has a diameter of 0.04 mm and was cold-drawn to 80% strain. Figure 5.10 presents the infrared spectrum for the nonprogrammed and programmed fibers, which was recorded from 600 cm^{-1} to 4000 cm^{-1}. The overall results shown in Figure 5.10(a) are divided into three groups in Figure 5.10(b) to (d) to give a better view. From Figure 5.10(a), the absorbances for both fiber

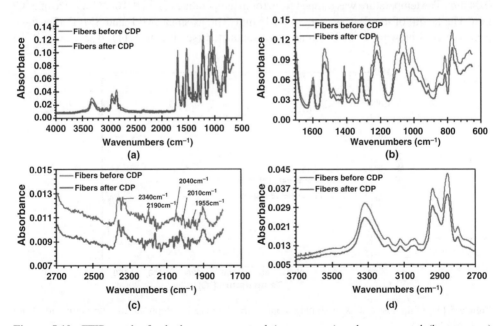

Figure 5.10 FTIR results for both nonprogrammed (upper curve) and programmed (lower curve) SMPFs. *Source:* [15] Reproduced with permission from John Wiley & Sons, Ltd

samples (cold-drawing programmed (CDP) and non-CDP) are mainly observed in the range from 600 to 1600 cm^{-1}, which is similar to the spectra of the PCL [32]. According to Figure 5.10(b) and (d), no changes in the ranges of 600 to 1700 cm^{-1} and 2700 to 4000 cm^{-1} are found. This suggests that the bonds of C—O—C and C=O are stabilized or fixed in their conformational configuration after the CDP process. The same observations are for groups of urethane, CH$_2$, and aromatics [33–35]. However, some peaks for the programmed fibers disappear as compared to those for nonprogrammed fibers, as shown in Figure 5.10(c). On the other hand, some new peaks form after the CDP process in Figure 5.10(c). As suggested by Wang *et al.* [33], the peaks disappeared and/or created are as a result of NCO group vibration and carbon dioxide vibration. This suggests that the CDP is a physical process; no chemical changes are involved.

As discussed previously, because it is the hard segment domain that determines the permanent shape and fixes the temporary shape, as well as mechanical properties of the SMPF, further alignment of the hard segment phase in the loading direction during cold-drawing programming may lead to an increase in stiffness during shape recovery and potentially higher recovery stress. The reason is that the soft segment domain is the switching phase in the shape recovery process which is either rubbery (if the glass transition is the switch temperature) or liquid (if the melt transition is the switch temperature). Therefore, the soft segment domain is very soft at the recovery temperature. However, the transition temperature of the soft segment phase is usually much lower than the transition temperature of the hard segment domain. Therefore, the recovery stress depends on the stiffness of the hard segment phase at the recovery temperature.

Direct evidence of the increase in stiffness after cold-drawing programming comes from the dynamic mechanical analyzer (DMA) test result. The DMA 2980 tester from TA instruments following the ASTM D 4092 standard was conducted. A dynamic load at 1 Hz was applied to SMPF bundles. Each fiber bundle contained 10 filaments. The diameter of the fiber was 0.04 mm. The temperature was ramped from room temperatures of 25 to 160 °C at a rate of 3 °C/min. The length of the fiber was set to be 15 mm. The as-spun SMPF and SMPF after three cycles of cold drawing to 250% strain were conducted (see Figure 5.11). It is clear that the

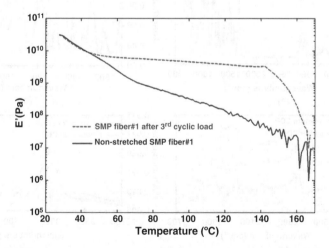

Figure 5.11 DMA test results of storage modulus E'. *Source:* [9] Reproduced with permission from the Royal Society

storage modulus of the cold-drawn SMPF is about an order higher than that of the as-spun SMPF at the recovery temperature of 80 °C. Because the recovery stress is in direct proportion to the stiffness, it is a direct indication of the increase in the constrained recovery stress of the cold-drawn SMPFs. This will be validated in the next section.

5.1.5   Summary

In summary, the SMPF was fabricated by melt spinning. SMPFs have a microphase segregated microstructure, which provides the key mechanisms for shape memory – a soft segment phase as the molecule switches and a hard segment phase as netpoints. It is found that cold-drawing programming, just like cold working on metals, changes the microstructures of the SMPF, particularly molecular alignment along the loading direction in both the soft segment and hard segment domains, and possible stress induced crystallization in the soft phase. As a result, the SMPF would increase the stiffness at temperatures above T_g of the soft segment phase, which is an indication of a potential increase in the recovery stress. It is also found that cold-drawing programming does not cause microphase separation of the soft segment from the hard segment, and also does not affect the microphase mixing temperature.

5.2   Characterization of Thermoplastic SMPFs

5.2.1   Thermomechanical Characterization

In this study, the SMPFs have a diameter of 0.06 mm. In order to show the effect of programming procedure on the thermomechanical behavior, both cold (within or below the soft–hard segment mixture T_g) and hot (above the T_g) drawing programming were conducted. SMPFs without programming were also used as control. Instead of testing one single filament, fiber bundles were used. Each bundle consisted of 100 filaments. For clarity, the nonstretched fiber bundle was named Sample 1. Sample 2 was hot-drawing programmed. The hot-drawing programming started by stretching the fiber bundle to 100% strain at a temperature of 100 °C, which was above the glass transition region of the soft–hard segment mixture (actually it is also above the melting temperature of the hard segment with a lower molecular weight and the microphase separation temperature), holding it at 100 °C for 1 hour, cooling down to room temperature while holding the strain constant, and unloading. Sample 3 was programmed at ambient temperature (about 20 °C), which was close to or even below the glass transition temperature zone of the mixture. Therefore, it is named cold-drawing programming. Cold-drawing programming started with stretching the bundle to 100% strain and holding at room temperature for 1 hour. After that the load was removed. For both cold and hot programmings, the MTS (Alliance RT/5, MTS Inc., USA) equipped with a 250 N load cell was used. The gauge length of the bundles was 10 mm. The cross-head speed was 200 mm/min during loading. In hot programming, the heating rate was 3 °C/min and the cooling rate was 0.5 °C/min.

After programming, both Samples 2 and 3 were relaxed at room temperature for 24 hours. During this period, a change in length was recorded to characterize the relaxation of fiber and determine the shape fixity. Then the stress recovery test was conducted with a fully constrained boundary condition. The programmed samples were heated from the ambient temperature to 150 °C. The heating rate was 0.5 °C/min. The temperature was monitored with an Omega XC-20-K-12 thermal couple by Yokogawa DC 100 data acquisition system. Once the temperature reached 150 °C, it was maintained for 10 hours until the recovery stress was stabilized. The recovered stress as a function of time was obtained. The peak recovered stresses and stabilized stresses of both the programmed samples were identified. It is noted that, in order to compensate for the thermal expansion effect, which makes the fiber bundle loose at the temperature below T_g because the fiber bundles could not take the compressive force, a pre-tension was applied to the bundle before the stress recovery test started. By a *trial and error* process, the pre-tension was determined as 4%, which ensured that the stress at the start of the stress recovery was about zero. Therefore, the pre-tension was a technique used to compensate for the inability for SMPF bundles to carry the compressive load. It did not affect the stress recovery test results.

Figure 5.12(a) shows the thermomechanical cycle of Sample 2 and Figure 5.12(b) shows the thermomechanical cycle of Sample 3. To demonstrate clearly the stress evolution with time, a two-dimensional figure illustrating the stress change with time is also presented for each sample (see Figure 5.12(c) for Sample 2 and (d) for Sample 3, respectively). It is noted that, for the stress relaxation after removing the programming stress, although 24 hours were monitored, it was found that the relaxation basically stabilized within about 10 minutes. Therefore, in Figure 5.12, the stress relaxation step lasted for 10 minutes instead of 24 hours. To compare the recovery stress programmed by both the hot- and cold-drawing programmings better, Figure 5.13 shows the recovery stress and temperature evolution with time.

For Sample 2, it can be seen that the stress increases linearly in step 1, due to the rubber elastic behavior of the soft phase above T_g. In step 2, the stress relaxes while holding the strain constant. In step 3, the stress decreases during cooling. It is noted that during cooling, two competing tendencies take effect. One is the stress increase due to thermal contraction and stiffening of the SMPF during cooling and the other is due to stress relaxation. The final stress shows a combined effect of these two tendencies. Depending on which tendency predominates, the stress may increase or decrease. In this test condition, we see a decrease in the stress, suggesting that stress relaxation exceeds thermal stress accumulation. In step 4, the stress is removed suddenly. Because the machine cannot record the unloading process, we use a vertical line to represents this step. In step 5, we see a viscoelastic springback. Although 24 hours were used to observe the springback with time, it was found that the springback was stabilized in about 10 minutes. In step 6, fully constrained stress recovery, the stress increases with temperature until a peak. After that, the stress relaxes until stabilization. Figure 5.13 highlights this step. Once the temperature reaches the T_g region of the soft–hard segment mixture, the shape memory is triggered and the fiber tends to shrink. Due to the constrained boundary condition, however, this tendency is not allowed. Therefore, the recovery stress is created. As the temperature rises further, the recovery stress accumulates until a peak is reached. It is noted that two competing mechanisms occur during the stress recovery. One is a stretch of molecules along the loading direction, which sees the stress increase, and the other is relaxation of the molecules, which leads to stress reduction. Therefore, once the rate of stress relaxation is equal to the rate of stress recovery, the peak recovery stress is reached. It is noted that there is another stress component in the fiber, that is, thermal expansion, or thermal stress. Because the thermal expansion has been technically removed in this study, we do not include the contribution of thermal stress to the observed recovery stress in the above discussion. As time goes by, stress relaxation predominates and we see a continuous reduction of stress until it is stabilized. This is because constrained stress recovery resembles the stress relaxation test.

For Sample 3, the stress increases linearly in step 1. While holding the strain constant, we see stress relaxation in step 2. Step 3 shows unloading instantly. In step 4, we see structural

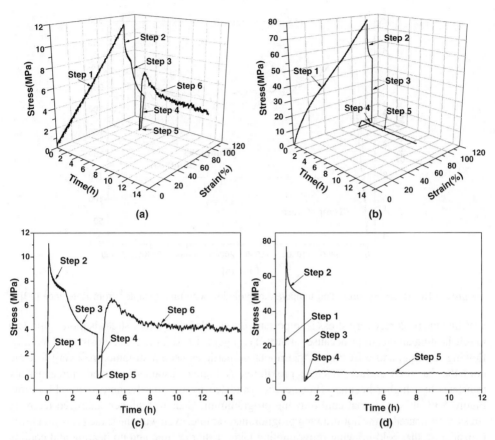

Figure 5.12 Thermomechanical behavior of SMPFs by both cold and hot tension programmings. (a) Stress–strain–time diagram for Sample 2. Steps 1 to 5 complete programming and Step 6 completes stress recovery, where step 1 is to stretch the fiber bundle to 100% strain at a rate of 200 mm/min at 100 °C; step 2 is to hold the strain constant for 1 hour; step 3 is to cool the fiber to room temperature slowly while holding the pre-strain constant; step 4 is to release the fiber bundle from the fixture (unloading); step 5 is to relax the fiber in the stress-free condition until the shape is fixed; and step 6 is to recover the fiber at 150 °C in the fully constrained condition (adapted from Reference [20]); (b) Stress–strain–time diagram for Sample 3. Steps 1–4 complete programming and step 5 completes stress recovery, where step 1 is to stretch the fiber bundle to 100% strain at a rate of 200 mm/min at room temperature; step 2 is to hold the strain constant for 1 hour; step 3 is to release the fiber bundle from fixtures (unloading); step 4 is to relax the fiber in the stress-free condition until the shape is fixed; and step 5 is to recover the fiber at 150 °C in the fully constrained condition (adapted from Reference [20]); (c) Stress evolution with time for Sample 2; (d) Stress evolution with time for Sample 3.

relaxation under zero stress, which is stabilized in about 10 minutes. In step 5, the fully constrained stress recovery step, the stress first increases due to the shape memory until the peak recovery stress. While holding the temperature constant with a zero recovery strain, we see a slight stress relaxation, that is, a small portion of the peak recovery stress is relaxed, similar to Sample 2.

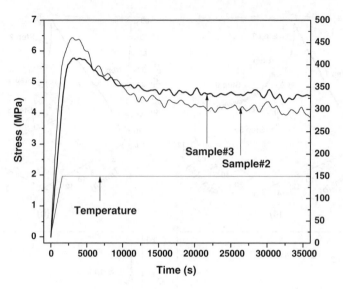

Figure 5.13 Recovery stress and temperature evolution with time (adapted from Reference [20])

While the peak recovery stress is important, the stabilized recovery stress is more important in self-healing applications because, in the CTH scheme [9–12], the crack closing and melting, flowing, wetting, and diffusing of thermoplastic particles need a sustained recovery force to hold the fractured surface in close proximity. A higher sustained recovery force usually translates to a higher healing efficiency, as validated by Li, Ajisafe, and Meng [11]. From Figure 5.13, it is clear that cold-drawing programming leads to a higher stabilized recovery stress as compared to the hot-drawing programmed sample, even with the same pre-strain level. Considering that cold-drawing programming takes a shorter time and no heating and cooling are needed, it is concluded that cold-drawing programming is a better way to train SMPFs for self-healing applications.

After programming and structural relaxation for Samples 2 and 3, some specimens were also tensile tested to determine their tensile strength, stiffness, and ductility. Sample 1 without programming was also tested as the control. The tensile properties of all three types of samples (Samples 1, 2, and 3) were tested using MTS (Alliance RT/5, MTS Inc., USA) equipped with a 250 N load cell. Sample clamps specified for fibers (Manual Bollard Grips, MTS Inc.) were used. The cross-head speed was 200 mm/min and the gauge length was 10 mm. Figure 5.14 shows the tensile test results.

As compared to the control Sample 1, it is clear that the programming procedure increases the stiffness (slope of the stress–strain curve) and the strength (stress at break) of the fibers, but reduces the ductility, regardless of hot or cold programming. This is simply because tension programming leads to the alignment of molecules along the loading direction. A further comparison of Samples 2 and 3 shows that hot programming results in a higher stiffness and strength than cold programming, but with lower ductility. The reason for this can be explained in terms of the microstructure change during programming. During cold-drawing programming, the soft amorphous phase molecules first align in the loading direction. However, this alignment is limited because of the strong resistance to moving in the glassy state. Further stretch leads to alignment of the hard segment along the loading direction. During hot-drawing

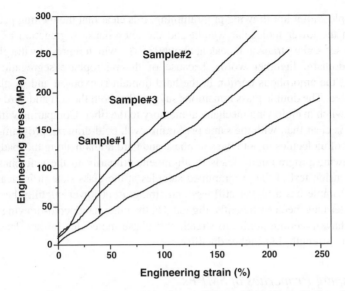

Figure 5.14 Tensile test results of the various fibers (adapted from Reference [20])

programming, the soft amorphous molecules can be sufficiently stretched along the loading direction because of the rubbery status at the programming temperature. Further, the programming temperature is higher than the microphase segregation temperature, leading to separation of the hard domain from the soft phase, which also provides an opportunity for stretching soft phase molecules along the loading direction. As a result, the alignment in the hard segment domain should be comparatively low. Figure 5.15 shows the first DSC cycle

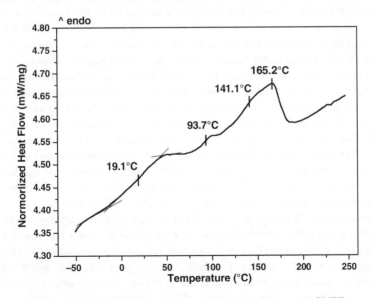

Figure 5.15 First DSC cycle of hot-drawing programmed SMPF

result of Sample 2 after hot-drawing programming. It is clear that the melting peak of the hard segment with the lower molecular weight and the microphase segregation peak disappear. However, we see a glass transition peak at about 93.7 °C, which represents the glass transition in the hard domain. In other words, because of the microphase segregation during hot programming, the amorphous portion in the hard domain is exposed and its glass transition can be identified. The double glass transitions suggest that both the soft and hard segments can be used as a switch in endowing mechanical memory to the fiber. Comparing with the result in Figure 5.8, it is clear that, with the same pre-strain level, cold-drawing programming leads to the alignment of molecules in both the soft phase and hard domain along the loading direction, while hot-drawing programming leads to alignment primarily in the soft phase. When we conduct the tension test of the programmed samples at the glass state, it is clear that the hot programmed sample has a higher stiffness and strength but lower ductility because the soft phase molecules have been sufficiently aligned. On the other hand, cold-drawing programmed samples still have a certain ability to stretch soft phase molecules further, leading to lower stiffness, lower strength, but higher ductility.

5.2.2　Damping Properties of SMPFs

Driven by the need for lighter and stronger engineering structures, fiber reinforced polymer (FRP) composite materials have emerged as a new class of engineering materials for load bearing structures. FRP composite represents a tough competitor for traditional materials such as metal, concrete, wood, etc. This is due to the high specific strength, stiffness, corrosion resistance, and tailorability of FRP composites. FRP composites in the form of laminate, sandwich, grid stiffened, 3-D woven fabric reinforced, Z-pinned, etc., have been manufactured. FRP composites have been widely used in many man-made structures, particularly in high tech and high value structures, including, but not limited to, aerospace (fixed wing aircraft, helicopters, etc.), defense (tank, armor, etc.), energy production, storage, and transportation (wind turbine blade, pipe, on-board and off-board storage tanks for natural gas or hydrogen, etc.), vehicles (car, truck, train, etc.), electrical and electronic components (rods, tubes, molded parts, electrical housings, etc.), construction items (bathtubs, decks, swimming pools, utility poles, bridge decks, railings, and repair, rehabilitation, reinforcement, and reconstruction of concrete structures, etc.), marine parts (ship hulls, decks, bulkheads, railings, offshore oil platforms, etc.), and consumer products (golf clubs, bicycles, fishing rods, skis, tennis rackets, snowmobiles, mobile campers, etc.). Also, cables made of fibers have been used as safety ropes or in cable stayed bridges. Owing to the light weight of FRPs, however, they are very easy to be shaken with less damping, leading to resonance and structural failure at loads well below the design level. Vibration induced structural failure is not uncommon. Collapse of buildings and bridges during an earthquake or hurricane provide typical examples of vibration induced catastrophic structural failure. In this section, we report on the damping properties of SMPFs, which are comparable to spider silk due to their similar microstructures.

Spider silks possess amazing properties such as high strength, toughness, and adaptive properties, which make them a perfect material for spiders to catch prey and also as the material for nests [36,37]. A falling spider hanging by a single silk can stop immediately and exactly where it wants, suggesting superior damping capabilities of the spider silk. Also, spider silk possesses super contraction, which is triggered by moisture. However, farming spider silk is very difficult because of the aggressive nature of spiders. Therefore, manufacturing artificial fibers with

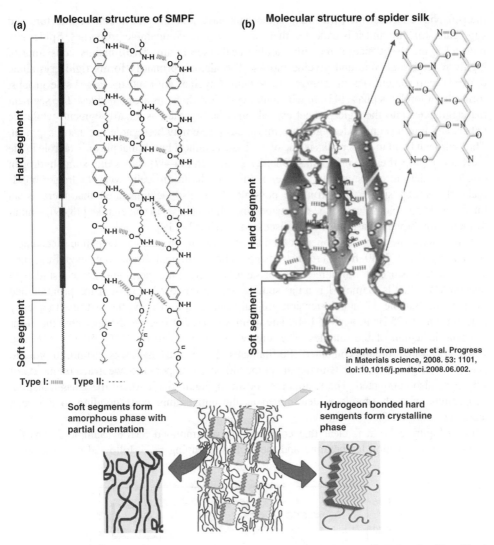

Figure 5.16 Similar molecular structure and aggregate structure between SMPF and spider silk. A hydrogen-bonded hard segment and soft segment form a microphase separated microstructure with a crystalline hard domain and an amorphous soft phase. The amorphous phase and crystalline phase are partially oriented. *Source:* [10] Reproduced with permission from the Royal Society

comparable properties has been a research focus for years. This type of research has focused mainly on producing spider silk-like proteins through recombinant biotechnology [38–40] and fabricating fibers from protein solutions by a wet spinning method [41,42]. Here, we would like to show that the melt spinning manufactured SMPFs have damping properties similar to spider silk due to their similar microphase segregated morphology (see Figure 5.16).

An in-depth study shows that spider silks are made of two-phase semicrystalline proteins with hydrogen bonded crystalline molecules as the hard domain and amorphous molecules as the soft phase. Hence, the SMPF and spider silk have a similar molecular architecture and aggregate structure and it is expected that they may have comparable properties. Spider silk molecules are a sequence of repetitive amino acids (see Figure 5.16), mainly consisting of highly repetitive alanine and glycine blocks. The alanine primarily forms rigid crystalline domains with dimensions of $2\,nm \times 5\,nm \times 7\,nm$. Glycine, with mostly larger side groups, forms the amorphous phase [43], in which the crystalline domains are embedded. The segment that aggregates into the rigid β sheet crystals may be regarded as a hard segment, while the helical and β turn segment forming the amorphous phase may be regarded as a soft segment. The hydrogen bond in spider silks is one of the basic chemical bonds that play a vital role in the mechanical properties of spider silks. Buehler's research [44–47] shows that some hydrogen bonds may deform cooperatively. This explains why the intrinsically weak hydrogen bonds make the strong spider silk structure possible. It is also demonstrated that there is an intermediate phase, also called the orientated amorphous phase, in spider silks [48,49], similar to the mesomorphic phase in SMPFs discussed previously [19].

In this study, the damping property of the SMPFs was investigated by Dynamic Mechanical Analyzer (DMA) 2980 from TA instruments. The isothermal strain-controlled frequency sweep mode was used for all three types of samples used in the thermomechanical studies in Section 5.2.1; that is, Sample 1 is an as-spun sample, Sample 2 is a hot-tension programmed sample, and Sample 3 is a cold-tension programmed sample. During the test, the frequency ranged from 1 to 25 Hz in steps of 1 Hz. The strain was fixed at 1%, which was within the linear viscoelastic region of the samples. The temperature was set to be 25 °C, 50 °C, 75 °C and 100 °C. Each fiber bundle contained 100 filaments. The gauge length was 15 mm. In order to determine the glass transition temperature of the SMPFs, a temperature sweep at a frequency of 1 Hz was also conducted. The results are shown in Figure 5.17, where the glass transition temperature of each sample is determined as the temperature corresponding to the peak damping ratio (tan δ).

From Figure 5.17, it is clear that cold-drawing programmed SMPF (Sample 3) has the highest glass transition temperature and the highest damping ratio, followed by the control

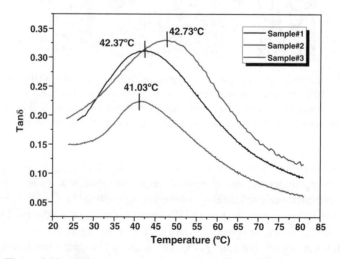

Figure 5.17 DMA temperature ramp test results of the three samples

sample without programming (Sample 1), and the lowest by the hot-drawing programmed sample (Sample 2). These changes in glass transition temperature and peak damping ratio can be explained by the change in morphology. Glass transition is the temperature region where amorphous materials change from the glassy state to the rubbery state upon heating, or vice versa if cooling. Below the glass transition zone, the entangled molecular chains are frozen and can hardly move. As the temperature increases, driven by thermodynamic forces, they are able to overcome the segmental rotation resistance and the internal friction, and finally enter the pure rubber-elastic state. Hence, the glass transition temperature is indicative of the internal restriction level. In other words, a higher internal constraint level leads to a higher glass transition temperature. As discussed previously (Figures 5.4(a), 5.8, and 5.15), the soft segment and short hard segment have been separated during hot-tension programming. Therefore, the resistance to various types of segmental motion in a soft segment is reduced, leading to the lowest glass transition in Sample 2. Under cold-drawing programming (Sample 3), both the soft segment and the short hard segment, particularly the soft segment, have been aligned along the loading direction. The alignment leads to a certain order in the molecules, similar to stress induced crystallization (SIC). As a result, a higher temperature is needed for the soft segment to gain sufficient mobility and to achieve a rubbery state, leading to an increase in the glass transition temperature. Of course, Sample 1 has a glass transition temperature in between those of Sample 2 and Sample 3 because its morphology is similar to Sample 3 but with a less soft segment alignment. Similar to the change in the glass transition temperature, the change in damping ratio can be understood. Damping is a macroscopic behavior of the internal resistance or barrier to molecular motion. In Sample 2, the soft segment and the short hard segment have been separated after hot-tension programming. Therefore, the segments are easier to move, leading to the lowest damping capability. In Sample 3, the soft segment and short hard segment are still entangled together after cold-drawing programming. Hence, the entanglement provides a higher resistance to segmental motion, resulting in the highest damping capability. In Sample 1, segmental motion is comparatively easier than that in Sample 3 because the segment is less aligned, leading to a damping capability slightly lower than that in Sample 3.

As compared to the DSC test results, it is seen that the DMA test only yields one individual glass transition, instead of several refined peaks as in the DSC results. This suggests that the DMA test is more suitable for determining the viscoelastic or macroscopic mechanical properties, such as damping ratio, while the DSC test is more insightful in revealing the microscopic details, such as multiple glass transitions and/or melting transitions. In other words, we can treat the broad glass transition in Figure 5.17 as a lump sum of several transitions.

Figure 5.18(a) to (d) shows the frequency sweep test results at sweeping temperatures of 25 °C, 50 °C, 75 °C, and 100 °C, respectively. It is seen that cold-drawing programmed SMPFs have the highest damping ratios, followed by the as-spun SMPFs, and the lowest by the hot-drawing programmed SMPFs, at all the testing temperatures except for 25 °C. We can understand the results from the point of view of energy dissipation. It is well known that a polymer exhibits the highest damping properties within the glass transition region. Because the test temperature of 50 °C is closest to the glass transition temperature of the cold-drawing programmed SMPFs, Sample 3 exhibits the highest damping ratio, followed by Sample 1, and the lowest by Sample 2. Another explanation is that both Sample 3 and Sample 1 have a soft and short hard segment mixture, while Sample 2 does not. The existence of the mixture provides entanglement and an additional barrier to the motion of molecules, leading to increased energy dissipation and damping ratio. At 75 °C, both Sample 3 and Sample 1 still

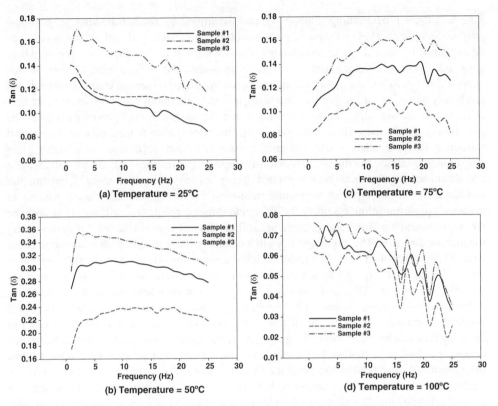

Figure 5.18 Damping ratio test results for the three samples at various temperatures and frequencies (adapted from Reference [20])

have a soft phase and hard domain mixture because 75 °C is lower than the microphase separation temperature (about 82.0 °C). In order to move the molecules in the mixture, more energy is needed due to the soft/hard segment interactions and barriers to overcome. For Sample 2, the microphase separation has occurred during the hot programming process. Therefore, energy dissipation due to the soft segment and hard segment mixture does not exist in Sample 2. Consequently, the same tendency at 50 °C is found for the damping properties at 75 °C. At 100 °C, although the soft segment for all three samples is similar (microphase separation has completed), the hard segment still has a certain difference. The hard segment has been aligned more in Sample 3 than in Sample 2, with Sample 1 in between them. Obviously, at 100 °C, the hard segment provides more resistance to movement of the soft segment in Sample 3 and Sample 1 than in Sample 2. Again, the same damping ranking is found. At 25 °C, Sample 2 consists of the largest amount of soft amorphous domain because the soft segment and hard segment separation has occurred during programming. Because 25 °C is within the glass transition region of the soft amorphous domain (Figures 5.4(a), 5.8, and 5.15), it is reasonable for Sample 2 to have the highest damping capability.

It is interesting to note that for spider dragline silk, the damping ratio is about 0.1. From Figure 5.18, it is seen that cold-drawing programmed SMPFs have a higher damping ratio in the frequency and temperature range examined, except for that at 100 °C. Because 100 °C is well above the working temperature for most engineering structures, it is safe to conclude that the SMPFs, particularly after cold-drawing programming, have a higher damping ratio than that of spider silk. However, the tensile strength and stiffness of the SMPFs are still considerably lower than those of the spider silk [10]. Efforts are needed to enhance the mechanical properties of the SMPFs while balancing its damping and stress memory capabilities.

5.2.3 Summary

In summary, SMPFs are very sensitive to their programming history. Even the as-spun SMPFs have been programmed during the manufacturing process. It has been demonstrated that different programmings lead to considerably different thermomechanical and damping properties, due to the significant impact of programming on microstructures. Cold-drawing programming leads to alignment in both the soft segment and the hard segment. As compared to hot-tension programming, cold-drawing programming is time and energy efficient. Also, cold-tension programming leads to a higher stabilized recovery stress, which is desired in self-healing applications. As compared to as-spun SMPFs, cold-drawing programmed SMPFs have a higher damping ratio as well as increased strength and stiffness, which are helpful in vibration control and load carrying. Therefore, we advocate cold-drawing programming for SMPFs in engineering structural applications. Finally, it is found that SMPFs, particularly cold-drawing programmed SMPFs, have a higher damping ratio than that of spider silks. Further endeavor should be towards enhancing the mechanical properties while maintaining the damping and shape memory capabilities of the SMPFs.

5.3 Constitutive Modeling of Semicrystalline SMPFs

As discussed above, SMPFs may have various types of morphologies or microstructures, and these microstructures are sensitive to composition, polymerization, manufacturing, temperature, time, and loading history. Therefore, various phenomenological models are needed to deal with the versatile microstructures. As an example, we will discuss constitutive modeling of semicrystalline SMPFs in this section. In particular, we will focus on semicrystalline SMPFs with a semicrystalline soft domain and a semicrystalline hard domain. We believe that this will yield a more general expression for SMPFs as other microstructures can be easily obtained using this general case. For example, if the crystalline volume fraction is zero, we have a pure amorphous phase, and vice versa.

The amorphous phase is produced when the randomly coiled molecules in the rubbery state are frozen by cooling. Crystallization can occur from the polymer melt by cooling, from solid polymer by stretching, and from solution. Like small molecule crystallization, polymer crystallization starts from an embryo, which is driven by random motion of the molecules. These embryos form and disappear dynamically, until the size of the embryo is large and stable enough to become a nucleus; this process is called nucleation. In addition to the thermal mechanism, nucleation is strongly affected by impurities, dyes, plasticizers, fillers, and other additives in the polymer. This is also referred to as heterogeneous nucleation. When the temperature drops below the melting temperature (T_m) but above the glass transition temperature (T_g), folded segments are added to the nucleus, leading to growth of the crystallite. It is noted that if the temperature is above the melting temperature, it destroys the molecular arrangement; if the temperature is below the glass transition temperature, the movement of

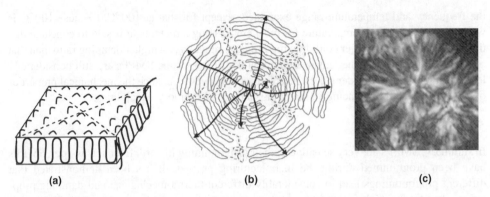

Figure 5.19 (a) Schematic of a lamella, (b) schematic of a spherulite, and (c) a polarized optical microscope image of a spherulite

molecular chains is frozen. Therefore, the favorable temperature range for crystal growth is between T_g and T_m. The growth of crystals depends on the temperature gradient because the growth of the crystalline region preferably occurs in the direction of the largest temperature gradient. If the gradient is very large in a certain direction, the crystal may have a unidirectional, dendritic character; if the temperature gradient is uniform in all directions, quasi-spherical crystal forms, which is spherulite. Usually, the thermodynamic force drives the lamellae to form spherical-shaped spherulites (see Figure 5.19). Crystallization can also occur by stretching. Stretching can be applied during manufacture, such as melt spinning for SMPFs, where tensile stress is applied to the polymer. Stretching can also be applied to polymers below T_g. The mechanism for stretching induced crystallization is the alignment of molecular segments along the loading direction. The aligned segments form crystallites, as discussed previously. In many cases the crystallization process is not fully developed chemically and the resulting microstructure contains 10 wt% up to 80 wt% crystalline segment. This partially ordered microstructure is called a semicrystalline polymer [50,51]. The properties of the semicrystalline polymers are determined not only by the degree of the crystallinity but also by the size and orientation of the crystalline and amorphous phases [52–55].

From Section 5.2, it was established that SMPFs have a microphase segregated structure. This segregated structure may motivate designers to incorporate a multiscale modeling approach for characterization of the semicrystalline polymer, in which the geometry of the microscale domains, together with their individual mechanical properties, can be utilized to obtain the associated macroscopic mechanical responses. In continuum mechanics, this needs knowledge of micromechanics, which is the key to assembling microsubphases into the macroscopic phase. Because this operation can be repeated from micro, meso, to macro, it provides a perfect platform for multiscale modeling.

5.3.1 Micromechanics Based Approaches

The microstructural configuration of heterogeneous materials can be correlated to the macroscopic constitutive relations within the micromechanics framework. In this approach the representative volume element (RVE) represents a specific arrangement of subphases, each of which has a specific geometry and mechanical properties. Selection of an RVE is extremely important. The RVE should be large enough to represent the microstructural morphology and sufficiently small to neglect the macroscopic mechanical gradients [56–58]. The RVE must be large enough because it needs to include all the microsubphases in it. Otherwise, it cannot represent the microstructural morphology of the material. On the other hand, the RVE must be sufficiently small so that, within the RVE, the physical and mechanical properties of each phase can be treated as independent of the material position.

In order to obtain the constitutive relation for a multiphase composite like an SMPF, in general, the micromechanics approaches are classified into the average field theory and the homogenization technique. The average field theory takes into account the average microstress and microstrain fields to estimate the macroscale mechanical response of the heterogeneous medium. This approach was proposed formerly by Eshelby [59] and later it has been developed to the Mori–Tanaka and self-consistent methods [56]. Homogenization micromechanical theory is applied to heterogeneous materials with a periodic microstructure, in which a multiscale perturbation analysis can be implemented to estimate the effective properties [60,61]. This method builds up the microscopic RVE through periodic unit cells and periodic boundary conditions, and the finite element (FE) or fast Fourier transforms are required to solve the problem.

In a multiscale analysis, the localization relations, that is, stress and strain concentration tenors, bridge the microscopic and macroscopic mechanical fields. When the medium behaves elastically, these relations are exact [56]. The main difficulty arises when nonlinearity is introduced in the mechanical behavior of the subphases, such as inelastic deformation or damage [56,62], which is the case for SMPFs. In general, three approaches within the micromechanics framework are available to establish the localized relations in the presence of such nonlinear processes.

5.3.1.1 Analytical Approach

The pioneering analytical solution by Eshelby [59], for an ellipsoidal inclusion embedded in an infinite elastic medium, has been extended to nonlinear cases in the literature. For example, the "secant" approach by Berveiller and Zaoui [63] and the self-consistent "tangent" method by Hill [64] and Hutchinson [65] are generalizations of this method for elastoplastic problems. The limitation of these analytical methods persists in their inability to simulate complex material structures, which result in inelastic responses that are too stiff [62,66]. Also, accurate stress redistribution in an inelastic analysis cannot be captured by these models [67]. Several models have been developed to resolve these issues in the literature, such as the above-mentioned "tangent" [64,66,68,69], "secant" [63,70], and "affine" [67,71] methods.

5.3.1.2 Numerical Approach

A direct approach in a multiscale analysis is to incorporate finite element analysis (FEA) to model the real microstructural morphologies, which are captured by advanced tomography techniques such as the scanning electron microscope (SEM). Upon applying their respective material constitutive equations to each of these microphases, FEA can effectively simulate their interactions. While FEA is a robust tool, computational cost and lack of knowledge of the microsubphase constituent properties remain the main difficulties facing these methods [72,73]. For example, how to experimentally determine the constitutive behavior of a crystallite in the soft phase or hard domain of the SMPFs remains a challenge because the crystal size is usually small and prevents direct measurement of the constitutive relationship using conventional mechanical testing.

5.3.1.3 Sequential Approach

This approach lies in between the two presented multiscale approaches in which the RVE is discretized into several subdomains and the constitutive equations are established from a multiscale analysis performed on the discretized RVE. The term sequential refers to the fact that a series of analytical, for example, Eshelby solution, and numerical, for example, FEA, procedures are implemented to correlate the microconstituent and macroscopic constitutive relations. The transformation field analysis (TFA) sequential approach, developed by Dvorak *et al.* [74–77], discretizes the exact solution of the Lippman–Schwinger equation, while uniform mechanical fields are assumed for each of the subphases.

In a micromechanics analysis of composite materials, the rule-of-mixture method is quite popular. While the rule-of-mixture method cannot link the constitutive behavior of the microsubphases with the constitutive behavior of the macrophase, it is widely used in estimating elastic properties or strength of composites. The basic idea is that the properties of the composites are a weighted volume average of the properties of the constituents. Because the rule-of-mixture method cannot consider the size and configuration of the subphases, the estimation is rough. Therefore, the theory of elasticity based method has been used. However, the theory of elasticity based method faces a tremendous challenge when a multiphase composite is considered, particularly when the subphases have irregular shapes and varying sizes, that is, with gradation. Therefore, combination of the rule-of-mixture method, the theory of elasticity, and Eshelby's equivalent medium theorem is usually used. In this approach, the multiphase composite is considered as one unit cell (RVE) embedded in an infinite equivalent medium that has the unknown properties of the composite. The equivalent medium model is then solved using the theory of elasticity, from which the unknown property of the equivalent medium, for instance the bulk modulus, can be solved. After that, the rule-of-mixture method is used to weigh the contribution of the subphases by taking into account their volume fraction and subphase size and distribution, such as particle size and gradation. This approach has been used widely in modeling the elastic modulus and tensile strength of particulate composites, including cement concrete, asphalt concrete, and conventional particulate polymer composites [78–84], and in guiding development of new composite materials [85–87].

In Section 5.2, we treated the SMPF as a microphase segregated structure and, based on their chemical composition and shape memory mechanisms, these phases are classified as soft phase and hard domain, and both may be semicrystalline. From the point of view of micromechanics modeling, however, this is not convenient because the RVE, which is made of a semicrystalline soft phase and a semicrystalline hard domain, will be difficult to analyze. Therefore, in the following, we will treat the SMPF as a two-phase composite with a crystalline phase and an amorphous phase; that is, the RVE will be a two-phase element. Clearly, both the soft phase and hard domain contribute to the amorphous phase and crystalline phase. Selection of the RVE in such way makes it easier to utilize existing constitutive relations for both amorphous polymer and crystalline polymer.

In this book asymmetric mechanical properties are assumed for both the stretched and nonstretched SMPFs, while the microstructural changes are assumed to be stress induced

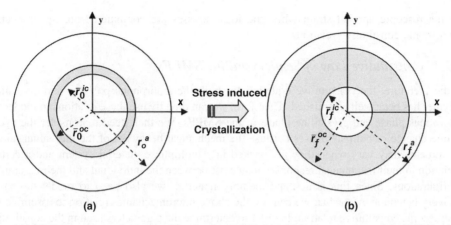

Figure 5.20 Microstructural representation for the SMPF: (a) initial microstructure of the SMPF, which contains some crystalline phase, (b) cold-drawn induced crystalline phase, which leads to growth in the crystalline phase. *Source:* [9] Reproduced with permission from the Royal Society

crystallization and morphological texture changes upon loading. Figure 5.20(a) shows the status of the microstructure for the as-spun SMPFs. Upon loading, the stress induced crystallization process results in a larger volume fraction of the oriented crystalline phase, as shown in Figure 5.20(b). In Figure 5.20, the initial inner, $\bar{r}_0^{ic}$, and outer, $\bar{r}_0^{oc}$, radii and the final inner, $\bar{r}_f^{ic}$, and outer, $\bar{r}_f^{oc}$, radii of the crystalline phase define the amorphous phase, where the superscript "a" means amorphous. In this book, the morphological texture updates for the amorphous and crystalline phases will be implicitly accounted for in their respective governing constitutive relations. In this way the simulations can monitor two significant microstructural changes, which are texture updates in the amorphous and crystalline phases and the stress induced crystallization process.

We have proved that stretching leads to alignment of both phases along the loading direction. Due to the fact that the manufactured fibers are assumed to be transversely isotropic, we may assume that the amorphous and crystalline phases in the SMPFs are assembled in a parallel configuration. While other micromechanics configurations, such as series (isostress) or rule-of-mixture's method [13] or transformation field analysis approach [14], are easily applicable without affecting the theory, the parallel configuration simplifies the numerical simulations significantly. This isostrain assumption suggests that

$$\varepsilon_z = \varepsilon_z^a = \varepsilon_z^c$$

and (5.2)

$$\Sigma_z = \sigma_z^a + \sigma_z^c$$

where ε_z and Σ_z are respectively the longitudinal macroscopic strain and stress and ε_z^a and ε_z^c are the microscale strain fields in the amorphous and crystalline phases, respectively; σ_z^a and σ_z^c are the local stresses in the amorphous and crystalline phases, respectively. We may assume that the loading condition is strain controlled; thus the strain in the two subphases are equal to the macroscopic applied strain while the local stresses are computed from the respective microscopic constitutive relations.

5.3.2 Constitutive Law of Semicrystalline SMPFs

In the literature, the constitutive equation for both the amorphous polymer and crystalline polymer has been well established. Therefore, we can directly use these relations to model the amorphous phase and crystalline phase of the SMPFs. We then need to consider the cyclic texture change of both subphases because the mechanical behaviors of the individual micro-constituents may vary when they are packed in a multiphase material system and a certain deviation in their mechanical responses may exist between the individual and their assembled configurations. Since this is a shape memory material, we also need to model the shape recovery behavior. After that, we can use the above micromechanics relation to assemble the macroscopic constitutive relation. In order to determine the parameters used in the constitutive model, we need to consider the kinematic relations under large deformation. Finally, we will discuss the numerical scheme to solve the coupled equations.

In order to keep matters simple, several basic assumptions were made in this study:

1. The SMPF consists of two phases – the amorphous phase and the crystalline phase.
2. The SMPF is considered to be an isostrain; that is, a uniform strain field assumption is held.
3. Heat transfer in the material is not considered. The temperature is then treated as uniform over the entire material body.
4. The structural relaxation and inelastic behavior of the material is assumed to be solely dependent on the temperature, time, and stress.
5. The material is assumed to be without any damage during the thermomechanical cycle.

The flowchart for the multiscale modeling of the constitutive behavior of the semicrystalline SMPFs is schematically shown in Figure 5.21.

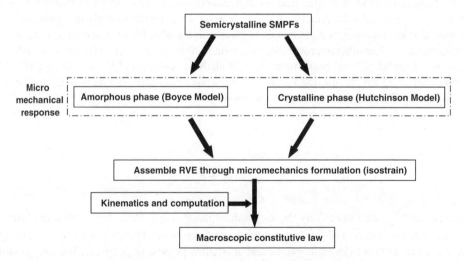

Figure 5.21 Flowchart of multiscale modeling of SMPFs

5.3.2.1 Amorphous Phase Constitutive Relation

In this book, the well-established Boyce model for the inelastic deformation of amorphous glassy polymers is assumed to be held for the amorphous phase of the semicrystalline SMPF. The inelastic deformation of glassy polymers initiates when the applied stress overcomes the intermolecular rotation resistance of a chain segment individually or in a cluster, leading to strain softening behavior. After initiation of the inelastic flow, the chain segments align in the direction of the principal inelastic deformation and result in a second source of resistance, which is termed network stretching resistance, leading to strain hardening behavior [9]. Amorphous inelastic deformation is affected by the strain rate, pressure and temperature effects. Argon [88] proposed a rate dependent micromechanical inelastic theory based on a double kink formation and the intermolecular resistance, which is the free energy barrier to chain segment rotation. Boyce and co-workers [89–91] extended Argon's viscoplastic element to a pressure, rate, and temperature dependent theory, which was based on modeling the intermolecular energy barriers. This theory considered a cubic unit cell consisting of eight polymeric chains. It is proven that, as compared to other models, such as the three-chain model or the five-chain model, the mechanical responses of this eight-chain unit cell to external loading represent the global deformation better [89]. Figure 5.22 shows a schematic of the eight-chain model. In comparison with phenomenological models, the main advantage of the Boyce model is that it incorporates physical parameters that are measurable from simple mechanical testing. The plastic multiplier in this model was introduced as follows [90]:

$$|\dot{\gamma}^p| = \dot{\gamma}_0^a \exp\left[-A\frac{s+ap}{\theta}\left(1-\left(\frac{|\tau|}{s+ap}\right)^{5/6}\right)\right] \tag{5.3}$$

where a is a material parameter, p is pressure, and s is the athermal shear strength. The material parameter A and the evolution law for s are defined as follows:

$$A = 39\frac{\pi\omega^2\bar{a}^3}{16k} \quad \text{and} \quad \dot{s} = h\left(1-\frac{s}{s_{ss}(\theta,|\dot{\gamma}^p|)}\right)|\dot{\gamma}^p| \tag{5.4}$$

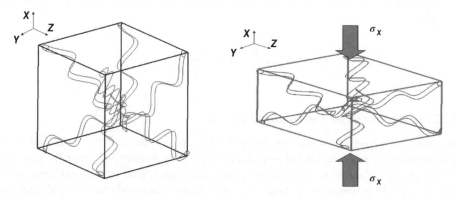

Figure 5.22 Schematic of the eight-chain model

where ω and $\bar{a}$ are material parameters and k is Boltzmann's constant [88]. The material parameter h shows the rate of the strain softening and s_{ss} represents the asymptotic preferred structure. The initial value of s for the annealed material is $s_0 = 0.077\mu/(1 - \nu)$, where μ is the elastic shear modulus and ν is Poisson's ratio. The phenomenological relation for athermal shear strength, s, takes into account the pressure, rate, and temperature dependency and the strain softening effect.

The material constant $\dot{\gamma}_0^a$ in Equation (5.3) is called the amorphous pre-exponential inelastic strain rate and $|\dot{\gamma}^p|$ is the effective equivalent inelastic deformation rate of a glassy polymer subjected to the effective equivalent shear stress, $|\tau|$, which is defined at the absolute temperature, θ, as follows:

$$|\tau| = \sqrt{\frac{1}{2}s_{ij}^* s_{ij}^*} \tag{5.5}$$

where $s_{ij}^* = s_{ij} - X_{ij}$ in which $s_{ij} = \sigma_{ij} - \frac{1}{3}\sigma_{kk}\delta_{ij}$ is the deviatoric Cauchy stress and $X_{ij} = \alpha_{ij} - \frac{1}{3}\alpha_{kk}\delta_{ij}$ is the deviatoric back stress tensor. The back stress tensor, α_{ij}, is defined as follows [89,91]:

$$\alpha_{ij} = nk\theta\frac{\lambda^L}{3}\left[\lambda_i^p L^{-1}\left(\zeta\frac{\lambda_j^p}{\lambda^L}\right) - \frac{1}{3}\lambda_k^p L^{-1}\left(\zeta\frac{\lambda_k^p}{\lambda^L}\right)\delta_{ij}\right] \tag{5.6}$$

where n is the number of chains per unit volume, k is Boltzmann's constant, λ^L is the limit of chain extensibility, and ζ is a viscoplastic related material constant that controls the magnitude of the hardening with respect to the inelastic stretches [92]. In the limit analysis the Langevin function, $L(\beta) = \coth(\beta) - 1/\beta$, is used extensively and imposes a limiting case in the evolution of the back stress tensor, which produces an extreme network resistance to the inelastic deformation and models the strain hardening. Once this limit is exceeded, the network structure fractures. Although $L^{-1}(\beta)$ can be obtained numerically, here the Padé analytical solution is utilized to obtain the inverse of the Langevin function as $L^{-1}(\beta) = \beta(3 - \beta^2)/(1 - \beta^2)$.

Equation (5.3) represents the magnitude of the amorphous inelastic strain rate, while the direction of the amorphous inelastic flow rate, $\dot{D}_{ij}^{ap}$, is governed by the deviatoric driving stress, s_{ij}^*. The following flow rule is proposed for the inelastic deformation in the amorphous phase [88]:

$$\dot{D}_{ij}^{ap} = |\dot{\gamma}^p|\frac{s_{ij}^*}{\sqrt{2}|\tau|} \tag{5.7}$$

5.3.2.2 Crystalline Phase Constitutive Relation

The crystalline phase follows a few independent slip systems in which classical crystal plasticity theories cannot be utilized to model them [93–95]. Similar to the metallic crystalline phases, inelastic deformation in crystalline polymeric systems follows three different mechanisms: (a) crystallographic slip, (b) twining, and (c) Martensite transformations [96]. All these mechanisms leave the crystallographic axis inextensible and provide less than five independent deformation modes to accommodate inelastic deformation [94]. In our study, the slippage mechanism is assumed to be the dominant influencing mechanism based on the work in References [93] and [94]. Basically, a crystalline structure consists of periodic unit cells where the same boundary condition is maintained for all unit cells. The elastic deformation of a crystalline solid is associated with distortion of the unit cells when the applied stress does not exceed the resistance of interatomic bonds in a crystallographic network. Once the loading breaks these interatomic bonds, permanent deformation is produced. In a crystalline polymeric network this permanent deformation changes the crystallographic texture. In order to represent the crystallographic texture change, a reference vector, c_i, is aligned with the crystallographic texture to show its evolution with deformation. Similar to skew dislocation and edge dislocation in metals, two slip mechanisms can be active in a slippage system of a polymeric network: (1) chain slip where the Burgers vector is aligned with c_i and (2) transverse slip where the Burgers vector is perpendicular to c_i [92]. There are only four linearly independent crystalline slip systems, which are represented by the unit vectors in the direction of the slip and normal to the slip planes, including: (a) chain slip, (100)[001] and (010)[001], and (b) transverse slip, (100)[010] and (010)[100] [93], where the round brackets stand for the Miller index of plane and the square bracket represents the Miller index of direction. Vector $s_i^{(\alpha)}$ denotes the slip direction and $n_i^{(\alpha)}$ represents the unit normal vector to the slip plane, where $\alpha = 1$ to 4 is the number of slip systems. The inelastic crystalline stretch rate tensor, $\dot{D}_{ij}^{cp}$, is introduced by the following relation [93]:

$$\dot{D}_{ij}^{cp} = \sum_{\alpha=1}^{K} \dot{\gamma}^{(\alpha)} R_{ij}^{(\alpha)} \tag{5.8}$$

where $R_{ij}^{(\alpha)}$ is the symmetric part of the Schmid tensor as defined by $R_{ij}^{(\alpha)} = \frac{1}{2}\left\{s_i^{(\alpha)} n_j^{(\alpha)} + s_j^{(\alpha)} n_i^{(\alpha)}\right\}$, and represents the αth crystalline slip plane. The inextensibility of the crystalline chain together with the incompressibility assumption results in the following constraint relation:

$$\dot{D}_{ij}^{cp} C_{ij}' = 0 \tag{5.9}$$

where $C_{ij}' = c_i c_j - \frac{1}{3} I_{ij}$ is the deviatoric part of dyadic $c_i c_j$.

The shear strain rate, $\dot{\gamma}^{(\alpha)}$, is defined as follows [65,93,97,98]:

$$\dot{\gamma}^{(\alpha)} = \dot{\gamma}_0^c \frac{\tau^{(\alpha)}}{g^{(\alpha)}} \left|\frac{\tau^{(\alpha)}}{g^{(\alpha)}}\right|^{n^c-1} \tag{5.10}$$

where $\dot{\gamma}_0^c$ is the crystalline reference inelastic strain rate, $g^{(\alpha)}$ is the shear strength for the αth slip system, and n^c is the rate sensitivity factor. The effective shear stress, $\tau^{(\alpha)}$, at the αth slippage system is given by [14]

$$\tau^{(\alpha)} = S_{ij}^{c*} R_{ij}^{(\alpha)} \tag{5.11}$$

where S_{ij}^{c*} is the projected deviatoric Cauchy stress, that is, $s_{ij} = \sigma_{ij} - \frac{1}{3} \sigma_{kk}\delta_{ij}$, in the direction of the deviatoric part of the dyadic $c_i c_j$, that is, $C_{ij}' = c_i c_j - \frac{1}{3} I_{ij}$, where I_{ij} is the unity second-order tensor. This constraint is enforced based on the inextensibility of the crystalline chain together with the incompressibility assumption [92]. The lattice spin that controls the rate of changes of the direction of c_i is introduced as follows [93,99]:

$$\dot{W}_{ij}^* = \dot{W}_{ij}^{cp} - \sum_{\alpha=1}^{K} \dot{\gamma}^{(\alpha)} A_{ij}^{(\alpha)} \tag{5.12}$$

where $A_{ij}^{(\alpha)} = \frac{1}{2}\left\{ s_i^{(\alpha)} n_j^{(\alpha)} - s_j^{(\alpha)} n_i^{(\alpha)} \right\}$ is the skew part of the Schmid tensor and $\dot{W}_{ij}^{cp}$ is the skew part of the velocity gradient L_{ij}^{cp} in which $L_{ij}^{cp} = \dot{D}_{ij}^{cp} + \dot{W}_{ij}^{cp}$.

5.3.2.3 Morphological Texture Update

It has been experimentally confirmed that the amorphous and crystalline polymers undergo morphological texture changes in their polymeric networks upon stretching [92–94]. The strain hardening phenomenon in the SMPF is influenced by texture changes in the crystalline and amorphous phases. In the following the governing relations for texture updates in each of these subphases are discussed.

(a) Crystalline Phase Texture Update

In the case of the crystalline phase, the crystallographic texture is updated by taking into account the changes in the crystallographic axes, which is given by the following rate equation [99]:

$$\dot{c}_i = W_{ij}^* c_j \tag{5.13}$$

Equation (5.13) captures the change in the crystallographic axis regarding lattice spins. The incremental form of Equation (5.13) is given by [93]

$$c_i(t + \Delta t) = \exp\left(W_{ij}^* \Delta t \right) c_j(t) \tag{5.14}$$

Based on the Cayley–Hamilton expression for an exponential term, one may find [14,92,93]

$$\exp\left(W_{ij}^* \Delta t \right) = I_{ij} + \frac{\sin w}{w} W_{ij}^* \Delta t + \left(\frac{1 - \cos w}{w^2} \right) W_{ik}^* W_{kj}^* \Delta t^2 \tag{5.15}$$

with $w^2 = -tr\left(W_{ik}^* W_{kj}^* \Delta t^2 \right)/2$.

(b) Amorphous Phase Texture Update

In the case of the amorphous phase one may only consider the influence of the inelastic deformation on molecular chain rotations and the subsequent strain hardening effects. The initial texture and pre-orientation effect can be considered by the segmental rotation resistance and network stretching resistance. It is common to neglect the texture effect in elastic deformation and consider it only during inelastic deformation. Dupaix and Boyce [100] relate the initial values of the back stress tensor, α_{ij}, athermal shear resistance, s, network stretch vector, Λ_i, and residual stress tensor, σ_{ij}, to the strain hardening effect in the amorphous phase. The pre-orientation effect may not be induced by inelastic deformations only, and spinning or other manufacturing processes may produce a preferential texture. Here the texture effect is modeled by introducing an initial inelastic deformation into the back stress evolution law. It is shown that setting these parameters can effectively monitor the hardening phenomenon in the amorphous phase.

(c) Morphological Texture Update

The RVE of SMPFs consists of both crystalline and amorphous subphases. Therefore, the texture updates in the two subphases lead to morphological texture change in the RVE. The morphological texture in the semicrystalline polymeric RVE is updated based on changes in the directions of the unit outer normal vectors to the interfaces in between the amorphous phase and the crystalline phase, n_i^I, which is computed according to the changes in the material coordinate system at the interface. Let the material coordinate system at the interface plane be represented by two infinitesimal independent vectors, δx_i and δx_j, at time $t = 0$. At time t, these vectors are transformed respectively to $F_{ij}(t)\delta x_i$ and $F_{ij}(t)\delta x_j$, where $F_{ij}(t)$ is the deformation gradient. The expression for the normal vector n_i^I is obtained as [13]

$$n_k^I(0) = \frac{\delta x_i^{(1)} \times \delta x_j^{(2)}}{|\delta x_i^{(1)} \times \delta x_j^{(2)}|} \tag{5.16}$$

$$n_k^I(t) = \frac{F_{ij}(t)\delta x_j^{(1)} \times F_{ij}(t)\delta x_i^{(2)}}{|F_{ij}(t)\delta x_i^{(1)} \times F_{ij}(t)\delta x_j^{(2)}|} \tag{5.17}$$

The deformation gradient is updated based on the rate equation, $\dot{F}_{ij} = L_{ik}F_{kj}$, which is converted into the following incremental form:

$$F_{ij}(t + \Delta t) = \exp(L_{ik}\Delta t)F_{kj}(t) \tag{5.18}$$

The deformation compatibility at the interface requires that the deformation gradient tensors of each phase be used in Equation (5.18). To take into account the molecular alignment in the amorphous phase, the deformation gradient of the amorphous phase is incorporated as follows:

$$F_{ij}^\alpha(t + \Delta t) = \exp(L_{ik}^\alpha \Delta t)F_{kj}^\alpha(t) \tag{5.19}$$

Basically the updated interfacial vector, n_i^I, renders the movement and rotation of the inclusion inside the RVE. This allows designers to update the morphology of the RVE based on the deformation gradient during a finite strain deformation process. Equations (5.18) and (5.19) will be utilized in coupling with the micromechanical formulations to investigate the effect of loading on the RVE configuration, where n_i^I is updated based on deformation gradients and the enforcing compatibility in the deformation field. Of course the bonding between the crystalline subphase and the amorphous subphase is assumed to be perfect.

5.3.2.4 Recoverable Stresses

Stress recovery is a unique feature for shape memory polymers. Particularly, as demonstrated in this book, the recovered stress of the SMPF has been utilized in the two-step biomimetic close-then-heal (CTH) self-healing scheme for closing wide-opened (millimeter scale) cracks. As discussed previously, strain hardening by cold-drawing programming is effective in increasing the recovery stress, and thus the healing efficiency. Basically, using the cold-drawing process, the semicrystalline SMPF undergoes the stress induced crystallization process (or alignment of molecules along the loading direction) and stores the applied deformations through entropy changes in the molecular network. Once the temperature exceeds the glass transition temperature of the semicrystalline polymer, the viscosity of the polymeric network drops and the frozen network is allowed to release the stored energy and achieve its minimum energy level. In other words, the stored energy in the polymer network is released upon heating and the polymeric network returns to its minimum level of internal energy. As shown in Reference [9], with a few cold-drawing cycles, the amount of the recoverable stresses is considerably increased. This suggests that one may relate the amount of recoverable stresses to the loading history and the stress induced crystallization process. In this book, the recoverable stress is assumed to be a function of the stress induced crystallization process and the accumulated inelastic strains in the amorphous phase. We focus on the amorphous phase because the amorphous phase is the molecular switch for shape memory in this SMPF. During stress recovery, a portion of these induced inelastic entropic and energetic changes in the SMPF molecular network is recoverable upon heating, where the viscosity of the frozen network drops and the recoverable inelastic strains are restored. Therefore, the proposed evolution law for the stress recovery should take into account the history of the loading, including the inelastic strains in the subphases. The kinematic and isotropic hardening relations in the classical continuum plasticity context provide a suitable governing equation form to explain the stress recovery process [101–103]. The stress recovery evolution relation is proposed as follows [14]:

$$\sigma^{\text{rec}} = R(1 - \exp(-\eta(T - T_{\text{room}})))$$ (5.20)

where η is a material constant that controls the rate of saturation of the recovery stress to its final value R, and T and T_{room} are, respectively, elevated and room temperatures. The parameter R takes into account the history of the loading, which includes the inelastic deformation, texture updates, and residual stresses due to the cyclic hardening. Taking the time derivative of Equation (5.20) results in the following incremental relation for the stress recovery computations [14]:

$$\dot{\sigma}^{\text{rec}} = \dot{T}R\eta(\exp(-\eta(T - T_{\text{room}})))$$ (5.21)

where $\dot{T}$ shows the rate of the heating process. The heating process is controlled by time integration of the heating rate as $T = \int \dot{T}\, dt + T_{\text{room}}$ and once the temperature reaches its final value during simulation, the heating rate is set to zero (holding the temperature constant for a stabilized stress recovery). The saturation limit, R, for the stress recovery is related to the loading history and the amount of the plastic strain by the following expression [14]:

$$\dot{R} = \omega(\Xi - R)|\dot{\varepsilon}^p|$$ (5.22)

where ω and Ξ are two material parameters for controlling the saturation parameter, R, and $|\dot{\varepsilon}^p| = \sqrt{\dot{D}_{ij}^{pc}\dot{D}_{ij}^{pc}}$ is the equivalent plastic strain rate for the crystalline phase. During each of the cyclic loadings the magnitude of R is updated incrementally and its final value is introduced in Equation (5.20) for the stress recovery calculation. Equation (5.22) represents a monotonically increasing value for the parameter R up to a certain saturation limit, which is enforced by Ξ. The underlying physics for Equations (5.20) and (5.22) is that the recoverable stress in SMPFs is a function of the recoverable microstructural changes created during the cold-drawing process and this stress recovery should saturate to a certain limit due to the limit in reversibility in these microstructural changes. In other words, a certain amount of microstructural changes is reversible in SMPFs, and after a certain limit these microstructural changes may result in failure of the polymeric networks and produce nonreversible defects. In this book, it is assumed that the recovery stress saturates to a certain limit as a function of the microstructural changes through Equations (5.20) and (5.22) and the physical parameter to control these changes is the accumulated inelastic strain and its rate.

5.3.2.5 Stress Induced Crystallization

The physics for stress induced crystallization has been discussed previously. For simplicity, a phenomenological constitutive relation for the stress induced crystallization process is introduced, as schematically shown in Figure 5.20. The RVE of the microstructure of the as-spun SMPF is shown in Figure 5.20(a), where a thin layer of initially formed crystalline phase acts in parallel with the amorphous phase. In Figure 5.20(b), the stress induced crystallization is represented by the enlarged cross-section of the crystalline phase. Therefore, we can assume that the microstructural changes are governed by the crystalline phase formation upon stretching. In order to take into account this fact in the governing relations of the semi-crystalline SMPFs, we propose the following relations, which relate the initial inner, $\bar{r}_0^{ic}$, and outer, $\bar{r}_0^{oc}$, radii and the final inner, $\bar{r}_f^{ic}$, and outer, $\bar{r}_f^{oc}$, radii of the crystalline phase to the magnitude of the inelastic strains [9]:

$$r^{oc} = \left(\bar{r}_f^{oc} - \bar{r}_0^{oc}\right)(1 - \exp(-q|\varepsilon^p|)) + \bar{r}_0^{oc}$$
and $\hspace{3cm}$ (5.23)
$$r^{ic} = \bar{r}_0^{ic}(\exp(-q'|\varepsilon^p|)) + \bar{r}_f^{ic}$$

where q and q' are material parameters and $|\varepsilon^p| = \sqrt{D_{ij}^{pc}D_{ij}^{pc}}$ are the effective accumulated crystalline plastic strains. The outer, r^{oc}, and inner, r^{ic}, crystalline radii start from their initial values and converge to their final values. Equations (5.23) provide a smooth transition between a highly amorphous to a highly crystalline microstructure where the history of loading is incorporated by formulating this process based on the inelastic strains. The crystalline inner and outer radii change rates are then given by the time derivative of Equations (5.23) as follows [9]:

$$\dot{r}^{oc} = q|\dot{\varepsilon}^p|\left(\bar{r}_f^{oc} - \bar{r}_0^{oc}\right)(\exp(-q|\varepsilon^p|))$$
and $\hspace{3cm}$ (5.24)
$$\dot{r}^{ic} = -q'|\dot{\varepsilon}^p|\bar{r}_0^{ic}\exp(-q'|\varepsilon^p|)$$

The volume fraction of the crystalline, c^{cry}, and amorphous, c^{amr}, phases are then given by the following relations [9]:

$$c^{\text{cry}} = \frac{\left(r^{oc} - \varepsilon_z \nu_{12}^c r^{oc}\right)^2 - \left(r^{ic} - \varepsilon_z \nu_{12}^c r^{ic}\right)^2}{\left(r_0^a - \varepsilon_z \nu_{12}^a r_0^a\right)^2} \quad \text{and} \quad c^{\text{amr}} = 1 - c^{\text{cry}} \tag{5.25}$$

where ε_z is the macroscopic strain in the fiber direction and ν_{12}^c and ν_{12}^a are, respectively, Poisson's ratio of the crystalline and amorphous phases. The fiber is assumed to be transversely isotropic, that is, $\nu_{12}^a = \nu_{13}^a$, $\nu_{12}^c = \nu_{13}^c$, where "a" stands for amorphous and "c" represents crystalline. The subscripts 1, 2, and 3 represent the principal material coordinate system.

5.3.2.6 Micromechanics

Once the constitutive equations and texture changes of the two subphases are established, we can assemble them together in the RVE by the micromechanics approach. Figure 5.23 shows a schematic of the RVE. The detailed discussion can be found in Section 5.3.1 and Equation (5.2).

5.3.3 Kinematics

Kinematics analysis is a very important step to relate the micromechanics result with the macroscopic constitutive behavior. The reason is that the material parameters involved in the subphases and RVE need to be determined through macroscopic testing. For example, a

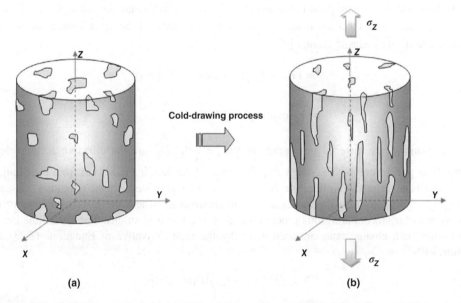

Figure 5.23 RVE: (a) a nonstretched SMPF and (b) a cold-stretched SMPF. *Source:* [13] Reproduced with permission from ASME

uniaxial tension test can determine the macroscopic stress, strain, and deformation gradient of the SMPF. Through multiplicative decomposition of the deformation gradient or additive decomposition of the strain, the macroscopic test results can be related to the deformation gradient or strain of each subphase or the elastic and plastic strain of the subphases. Therefore, kinematics analysis is a very important step in obtaining the macroscopic constitutive relation.

As discussed in Chapter 4, a strain description based on displacement gradients becomes nonlinear, which may result in computational difficulties for elastoplastic analysis. Therefore, the deformation gradient and multiplicative decomposition technique are usually used:

$$F_{ij} = F^e_{ik} F^p_{kj} \qquad (5.26)$$

where F_{ij} is the total deformation gradient, which correlates the material (undeformed) and deformed configurations, F^e_{ij} is the elastic deformation gradient, which is obtained by elastically unloading the deformed configuration to a stress-free status or the intermediate configuration, and F^p_{ij} is the plastic deformation gradient. Of course, as discussed in Chapter 4, the deformation gradient can also be multiplicatively decomposed into rotational and stretching tensors. Details can be found in Chapter 4.

5.3.4 Computational Aspects

The elastoplastic multiscale analysis requires several computational modules, including (1) a microscale computation module, which consists of a set of numerical solutions for the local constitutive equation of each subphase, (2) a micromechanical computation module, which provides numerical tools to link the mechanical properties of each of the local subphases to the macroscopic responses, and (3) a macroscale computation module, in which the continuum mechanics governing equations are enforced to simulate the overall mechanical response of the material and to identify the local loading conditions over the RVE. Each of these computational modules is discussed in the following. A flowchart of the multiscale analysis is shown in Figure 5.24.

5.3.4.1 Subphase Computation Module

The well-established elastic-predictor/plastic-corrector return mapping algorithm can be utilized to obtain the inelastic responses of the microscale amorphous and crystalline phases. Here, we only outline the steps to be used. A detailed description of this solution algorithm can be found in References [103] to [105]. The return mapping technique is capable of handling both associative and nonassociative flow rules with variant tangent stiffnesses and results in a consistent solution approach [105]. It is noted that this algorithm is applicable to the material, intermediate, or spatial formulations.

The return mapping techniques in inelastic solutions are a natural consequence of splitting the total strain into elastic and inelastic strains. Let tensor u_{ij}, an incremental field to describe the deformation, and its gradient, ∇u_{ij}, show the deformation rate. The solution is implemented by the following steps. Step 1 introduces a loading condition such as $F^{(n+1)}_{ij} = (I_{ik} + \nabla u_{ik})F^{(n)}_{kj}$, where I_{ij} is the unity second-rank tensor and the superscripts n and $n+1$ represent, respectively, the previous and current load steps. In step 2 the material is elastically stretched

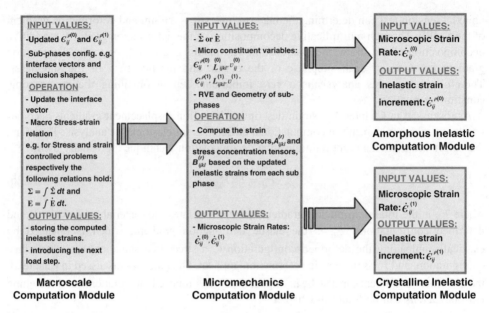

Figure 5.24 Computation algorithm for a micromechanics based multiscale analysis. *Source:* [9] Reproduced with permission from the Royal Society

by a "trial elastic" deformation gradient given by $F_{ij}^{e(n+1)\text{Trial}} = F_{ik}^{(n+1)} F_{kj}^{p(n)-1}$, where the super-scripts "e" and "p" represent "elastic" and "plastic" components, respectively, and $F_{ij}^{p(n)}$ shows the frozen inelastic deformation gradient. In step 3 the elastically stretched configuration is relaxed until the state of stress satisfies the yield condition ($\psi = 0$). The return mapping relaxes the stresses along the steepest descent path, which is defined based on either the yield function (associated flow rule) or the potential function (nonassociated flow rule). The proposed microscale formulations are based on the incompressibility assumption [89,93], which is represented by $\lambda_1\lambda_2\lambda_3 = 1$, where λ_i ($i = 1, 2, 3$) indicates the principal stretches. Simo and Ortiz [106] introduced the isochoric assumption into the kinematics of large deformation by splitting the deformation gradient into volume preserving and nonpreserving parts, as shown in the following:

$$\hat{F}_{ij} = J^{-1/3} F_{ij} \tag{5.27}$$

where the hat symbol, $\hat{}$, shows the isochoric components. Therefore, $\hat{F}_{ij}$ is the volume-preserving deformation gradient and $J = \det(F_{ij})$. This decomposition also holds for the elastic deformation, $\hat{F}_{ij}^e = J^{-1/3} F_{ij}^e$. The proposed kinematic decomposition by Reference [106] ensures that the computed inelastic deformation is isochoric, that is, $\det(F_{ij}^p) \equiv 1$. The volume preservative right, $C_{ij} = F_{ik}^T F_{kj}$, and left, $b_{ij} = F_{ik}^T F_{kj}$, Cauchy–Green tensors are then obtained as follows:

$$\hat{C}_{ij} = J^{-2/3} C_{ij} \quad \text{and} \quad \hat{b}_{ij} = J^{-2/3} b_{ij} \tag{5.28}$$

To hold the incompressibility assumption the volume preserving part of the applied deformation gradient needs to be utilized in the trial elastic part, with the left and right Cauchy–Green tensors given by [14]

$$\hat{b}_{ij}^{e(n+1)^{\text{Trial}}} = \hat{F}_i^{e(n+1)^{\text{Trial}}} \left(\hat{F}_j^{e(n+1)^{\text{Trial}}} \right)^T = J^{(n+1)^{-2/3}} b_{ij}^{e(n+1)^{\text{Trial}}}$$

$$\hat{C}_{ij}^{e^{\text{Trial}}} = J^{(n+1)^{-2/3}} C_{ij}^e \tag{5.29}$$

where $J^{(n+1)} = \det\left(F_{ij}^{(n+1)} \right)$.

There are a number of stress–strain relations to relate the stress and strain states in a finite deformation problem [107]. A relation between the Cauchy stress and Hencky strain is used [108]:

$$\sigma_{ij}^{(n+1)^{\text{Trial}}} = \frac{1}{J^{(n+1)}} L_{ijkl}^e \ln\left(U_{kl}^{e(n+1)^{\text{Trial}}} \right) \tag{5.30}$$

where $U_{ij}^{e(n+1)^{\text{Trial}}} = \sqrt{\hat{C}_{ij}^{e^{\text{Trial}}}}$.

The computed trial elastic stress is relaxed by the plastic-corrector method, as described in Table 5.1. In Table 5.1, Q_{ij} is the hardening tensor, $|\dot{\lambda}|$ is the inelastic multiplier, which shows the magnitude of inelastic deformation, and μ is the shear modulus. The second-order tensor $\hat{n}_{ij}$ defines the direction of the inelastic flow; for example, in the case of the associated formulation it is $\hat{n}_{ij} = \partial \psi / \partial \sigma_{ij}$, where the yield surface is $\psi = 0$. The deviatoric stress and back stress tensors are, respectively, identified by $s_{ij} = \sigma_{ij} - \frac{1}{3} \sigma_{mm} I_{ij}$ and $\alpha_{ij} = X_{ij} - \frac{1}{3} X_{mm} I_{ij}$.

5.3.4.2 Micromechanical Computation Module

One of the most vital decisions in a micromechanical based analysis is the configuration of the RVE. An RVE should contain the most dominant phases with direct impact on overall mechanical responses, while it should result in a simple model. In other words, it cannot be too large and at the same time it cannot be too small. This configuration is obtained through examination of physical facts and experimental results. During large deformation, the microstructure of an RVE may change, such as stress induced crystallization in SMPFs. Therefore, an incremental formulation is required to capture these changes. It is noted that the RVE can be further subdivided into sub-RVEs. This process can continue until the fundamental phase or length scale is reached. In other words, micromechanics can be used in RVE and sub-RVEs. The choice should be a balance between the computation cost and accuracy. In this computational module the micromechanics governing relations are enforced to obtain the local overall relations.

5.3.4.3 Macroscale Computation Module

The micromechanical based analysis may be utilized to find the overall behavior of the macroscale RVE as a function of the predefined incremental surface boundary conditions on

Table 5.1 Iterative algorithm for an elastic-predictor/return mapping solution

---

Step 1: Geometric update for a deformation controlled loading

$$u_{ij}; F_{ij}^u = I_{ij} + \nabla u_{ij}; F_{ij}^{(n+1)} = F_{ik}^u F_{kj}^{(n)}$$

Step 2: Elastic predictor

$$F_{ij}^{p(n+1)} = F_{ij}^{p(n)}; F_{ij}^{e(n+1)} = F_{ij}^{(n+1)} F_{ij}^{p(n)-1}; \psi = \psi\left(E_{ij}^{e(n+1)}, E_{ij}^{p(n)}, Q_{ij}\right);$$

$$\sigma_{ij}^{(n+1)\text{Trial}}$$

Step 3: Yield condition

$$\Psi\left(\sigma_{ij}^{(n+1)}, Q_{ij}^{(n+1)}\right) \leq 0?$$

Yes: go to step 1
No: continue to step 4

Step 4: Plastic-corrector

$$\sigma_{ij}^{(n+1)} = \sigma_{ij}^{(n+1)\text{Trial}} - 2\mu\left|\dot{\lambda}^{(n+1,\,\Xi)}\right|\hat{n}_i^{(n+1)} \text{ and } \dot{Q}_{ij} = \dot{\lambda}^{(n+1,\,\Xi)} H_{ij}^{(n+1)}; \hat{n}_{ij}^{(n+1)} = \frac{\zeta_{ij}^{(n+1)}}{\zeta_{ij}^{(n+1)}}; \zeta_{ij}^{(n+1)} = s_{ij}^{(n+1)} - \alpha_{ij}^{(n+1)}$$

where Ξ shows the iteration number and $\dot{\lambda}^{(n+1,\,\Xi)}$ is a primary value for the plastic multiplier and

$\dot{\lambda}^{(n+1,\,\Xi+1)}$ is its updated value. The updated value is obtained by enforcing the consistency condition
(inviscid flow) or viscoplastic potential (viscid flow):

$$\dot{\varepsilon}_{ij}^{p(n+1,\Xi+1)} = \sqrt{\frac{2}{3}}\left|\dot{\lambda}^{(n+1,\Xi+1)}\right|\hat{n}_{ij}^{(n+1)}$$

Step 5: Convergence check

$$\text{Eerror} = \frac{\left|\dot{\lambda}^{(n+1,\Xi+1)} - \dot{\lambda}^{(n+1,\Xi)}\right|}{\dot{\lambda}^{(n+1,\Xi)}} < \text{Total}?$$

Yes:

(a) store $\sigma_{ij}^{(n+1)}$, (b) compute $F_{ij}^{e(n+1)}$ from the updated stress state $\sigma_{ij}^{(n+1)}$, (c) $F_{ij}^{p(n+1)} = F_{ij}^{(n+1)} F_{ij}^{e(n+1)-1}$,

(d) store $Q_{ij}^{(n+1)}$, (e) $\varepsilon_{ij}^{p(n+1)} = \varepsilon_{ij}^{p(n)} + \dot{\varepsilon}_{ij}^{p(n+1)}$, (f) $\alpha_{ij}^{(n+1)} = \alpha_{ij}^{(n)} + \dot{\alpha}_{ij}^{(n+1)}$, (g) $\dot{\lambda}^{(n+1,\Xi)} = \dot{\lambda}^{(n+1,\Xi+1)}$
and go to the next load step

No:
go to Step 4 with $\dot{\lambda}^{(n+1,\Xi)} = \dot{\lambda}^{(n+1,\Xi+1)}$

Source: [14] Reproduced with permission from Elsevier.

the microscale subphases or, vice versa, it can be used to establish the local mechanical behaviors based on a predefined macroscale loading condition on the RVE surface. The microscale surface tractions should result in a self-equilibrium traction field on the boundary of the macroscopic RVE or microscale surface deformations should yield a self-compatible displacement field in the macroscopic RVE. The macroscale stress and strain states in a

Table 5.2 Material parameters for the stress recovery and cyclic hardening simulations

η	$T_{room}(°C)$	$T_{final}(°C)$	ω	$\aleph$ (MPa)	q	q'	$\bar{r}_f^{oc}$	$\bar{r}_0^{oc}$	$\bar{r}_0^{ic}$
0.0294	25	90	1	200	0.2	0.1	0.65	0.5	0.2

Source: [9] Reproduced with permission from the Royal Society.

continuum medium are respectively denoted by the tensors Σ_{ij} and E_{ij}. In micromechanics it is assumed that the averaged uniform stress, $\bar{\sigma}_{ij}$, and strain, $\bar{\varepsilon}_{ij}$, tensors in an RVE are equal to those applied on the continuum media, $\Sigma_{ij} = \bar{\sigma}_{ij}$ and $E_{ij} = \bar{\varepsilon}_{ij}$. Then $\bar{\sigma}_{ij}$ and $\bar{\varepsilon}_{ij}$ should satisfy the continuum equilibrium equations and strain displacement relations. Consequently, the computed microscale mechanical responses are enforced (1) to build the macroscale constitutive behavior and (2) to ensure the compatibility between all subphases.

5.3.5 Results and Discussion

In order to demonstrate the predictability of the above model, the macroscopic behavior of the SMPF is estimated and the prediction is compared with the test result. In the following, three such predictions and/or comparisons are made. The first one investigates the effect of the heating rate on fully constrained stress recovery. The second one evaluates the effect of the amount of the amorphous phase and the crystalline phase on the stress–strain behavior under cyclic tension. The last one examines the growth of the crystalline phase due to stress induced crystallization. In all cases, the SMPF has a diameter of 0.04 mm. The material parameters for the stress recovery and strain hardening modeling as well as for the amorphous and crystalline subphase modules and for crystalline phase slip systems are summarized in Tables 5.2 to 5.4, respectively.

Figure 5.25 shows the effect of the heating rate on stress recovery. Several observations can be made. (1) The predicted result and the test result are pretty close when the heating rate is 0.35 °C/min. This demonstrates the predictability of the stress recovery model. (2) The stress recovers faster as the heating rate increases. When the heating rate is at its highest (0.7 °C/min), the recovery stress stabilizes (plateaus) at about 127 s, while it is about 255 s when the heating rate becomes 0.35 °C/min. When the heating rate is 0.17 °C/min, the stress does not plateau in

Table 5.3 Material parameters for crystalline and amorphous computational modules

Amorphous computational module (soft segment)									
nkT (MPa)	$\dot{\gamma}_0$ (s^{-1})	A	ζ	λ_L	k	T (K)	σ_y (MPa)	E_z (MPa)	E_t (MPa)
0.5	0.03	3.31×10^{-27}	0.1	0.1	1.38×10^{-23}	298	20	100	70
Crystalline computational module (hard segment)									
n	$\dot{\gamma}_0$ (s^{-1})	Reference crystalline axes		Slippage systems		σ_y (MPa)	E_z (MPa)	E_t (MPa)	
5	0.05	(0.5, 0.5, 0.5)		See Table 5.4		100	70	70	

Source: [9] Reproduced with permission from the Royal Society.

Table 5.4 Hypothetical crystalline slippage systems

Slippage type	Indicial notation	Normalized resistance (g^{α}/τ_0)
Chain slip	(100)[001]	1
	(010)[001]	2.5
	{110}[001]	2.5
Transverse slip	(100)[010]	1.6
	(010)[100]	2.5
	{110}⟨1$\bar{1}$0⟩	2.5

Source: [9] Reproduced with permission from the Royal Society.

the range of computation. The reason for this is that the stress recovery is controlled by the glass transition temperature of the amorphous phase. A higher heating rate needs less time to reach the glass transition temperature and thus the recovery stress plateaus earlier. (3) It seems that the heating rate does not affect the stabilized recovery stress. From Figure 5.25, the stabilized recovery stress with heating rates of 0.7 °C/min and 0.35 °C/min is the same. Therefore, the heating rate affects the stress recovery profile, but not the stabilized recovery stress. This type of behavior has also been found in cold-compression programmed amorphous thermosetting SMPs [109].

It is noted that, in this analysis, we assumed that the temperature distribution is uniform throughout the SMPF. In actual specimens, it needs time for the heat to transfer from the outer surface to the center of the fiber. In other words, a heat transfer analysis is needed. Due to the thin diameter of the fibers, however, we did not conduct a heat transfer analysis. Instead, we assumed a uniform temperature distribution.

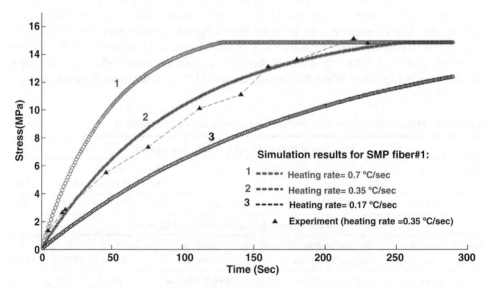

Figure 5.25 Simulation and experimental results for stress recovery of the SMPF with different heating rates. *Source:* [9] Reproduced with permission from the Royal Society

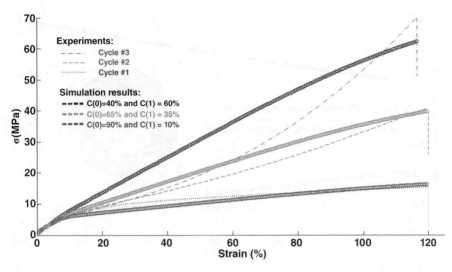

Figure 5.26   Simulation results for various volume fractions of the crystalline and amorphous phases together with cyclic tension test results for the SMPF up to the 120% strain level with a loading rate of 50.8 mm/min. *Source:* [9] Reproduced with permission from the Royal Society

Because the proposed multiscale viscoplastic model incorporates the individual constitutive equation of each subphase, it can easily capture a vast variety of microstructural configurations, such as 100% amorphous or crystalline fibers or any microstructure configuration that lies between these two extreme cases. Now we would like to demonstrate the capability of the model in predicting strain hardening behavior with various amorphous and crystalline phase loadings. Figure 5.26 shows three cyclic tensile test results and model predictions with various amorphous volume fractions C(0) and crystalline volume fractions C(1). Clearly, the prediction with C(1) = 10% by volume is very close to the test result in the first strain hardening cycle. In the second and third cyclic tensions, the SMPF is work hardened and shows stiffer mechanical responses. With C(1) = 35% by volume, the proposed theory can capture the second cycle of strain hardening. While in the third cycle, the prediction shows some deviation from the experiments, regardless of the crystalline volume fraction adopted. It is expected that the prediction may be better if true stress and true strain are used in describing the test results. This is because the large axial deformation leads to considerable shrinkage in the transverse direction due to Poisson's ratio effect. Owing to the small diameter of the SMPF, however, direct measurement using traditional methods such as a strain gauge is not realistic. Other noncontact methods such as imaging using a charge coupled device (CCD) camera together with the image processing technique such as those used in measuring microcrack initiation and propagation in an adhesively bonded joint may be used [110].

Finally, the stress induced crystallization process is also modeled. The result is shown in Figure 5.27. The inner radius of the crystalline annulus is shrinking while the outer radius of the crystalline annulus is expanding as the tensile strain increases during the strain hardening test by cold-drawing programming. Clearly, this shows that the area of the crystalline annulus is increasing and thus the volume fraction of the crystalline phase is growing. At the same time, the area for the amorphous annulus is decreasing and the volume fraction of the amorphous

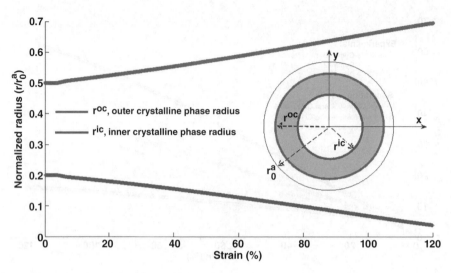

Figure 5.27 Simulation of the stress induced crystallization process in which the inner (bottom curve) and outer (top curve) radii of the crystalline phase evolve with the loading process. *Source:* [9] Reproduced with permission from the Royal Society

phase is reducing. In other words, stress induced crystallization suggests a transformation of the amorphous phase to the crystalline phase. Owing to the ordered microstructure of the crystalline phase, stress induced crystallization leads to a stiffer and stronger SMPF.

5.3.6 Summary

In general, SMPF is perceived as a two-phase composite material with a crystalline phase mixed with an amorphous phase. A multiscale viscoplasticity theory is developed. The amorphous phase is modeled using the Boyce model, while the crystalline phase is modeled using the Hutchinson model. Under an isostrain assumption, the micromechanics approach is used to assemble the microscale RVE. The kinematic relation is used to link the micromechanics constitutive relation to the macroscopic constitutive law. The proposed theory takes into account the stress induced crystallization process and the initial morphological texture, while the polymeric texture is updated based on the applied stresses. The related computational issue is discussed. The predictability of the model is validated by comparison with test results. It is expected that more accurate measurement of the stress and strain in the SMPF with large deformation may further enhance the predictability of the developed model. It is also desired to reduce the number of material parameters in the model. In other words, a deeper understanding and physics based theoretical modeling are needed.

5.4 Stress Memory versus Strain Memory

5.4.1 Stress–Strain Decomposition during Thermomechanical Cycle

While shape (strain) memory has been traditionally used as a parameter to characterize shape memory capabilities of shape memory polymers, stress memory is more important for some applications, such as serving as actuators or providing the crack closure force in close-then-heal (CTH) self-healing applications. It is known that both stress recovery and strain recovery depend on the boundary conditions of the specimens. Maximum stress recovery occurs under fully constrained boundary conditions (zero recovery strain) and maximum strain recovery occurs under free boundary conditions (zero recovery stress). Therefore, when we talk about the stress recovery ratio and strain recovery ratio, we are referring to these two extreme cases. Corresponding to the terminology used for shape memory, stress memory can also have pre-stress, temporary stress, or "fixed" stress and stress fixity ratio, as well as recovered stress and stress recovery ratio. In order to define the stress fixity ratio and the stress recovery ratio, we need first to define the pre-stress, fixed stress, and recovered stress.

Let us first define the recovery stress. From Figures 5.12 and 5.13, it is clear that the recovery stress has two characteristic values: one is the peak recovery stress and the other is the stabilized recovery stress. As discussed previously, while the peak recovery stress is important, the stabilized recovery stress is more important in self-healing applications because crack closing, as well as melting, flowing, wetting, and diffusing of thermoplastic particles, need a sustained recovery force to hold the fractured surface in close proximity. Therefore, for practical applications, the stabilized recovery stress is more meaningful. In this book, we define the stabilized recovery stress as the recovery stress to be used in the calculation of the stress recovery ratio.

In order to calculate the stress fixity ratio and stress recovery ratio, we also need to define the pre-stress and the fixed temporary stress. Taking the stress–strain behavior during cold-drawing programming as an example, both the stress and strain can be divided into several components, as shown in Figure 5.28. Similar to the definition of pre-strain, we can define the pre-stress as the peak applied stress, that is, stress at the end of the loading step (step 1), σ_p. The challenge is how to define the fixed stress.

It is well known that we cannot see the stress, but we can tell if the stress is applied or not based on its effect, that is, displacement or deformation or strain. Programming is a process of deforming the molecules from a coiled configuration to an aligned configuration. Therefore, any stress related to such molecular deformation should be treated as a fixed stress. In other words, locking programming stress is in the form of inelastic molecular deformation. Here, we emphasize inelastic deformation because the elastic strain cannot be fixed or memorized (released after unloading) and thus cannot lock stress. Based on Figure 5.28, four steps are used in the programming process. Let us analyze each step to see whether or not it contributes to stress storage. Step 1, the loading step, can cause inelastic molecular deformation. As indicated by Li and Xu [109], as long as the loading in step 1 is beyond yielding, a certain pseudo-plastic deformation can be fixed and the SMP is programmed, even without the stress relaxation step (step 2). In other words, if unloading immediately at the end of step 1, we should see a certain pseudo-plastic deformation if the SMP is deformed beyond yielding, suggesting that a certain inelastic strain is fixed. In conclusion, step 1 contributes to stress locking if the SMP is deformed beyond yielding (the unloading stress–strain curve will not return to the origin). However, not all the stress applied in step 1 can be locked. Some stress causes elastic deformation, which will be released by unloading to zero stress. Later, we will see that this portion of stress can be determined in steps 3 and 4 of Figure 5.28. Step 2 is the stress relaxation step. The reason for stress relaxation is due to the deformation of molecules – molecules adjust their configuration to align along the loading direction or to relax. Therefore, the stress that disappeared in step 2, or the stress relaxation step, is indeed the stress locked in the polymer. In

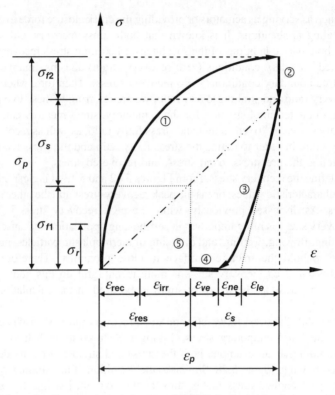

Figure 5.28 Schematic of stress–strain behavior of SMPFs in the entire thermomechanical cycle by cold-drawing programming (step 1, loading; step 2, holding the strain constant for a while (stress relaxation); step 3, unloading; step 4, structural relaxation; step 5, fully constrained stress recovery)

order words, the relaxed stress should also be a portion of the stress fixed because it causes inelastic deformation of molecules. Direct evidence is that, during cold-compression programming of amorphous SMPs, the longer the stress relaxation step, the greater are the shape fixed ratio and the shape recovery ratio [109]. Step 3 is the unloading step. It is accompanied by elastic springback. Therefore, the stress corresponding to step 3 does not contribute to stress locking. Instead, this portion of stress should be removed from the pre-stress in order to find the fixed stress. Step 4 is the structural relaxation step, representing viscoelastic springback. Therefore, the stress corresponding to this step also needs to be removed. Based on the above analyses, we have found that the programming stress locked or fixed should be equal to the pre-stress (stress at the end of step 1) minus the stress corresponding to springback in steps 3 and 4, which is the stress locked in step 1 plus the stress stored in step 2.

From Figure 5.28, the total programming stress, σ_p, can be decomposed into fixed stress, σ_f (σ_{f1} in step 1 and σ_{f2} in step 2), and stress corresponding to springback (linear elastic σ_{le}, nonlinear elastic σ_{ne}, and viscoelastic σ_{ve}), σ_s, which is reversible and can be determined from the unloading stress–strain curve, as shown in Figure 5.28. From this definition, it is clear that we define fixed stress as subtracting the stress corresponding to springback from the peak applied stress. This makes sense because the fixed stress should be the portion of stress that causes fixed strain. Springback stress just causes elastic deformation, which is reversible. Therefore, removing springback stress from the programming stress is reasonable in order to determine the fixed stress. Of course, as shown in Figure 5.28, we have simplified the determination of the springback stress by an equivalent linear elastic relationship of the unloading stress–strain behavior (the dot–dot–dash line), just like a linear elastic spring when the applied load is removed. More studies are needed to represent the springback stress better.

Similarly, the total programming strain, ε_p, can be divided into springback strain, ε_s, and residual viscoplastic damage strain, ε_{res}. The springback strain ε_s is further decomposed into linear elastic strain, ε_{le}, nonlinear elastic strain, ε_{ne}, and viscoelastic strain, ε_{ve}. It is noted that the linear and nonlinear elastic strains are independent of time during the unloading process, while the viscoelastic strain is time dependent after full removal of the applied stress. It usually takes from minutes to hours to stabilize after the removal of the applied load. As shown in Figure 5.12, the SMPFs took about 10 minutes to stabilize. The residual viscoplastic damage strain, ε_{res}, is further divided into irrecoverable strain, ε_{irr}, and recoverable strain, ε_{rec}. More detailed decomposition can divide the irrecoverable strain further into irreversible plastic strain and damage. Sometimes, we can also call the recoverable strain, which is the portion that can be recovered due to shape memory, as pseudo-plastic strain, meaning a type of plastic strain that can be recovered, in order to differentiate from the classical definition of plastic strain, which is permanent strain and cannot be recovered.

From Figure 5.28, the shape fixity ratio R_f can be defined as the residual strain over the pre-strain:

$$R_f = \frac{\varepsilon_{res}}{\varepsilon_p} \tag{5.31}$$

The shape recovery ratio R_r is defined as the recovered strain over the fixed strain:

$$R_r = \frac{\varepsilon_{rec}}{\varepsilon_{res}} \tag{5.32}$$

It is noted that this definition for the shape recovery ratio is different from the definition by Behl and Lendlein [111]. In their definition, the shape recovery ratio for the first thermomechanical cycle is

$$R_r = \frac{\varepsilon_p - \varepsilon_{irr}}{\varepsilon_p} = \frac{\varepsilon_s + \varepsilon_{rec}}{\varepsilon_s + \varepsilon_{res}} \tag{5.33}$$

In other words, the shape recovery ratio is defined as the ratio of the recovered strain (springback strain and shape memory strain) over the programming strain. Obviously, if the springback strain (ε_s) is small, which is usually the case for hot programming, the two definitions yield similar results; if the springback strain is large, such as in cold-drawing programming, the two definitions yield considerably different results. Because springback does not represent shape memory, we advocate Equation (5.32).

In a manner similar to the shape fixity and the shape recovery ratio, we can define the stress fixity and the stress recovery ratio. From Figure 5.28, the temporarily fixed stress can be obtained by subtracting the springback stress from the total stress (peak stress). Once the pre-stress and fixed stress are obtained, we can define the stress fixity ratio S_f as the temporarily fixed stress over the pre-stress:

$$S_f = \frac{\sigma_{f1} + \sigma_{f2}}{\sigma_p} \tag{5.34}$$

We can also define the stress recovery ratio S_r as the recovered stress over the fixed stress:

$$S_r = \frac{\sigma_r}{\sigma_{f1} + \sigma_{f2}} \tag{5.35}$$

where σ_r is the recovered stress.

Based on Figure 5.12, the stress–strain behavior of Sample 2 and Sample 3 during the entire thermomechanical cycles is redrawn, as shown in Figure 5.29(a) and (b), respectively. Because the unloading stress–strain curve cannot be recorded by the MTS machine, we use a straight line to represent the elastic unloading step. In Figure 5.29, the dotted auxiliary lines are drawn to help identify the peak stress, the springback stress, and the recovered stress. From Figure 5.29(a), the pre-stress is $\sigma_p = 11.2\,\text{MPa}$, the recovered stress is $\sigma_r = 4.2\,\text{MPa}$, the springback stress is $\sigma_s = 0.3\,\text{MPa}$, and the fixed stress $(\sigma_{f1} + \sigma_{f2}) = (\sigma_p - \sigma_s)$ is $(11.2–0.3\,\text{MPa}) = 10.9\,\text{MPa}$. Using Equations (5.34) and (5.35), the stress fixity ratio is $S_f = 10.9/11.2 = 97.3\%$ and the stress recovery ratio is $S_r = 4.2/10.9 = 38.5\%$. We have conducted a free shape recovery test for Sample 2. The shape recovery ratio is about 100%. From Figure 5.29(b), the pre-stress $\sigma_p = 77.5\,\text{MPa}$, the recovered stress is $\sigma_r = 4.7\,\text{MPa}$, the springback stress is $\sigma_s = 10.1\,\text{MPa}$, and the fixed stress $(\sigma_{f1} + \sigma_{f2}) = (\sigma_p - \sigma_s)$ is $(77.5–10.1\,\text{MPa}) = 67.4\,\text{MPa}$. Using Equations (5.34) and (5.35), the stress fixity ratio is $S_f = 67.4/77.5 = 86.9\%$ and the stress recovery ratio is $S_r = 4.7/67.4 = 7.0\%$. Again, we have conducted a free shape recovery test for Sample 3. The shape recovery ratio is about 100%.

Obviously, the stress recovery ratio is much lower than the strain recovery ratio. The difference is even bigger for cold-drawing programming. In other words, the SMPF has a good memory of strain or shape, but a poor memory of stress or load. Actually, this is true not only for SMPFs but also for almost all shape memory polymers. On the one hand, this shows the limitations of SMPs in certain applications, such as those that need a large recovery force. On the other hand, this also provides a unique opportunity for researchers to develop SMPs with a higher recovery stress and a higher stress recovery ratio.

Explanation of the different memories in stress and strain needs in-depth understanding of the mechanism and physics on locking stress and storing strain. From the macroscopic point of view, strain or shape recovery is a global behavior of the materials. Stress recovery depends on the level of recovery strain and also on the stiffness of the materials at the recovery temperature. Because the switching phase is either an amorphous phase by the glass transition or crystalline phase by the melting transition, the stiffness at the recovery temperature is usually very low (rubbery or liquid). Therefore, the recovery stress is very low although the recovery strain may be very high. Increasing the stiffness of the materials at the recovery temperature is the key to increase the recovery stress.

From the microscopic point of view, strain recovery in a stress free status is driven by the conformational entropy; that is, molecules recoil from the ordered configuration to the

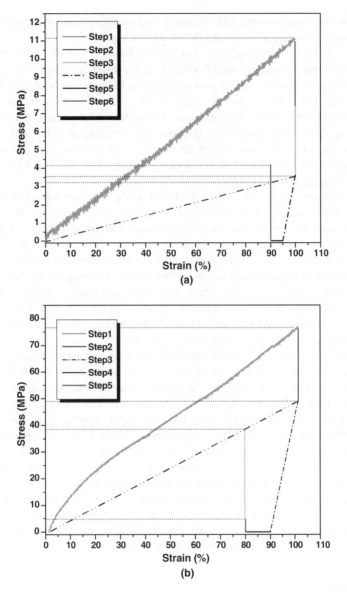

Figure 5.29 Stress–strain behavior of SMPFs during the entire thermomechanical cycle. (a) Sample 2 (hot-drawing programmed sample): step 1, stretch the fiber bundle to 100% strain at a rate of 200 mm/min at 100 °C; step 2, hold the strain constant for 1 hour; step 3, cool the fiber to room temperature slowly while holding the pre-strain constant; step 4, release the fiber bundle from the fixture (unloading); step 5, relax the fiber in the stress free condition until the shape is fixed; step 6, recover the fiber at 150 °C in a fully constrained condition; (b) Sample 3 (cold-drawing programmed sample): step 1, stretch the fiber bundle to 100% strain at a rate of 200 mm/min at room temperature; step 2, hold the strain constant for 1 hour; step 3, release the fiber bundle from fixtures (unloading); step 4, relax the fiber in a stress free condition until the shape is fixed; step 5, recover the fiber at 150 °C in a fully constrained condition. (Because the curves are black and white, please refer to Figure 5.12 to better differentiate each step).

disordered random configuration. This of course leads to appreciable strain recovery because there is not much resistance to the recoil of the molecules. For the stress recovery, two competing events occur simultaneously. On the one hand, the recoil of molecules leads to shortening of the tension programmed fiber. Due to the fixed boundary condition, however, this shortening is not allowed, leading to recovery stress. On the other hand, under the recovery stress, molecules start to align along the recovery stress direction, which counterbalances the tendency for molecules to recoil. In other words, alignments of the molecules tend to reduce the recovery stress. These two tendencies continue until a balance is reached. Additionally, stressed molecules tend to relax, which also leads to stress reduction (stress relaxation). Consequently, the recovery stress is not high enough because the molecules cannot fully recover their original random configuration.

These analyses can also be validated by DSC test results after free and constrained recovery of Samples 2 and 3, discussed in Section 5.2. Under free shape recovery, the SMPF can be considered as annealing at the recovery temperature, while under fully constrained stress recovery, the SMPF can be considered as another round of hot programming. Because the annealing temperature of 150 °C is higher than the glass transition and melting temperature of the soft and short hard segments and the microphase separation temperature, but lower than the microphase mixing temperature, it is expected that the broad melting temperature of the short hard segment and the microphase separation temperature of the DSC after free shape recovery would disappear because the short-range ordered hard segments have melted and microphase mixing has not occurred. Under constrained stress recovery, however, it is expected that the molecule alignments in both the soft and hard segments are maintained at a certain level after stress recovery. Therefore, the DSC result should show a fingerprint of this behavior. From Figure 5.30, this is exactly the case. As compared to the free recovered specimen, the stress recovered specimen shows broadened transition in the soft–hard segment mixture and broadened melting transition in the long hard segments, regardless of the programming approach (hot or cold drawing). This suggests that, during stress recovery, the molecules are not able to fully display their memory effect or recoil to their relaxed status, leading to a lower recovered stress or stress recovery ratio. On the other hand, free shape recovery allows the memory effect to be sufficiently displayed, as evidenced by the disappearance of the hard segment broad melting transition and narrowing of the soft–hard segment mixture broad transition, leading to a higher recovery strain or shape recovery ratio. A schematic illustration of the fully constrained stress recovery and fully free strain recovery of a tension programmed cross-linked amorphous SMP is shown in Figure 5.31.

5.4.2 Summary

In general, SMPFs have a good memory of strain or shape, but a poor memory of stress or load. It is expected that, if other properties such as conductivity, color, density, etc., were used to evaluate the memory capability of SMPs, different recovery ratios would be obtained. Therefore, the memory capability depends on the parameter or criterion used. Although the current dominant criterion used to evaluate memory capability is the shape or strain recovery ratio, it is expected that other indicators such as the stress recovery ratio would be more and more used because SMPs will find new applications in various areas. Most importantly, we need to better understand the underlying physics for each parameter because different parameters may not be related linearly or proportionally. It is expected that this opens

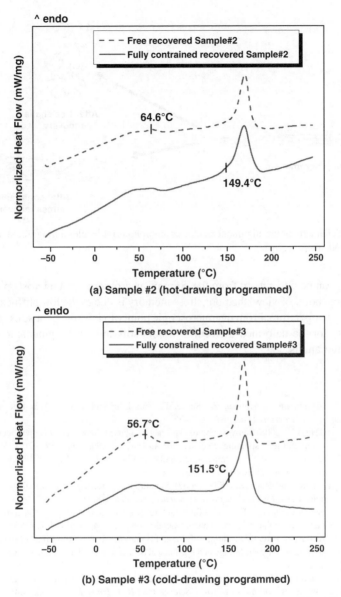

Figure 5.30 (a) DSC curves of recovered Sample 2; (b) DSC curves of recovered Sample 3. The samples after constrained stress recovery show a broadened transition temperature in the soft–hard segment mixture and a broadened melting transition in the long hard segments, signifying an inability for the deformed molecules to return to their original configuration during constrained stress recovery.

up new opportunities and there will be foreseeable research activities in the near future. It is also expected that the general name of the shape memory polymer may be changed to more specific names, such as stress memory polymer, conductivity memory polymer, color memory polymer, density memory polymer, etc. As for the difference between strain memory and stress

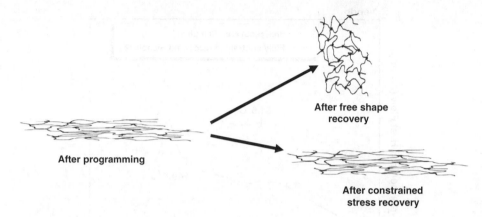

Figure 5.31 Schematic of the alignment change of amorphous molecules during free shape recovery and fully constrained stress recovery.

memory, they can be understood from the point of macroscopic and microscopic views. From the macroscopic point of view, the poor stress memory is due to the low stiffness of the SMPF at the recovery temperature. From the microscopic point of view, the poor stress memory is due to the inability for the deformed molecules in both soft and hard segments to return to their original configurations.

References

1. Nangrejo, M., Bragman, F., Ahmad, Z., Stride, E., and Edirisinghe, M. (2012) Hot electrospinning of polyurethane fibers. *Materials Letter*, **68**, 482–485.
2. Meng, Q.H. and Hu, J.L. (2008) Study on poly(e-caprolactone)-based shape memory copolymer fiber prepared by bulk polymerization and melt spinning. *Polymers for Advanced Technologies*, **19**, 131–136.
3. Meng, Q.H. and Hu, J.L. (2008) A temperature-regulating fiber made of PEG-based smart copolymer. *Solar Energy Materials and Solar Cells*, **9**, 1245–1252.
4. Chattopadhyay, D.K. and Raju, K.V.S.N. (2007) Structural engineering of polyurethane coatings for high performance applications. *Progress in Polymer Science*, **32**, 352–418.
5. Skarja, G.A. and Woodhouse, K.A. (2000) Structure–property relationships of degradable polyurethane elastomers containing an amino acid-based chain extender. *Journal of Applied Polymer Science*, **75**, 1522–1534.
6. LaShanda, T., Korley, J., Pate, B.D., Thomas, E.L., and Hammond, P.T. (2006) Effect of the degree of soft and hard segment ordering on the morphology and mechanical behavior of semicrystalline segmented polyurethanes. *Polymer*, **47**, 3073–3082.
7. Lee, H.S. and Hsu, S.L. (1994) Structural changes and chain orientational behavior during tensile deformation of segmented polyurethanes. *Journal of Polymer Science Part B: Polymer Physics*, **32**, 2085–2098.
8. Hu, J., Meng, H., Li, G., and Ibekwe, S. (2012) A review of stimuli-responsive polymers for smart textile applications. *Smart Materials and Structures*, **21**, paper 053001 (23 pages).
9. Li, G. and Shojaei, A. (2012) A viscoplastic theory of shape memory polymer fibers with application to self-healing materials. *Proceedings of the Royal Society A – Mathematical Physical and Engineering Sciences*, **468**, 2319–2346.
10. Li, G., Meng, H., and Hu, J. (2012) Healable thermoset polymer composite embedded with stimuli-responsive fibers. *Journal of the Royal Society Interface*, **9**, 3279–3287.
11. Li, G., Ajisafe, O., and Meng, M. (2013) Effect of strain hardening of shape memory polymer fibers on healing efficiency of thermosetting polymer composites. *Polymer*, **54**, 920–928.
12. Li, G. and Zhang, P. (2013) A self-healing particulate composite reinforced with strain hardened shape memory polymer fiber. *Polymer*, **54**, 5075–5086.

13. Shojaei, A., Li, G., and Voyiadjis, G.Z. (2013) Cyclic viscoplastic–viscodamage analysis of shape memory polymer fibers with application to self-healing smart materials. *ASME Journal of Applied Mechanics*, **80**, paper 011014 (15 pages).

14. Shojaei, A. and Li, G. (2013) Viscoplasticity analysis of semicrystalline polymers: a multiscale approach within micromechanics framework. *International Journal of Plasticity*, **42**, 31–49.

15. Zhang, P. and Li, G. (2013) Structural relaxation behavior of strain hardened shape memory polymer fibers for self-healing applications. *Journal of Polymer Science Part B: Polymer Physics*, **51**, 966–977.

16. Meng, Q.H., Hu, J.L., Shen, L., Hu, Y., and Han, J. (2009) A smart hollow filament with thermal sensitive internal diameter. *Journal of Applied Polymer Science*, **113**, 2440–2449.

17. Meng, Q.H. and Hu, J.L. (2008) The influence of heat treatment on properties of shape memory fibers: I. Crystallinity, hydrogen bonding and shape memory effect. *Journal of Applied Polymer Science*, **109**, 2616–2623.

18. Meng, Q.H. and Hu, J.L. (2008) Self-organizing alignment of carbon nanotube in shape memory segmented fiber prepared by polymerization and melt spinning. *Composites Part A: Applied Science and Manufacturing*, **39**, 314–321.

19. Stoclet, G., Seguela, R., Lefebvre, J.M., Elkoun, S., and Vanmansart, C. (2010) Strain-induced molecular ordering in polylactide upon uniaxial stretching. *Macromolecules*, **43**, 1488–1498.

20. Yang, Q. and Li, G. (2014) A spider silk like fiber for vibration damping. *Smart Materials and Structures* (accepted).

21. Saiani, A., Daunch, W.A., Verbeke, H., Leenslag, J.-W., and Higgins, J.S. (2001) Origin of multiple melting endotherms in a high hard block content polyurethane. 1. Thermodynamic investigation. *Macromolecules*, **34**, 9059–9068.

22. Zhou, D., Zhang, G.G., Law, D., Grant, D.J., and Schmitt, E.A. (2002) Physical stability of amorphous pharmaceuticals: importance of configurational thermodynamic quantities and molecular mobility. *Journal of Pharmaceutical Science*, **91**, 1863–1872.

23. Hu, J.L. (2007) *Shape Memory Polymers and Textiles*, Woodhead Publishing Limited.

24. Koerner, H., Price, G., Pearce, N.A., Alexander, M., and Vaia, R.A. (2004) Remotely actuated polymer nanocomposites – stress-recovery of carbon-nanotube-filled thermoplastic elastomers. *Nature Materials*, **3**, 115–120.

25. Ping, P., Wang, W., Chen, X., and Jing, X. (2005) Poly(ε-caprolactone) polyurethane and its shape-memory property. *Biomacromolecules*, **6**, 587–592.

26. Lendlein, A., and Langer, R. (2002) Shape memory polymers. *Angewandte Chemie International Edition*, **41**, 2034–2057.

27. Rabani, G., Luftmann, H., and Kraft, A. (2006) Synthesis and characterization of two shape-memory polymers containing short aramid hard segments and poly(ε-caprolactone) soft segments. *Polymer*, **47**, 4251–4260.

28. Wang, W., Jin, Y., Ping, P., Chen, X., Jing, X., and Su, Z. (2010) Structure evolution in segmented poly(ester urethane) in shape-memory process. *Macromolecules*, **43**, 2942–2947.

29. Zotzmann, J., Behl, M., Feng, Y., and Lendlein, A. (2010) Copolymer networks based on poly(ω-pentadeca-lactone) and poly(ε-caprolactone) segments as a versatile triple-shape polymer system. *Advanced Functional Materials*, **20**, 3583–3594.

30. Siesler, H.W. and Holland-Moritz, K. (1980) *Infrared and Raman Spectroscopy of Polymers*, Marcel Dekker Inc., New York.

31. Estes, G.M., Seymour, R.W., and Cooper, S.L. (1971) Infrared studies of segmented polyurethane elastomers. II. Infrared dichroism. *Macromolecules*, **4**, 452–457.

32. Elzein, T., Nasser-Eddine, M., Delaite, C., Bistac, S., and Dumas, P. (2004) FTIR study of polycaprolactone chain organization at interfaces. *Journal of Colloid and Interface Science*, **273**, 381–387.

33. Wang, F.C., Feve, M., Lam, T.M., and Pascault, J.P. (1994) FTIR analysis of hydrogen bonding in amorphous linear aromatic polyurethanes. I. Influence of temperature. *Journal of Polymer Science Part B: Polymer Physics*, **32**, 1305–1313.

34. Christenson, E.M., Anderson, J.M., Hiltner, A., and Baer, E. (2005) Relationship between nanoscale deformation processes and elastic behavior of polyurethane elastomers. *Polymer*, **46**, 11744–11754.

35. Zhu, Y., Hu, J., and Yeung, K. (2009) Effect of soft segment crystallization and hard segment physical crosslink on shape memory function in antibacterial segmented polyurethane ionomers. *Acta Biomaterialia*, **5**, 3346–3357.

36. Eisoldt, L., Smith, A., and Scheibel, T. (2011) Decoding the secrets of spider silk. *Materials Today*, **14**, 80–86.

37. Buehler, M.J., Keten, S., and Ackbarow, T. (2008) Theoretical and computational hierarchical nanomechanics of protein materials: deformation and fracture. *Progress in Materials Science*, **53**, 1101–1241.

38. Scheller, J., Ghrs, K.H., Grosse, F., and Conrad, U. (2001) Production of spider silk proteins in tobacco and potato. *Nature Biotechnology*, **19**, 573–577.

39. Lazaris, A., Arcidiacono, S., Huang, Y., Zhou, J.F., Duguay, F., Chretien, N., Welsh, E.A., Soares, J.W., and Karatzas, C.N. (2002) Spider silk fibers spun from soluble recombinant silk produced in mammalian cells. *Science*, **295**, 472–476.

40. Huemmerich, D., Scheibel, T., Vollrath, F., Cohen, S., Gat, U., and Ittah, S. (2004) Novel assembly properties of recombinant spider dragline silk proteins. *Current Biology*, **14**, 2070–2074.

41. Zhou, G., Shao, Z., Knight, D.P., Yan, J., and Chen, X. Silk (2009) Fibers extruded artificially from aqueous solutions of regenerated Bombyx Mori silk fibroin are tougher than their natural counterparts. *Advanced Materials*, **21**, 366–370.

42. Yan, J., Zhou, G., Knight, D.P., Shao, Z., and Chen, X. (2010) Wet-spinning of regenerated silk fiber from aqueous silk fibroin solution: discussion of spinning parameters. *Biomacromolecules*, **11**, 1–5.

43. Gatesy, J., Hayashi, C., Motriuk, D., Woods, J., and Lewis, R. (2001) Extreme diversity, conservation, and convergence of spider silk fibroin sequences. *Science*, **291**, 2603–2605.

44. Qin, Z. and Buehler, M.J. (2010) Cooperative deformation of hydrogen bonds in beta-strands and beta-sheet nanocrystals. *Physical Review E*, **82**, 061906.

45. Keten, S., Xu, Z., Ihle, B., and Buehler, M.J. (2010) Nanoconfinement controls stiffness, strength and mechanical toughness of beta-sheet crystals in silk. *Nature Materials*, **9**, 359–367.

46. Keten, S. and Buehler, M.J. (2008) Geometric confinement governs the rupture strength of H-bond assemblies at a critical length scale. *Nano Letters*, **8**, 743–748.

47. Keten, S. and Buehler, M.J. (2010) Nanostructure and molecular mechanics of spider dragline silk protein assemblies. *Journal of the Royal Society Interface*, **7**, 1709–1721.

48. Glisovic, A., Vehoff, T., Davies, R.J., and Salditt, T. (2008) Strain dependent structural changes of spider dragline silk. *Macromolecules*, **41**, 390–398.

49. Blackledge, T.A., Swindeman, J.E., and Hayashi, C.Y. (2005) Quasistatic and continuous dynamic characterization of the mechanical properties of silk from the cobweb of the black widow spider *Latrodectus hesperus*. *Journal of Experimental Biology*, **208**, 1937–1949.

50. Brunette, C.M., Hsu, S.L., and MacKnight, W.J. (1982) Hydrogen-bonding properties of hard-segment model compounds in polyurethane block copolymers. *Macromolecules*, **15**, 71–77.

51. Seefried, C.G., Koleske, J.V., and Critchfield, F.E. (1975) Thermoplastic urethane elastomers. II. Effects of variations in hard-segment concentration. *Journal of Applied Polymer Science*, **19**, 2503–2513.

52. Kim, B.K., Lee, S.Y., and Xu, M. (1996) Polyurethanes having shape memory effects. *Polymer*, **37**, 5781–5793.

53. Niesten, M.C.E.J., Feijen, J., and Gaymans, R.J. (2000) Synthesis and properties of segmented copolymers having aramid units of uniform length. *Polymer*, **41**, 8487–8500.

54. Niesten, M.C.E.J., Harkema, S., van der Heide, E., and Gaymans, R.J. (2001) Structural changes of segmented copolyetheresteramides with uniform aramid units induced by melting and deformation. *Polymer*, **42**, 1131–1142.

55. Tobushi, H., Hashimoto, T., Hayashi, S., and Yamada, E. (1997) Thermomechanical constitutive modeling in shape memory polymer of polyurethane series. *Journal of Intelligent Material Systems and Structures*, **8**, 711–718.

56. Nemat-Nasser, S. and Hori, M. (1993) *Micromechanics: Overall Properties of Heterogeneous Materials*, Elsevier, Amsterdam.

57. Castañeda, P.P. (2002) Second-order homogenization estimates for nonlinear composites incorporating field fluctuations: I – Theory. *Journal of the Mechanics and Physics of Solids*, **50**, 737–757.

58. Castañeda, P.P. (2002) Second-order homogenization estimates for nonlinear composites incorporating field fluctuations: II – Applications. *Journal of the Mechanics and Physics of Solids*, **50**, 759–782.

59. Eshelby, J.D. (1957) The determination of the elastic field of an ellipsoidal inclusion, and related problems. *Proceedings of the Royal Society of London. Series A. Mathematical and Physical Sciences*, **241**, 376–396.

60. Fotiu, P.A. and Nemat-Nasser, S. (1996) Overall properties of elastic–viscoplastic periodic composites. *International Journal of Plasticity*, **12**, 163–190.

61. Hori, M. and Nemat-Nasser, S. (1999) On two micromechanics theories for determining micro–macro relations in heterogeneous solids. *Mechanics of Materials*, **31**, 667–682.

62. Chaboche, J.L., Kruch, S., Maire, J.F., and Pottier, T. (2001) Towards a micromechanics based inelastic and damage modeling of composites. *International Journal of Plasticity*, **17**, 411–439.

63. Berveiller, M. and Zaoui, A. (1979) An extension of the self-consistent scheme to plastically-flowing polycrystals. *Journal of the Mechanics and Physics of Solids*, **26**, 325–344.

64. Hill, R. (1965) Continuum micro-mechanics of elastoplastic polycrystals. *Journal of the Mechanics and Physics of Solids*, **13**, 89–101.

65. Hutchinson, J.W. (1976) Bounds and self-consistent estimates for creep of polycrystalline materials. *Proceedings of the Royal Society of London, Series A: Mathematical and Physical Sciences*, **348**, 101–127.

66. Chaboche, J.L., Kanoute, P., and Roos, A. (2005) On the capabilities of mean-filed approaches for the description of plasticity in metal matrix composites. *International Journal of Plasticity*, **21**, 1409–1434.

67. Masson, R. and Zaoui, A. (1999) Self-consistent estimates for the rate-dependent elastoplastic behaviour of polycrystalline materials. *Journal of the Mechanics and Physics of Solids*, **47**, 1543–1568.

68. Doghri, I., Adam, L., and Bilger, N. (2010) Mean-field homogenization of elasto-viscoplastic composites based on a general incrementally affine linearization method. *International Journal of Plasticity*, **26**, 219–238.

69. Doghri, I., Brassart, L., Adam, L., and Gérard, J.S. (2011) A second-moment incremental formulation for the mean-field homogenization of elasto-plastic composites. *International Journal of Plasticity*, **27**, 352–371.

70. Chu, T.Y. and Hashin, Z. (1971) Plastic behavior of composites and porous media under isotropic stress. *International Journal of Engineering Science*, **9**, 971–994.

71. Masson, R., Bornert, M., Suquet, P., and Zaoui, A. (2000) An affine formulation for the prediction of the effective properties of nonlinear composites and polycrystals. *Journal of the Mechanics and Physics of Solids*, **48**, 1203–1227.

72. Feyel, F. and Chaboche, J.L. (2000) Fe^2 multi-scale approach for modeling the elasto-viscoplastic behaviour of long fiber SiC/Ti composite materials. *Computer Methods in Applied Mechanics and Engineering*, **183**, 309–330.

73. Ghosh, S. (2010) *Micromechanical Analysis and Multi-scale Modeling Using the Voronoi Cell Finite Element Methods*, CRC Press, Boca Raton, Florida.

74. Dvorak, G.J. (1990) On uniform fields in heterogeneous media. *Proceedings of the Royal Society of London, Series A: Mathematical and Physical Sciences*, **431**, 89–110.

75. Dvorak, G.J. (1992) Transformation field analysis of inelastic composite materials. *Proceedings of the Royal Society of London, Series A: Mathematical and Physical Sciences*, **437**, 311–327.

76. Dvorak, G.J. and Benveniste, Y. On transformation strains and uniform fields in multiphase elastic media. *Proceedings of the Royal Society of London, Series A: Mathematical and Physical Sciences*, **437**, 291–310.

77. Dvorak, G.J. and Zhang, J. (2001) Transformation field analysis of damage evolution in composite materials. *Journal of the Mechanics and Physics of Solids*, **49**, 2517–2541.

78. Li, G., Zhao, Y., and Pang, S.S. (1998) A three-layer built-in analytical modeling of concrete. *Cement and Concrete Research*, **28**, 1057–1070.

79. Li, G., Zhao, Y., and Pang, S.S. (1999) Four-phase sphere modeling of effective bulk modulus of concrete. *Cement and Concrete Research*, **29**, 839–845.

80. Li, G., Li, Y., Metcalf, J.B., and Pang, S.S. (1999) Elastic modulus prediction of asphalt concrete. *ASCE Journal of Materials in Civil Engineering*, **11**, 236–241.

81. Li, G., Zhao, Y., Pang, S.S., and Li, Y. (1999) Effective Young's modulus estimation of concrete. *Cement and Concrete Research*, **29**, 1455–1462.

82. Li, G., Zhao, Y., and Pang, S.S. (1999) Analytical modeling of particle size and cluster effects on particulate-filled composite. *Materials Science and Engineering A*, **271**, 43–52.

83. Li, G., Helms, J.E., Pang, S.S., and Schulz, K.C. (2001) Analytical modeling of tensile strength of particulate filled composites. *Polymer Composites*, **22**, 593–603.

84. Huang, B., Li, G., and Mohammad, L.N. (2003) Analytical modeling and experimental study of tensile strength of asphalt concrete composite at low temperatures. *Composites Part B: Engineering*, **34**, 705–714.

85. Li, G. and Nji, J. (2007) Development of rubberized syntactic foam. *Composites Part A: Applied Science and Manufacturing*, **38**, 1483–1492.

86. Li, G. and John, M. (2008) A crumb rubber modified syntactic foam. *Materials Science and Engineering A*, **474**, 390–399.

87. Nji, J. and Li, G. (2008) A CaO enhanced rubberized syntactic foam. *Composites Part A: Applied Science and Manufacturing*, **39**, 1404–1411.

88. Argon, A.S. (1963) A theory for the low-temperature plastic deformation of glassy polymers. *Philosophical Magazine*, **28**, 839–865.

89. Arruda, E.M. and Boyce, M.C. (1993) A three-dimensional constitutive model for the large stretch behavior of rubber elastic materials. *Journal of the Mechanics and Physics of Solids*, **41**, 389–412.

90. Boyce, M.C., Parks, D.M., and Argon, A.S. (1989) Plastic flow in oriented glassy polymers. *International Journal of Plasticity*, **5**, 593–615.

91. Arruda, E.M., Boyce, M.C., and Quintus-Bosz, H. (1993) Effects of initial anisotropy on the finite strain deformation behavior of glassy polymers. *International Journal of Plasticity*, **9**, 783–811.

92. Ahzi, S., Park, D.M., and Argon, A.S. (1995) Estimates of the overall elastic properties in semi-crystalline polymers. *AMD – Current Research in the Thermo-Mechanics of Polymers in the Rubbery-Glassy Range*, **203**, 31–44.

93. Lee, B.J., Parks, D.M., and Ahzi, S. (1993) Micromechanical modeling of large plastic deformation and texture evolution in semi-crystalline polymers. *Journal of the Mechanics and Physics of Solids*, **41**, 1651–1687.

94. Parks, D.M. and Ahzi, S. (1990) Polycrystalline plastic deformation and texture evolution for crystals lacking five independent slip systems. *Journal of the Mechanics and Physics of Solids*, **38**, 701–724.

95. Schoenfeld, S.E., Ahzi, S., and Asaro, R.J. (1995) Elastic–plastic crystal mechanics for low symmetry crystals. *Journal of the Mechanics and Physics of Solids*, **43**, 415–446.

96. Stevenson, A. (1995) Effect of crystallization on the mechanical properties of elastomers under large deformation, in Current Research in the Thermo-Mechanics of Polymers in the Rubbery–Glassy Range, ASME.

97. Asaro, R.J. (1979) Geometrical effects in the inhomogeneous deformation of ductile single crystals. *Acta Metallurgica*, **27**, 445–453.

98. Asaro, R.J. and Needleman, A. (1985) Overview no. 42, Texture development and strain hardening in rate dependent polycrystals. *Acta Metallurgica*, **33**, 923–953.

99. Asaro, R.J. and Rice, J.R. (1977) Strain localization in ductile single crystals. *Journal of the Mechanics and Physics of Solids*, **25**, 309–338.

100. Dupaix, R.B. and Boyce, M.C. (2007) Constitutive modeling of the finite strain behavior of amorphous polymers in and above the glass transition. *Mechanics of Materials*, **39**, 39–52.

101. Chaboche, J.L. (1991) On some modifications of kinematic hardening to improve the description of ratchetting effects. *International Journal of Plasticity*, **7**, 661–678.

102. Shojaei, A., Eslami, M., and Mahbadi, H. (2010) Cyclic loading of beams based on the chaboche model. *International Journal of Mechanics and Materials in Design*, **6**, 217–228.

103. Voyiadjis, G.Z., Shojaei, A., and Li, G. (2012) A generalized coupled viscoplastic– viscodamage–viscohealing theory for glassy polymers. *International Journal of Plasticity*, **28**, 21–45.

104. Simo, J.C. and Taylor, R.L. A return mapping algorithm for plane stress elastoplasticity. *International Journal for Numerical Methods in Engineering*, **22**, 649–670.

105. Simo, J.C. and Hughes, T.J.R. (1997) *Computational Inelasticity*, Springer, New York.

106. Simo, J.C. and Ortiz, M. (1985) A unified approach to finite deformation elastoplastic analysis based on the use of hyperelastic constitutive equations. *Computer Methods in Applied Mechanics and Engineering*, **49**, 221–245.

107. Nemat-Nasser, S. (2006) *Plasticity: A Treatise on Finite Deformation of Heterogeneous Inelastic Materials*, Cambridge University Press, Cambridge.

108. G'Sell, C. (2001) Dislocations in glassy polymers. Do they exist? Are they useful? *Materials Science and Engineering A*, **309–310**, 539–543.

109. Li, G. and Xu, W. (2011) Thermomechanical behavior of thermoset shape memory polymer programmed by cold-compression: testing and constitutive modeling. *Journal of the Mechanics and Physics of Solids*, **59**, 1231–1250.

110. Ji, G., Ouyang, Z., Li, G., Ibekwe, S.I., and Pang, S.S. (2010) Effects of adhesive thickness on global and local Mode-I interfacial fracture of bonded joints. *International Journal of Solids and Structures*, **47**, 2445–2458.

111. Behl, M. and Lendlein, A. (2007) Shape-memory polymers. *Materials Today*, **20**, 20–28.

6

Self-Healing with Shape Memory Polymer as Matrix

While all structural design takes a proactive measure to prevent premature failure, such as using safety factors greater than 1, premature failures do occur due to accidental loading, environmental attacks, design fault, and construction defects. Therefore, proactive maintenance by active intervention (repair) is needed to restore some or all of the lost structural capacity, which may eventually extend the service life of the structures. Hence, repair or self-healing is not just a way of passively restoring structural service but can actively manage the structural life by repair or healing at the right time and the right place. Figure 6.1 shows a schematic of active repairs on the extension of service life of structures.

Self-healing of structural damage has been a tremendous interest in the scientific community recently [1]. The ability to heal wounds is one of the truly remarkable properties of biological systems. A significant challenge facing the materials science community is to design smart synthetic systems that can mimic this behavior by not only sensing the presence of a wound or defect but also by actively re-establishing the continuity and integrity of the damaged area. Such self-healing materials would significantly extend the lifetime and utility of a vast array of manufactured structures [2].

Thermosetting polymer composites have been widely used in almost all man-made structures, particularly in high tech and high value structures such as aircrafts, ships, autos, trains, pressure vessels, pipelines, wind-turbine blades, bridges, buildings, etc. Owing to various types of loadings such as fatigue, impact, creep, earthquake, hurricane, etc., and environmental conditions such as corrosion, hygrothermal, ultraviolet radiation, fire, etc., no engineering structures last forever. Therefore, self-healing is a highly desired feature in engineering structures. With the fast advancement in materials science and engineering, as well as understanding of self-healing mechanisms and enhanced modelling capability, self-healing is evolving towards engineering practice instead of engineering dream. Self-healing is gradually changing the paradigm of structural design, which can be evidenced in a number of recent review articles and books [1–28].

Almost all types of polymers, including thermoplastic, thermoset, and elastomers, have either demonstrated a self-healing capability or have shown the potential to be healed. Various

Self-Healing Composites: Shape Memory Polymer-Based Structures, First Edition. Guoqiang Li.
© 2015 John Wiley & Sons, Ltd. Published 2015 by John Wiley & Sons, Ltd.

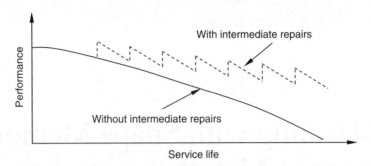

Figure 6.1 Schematic of intermediate interventions on the service life and performance of structures

applications or potential applications have been identified [1–28]. Basically, any polymer based structures or devices or coatings can be endowed with self-healing capabilities. While a number of self-healing approaches and strategies have been explored, and the amount of literature is exploding, healing can be divided into two categories. One is extrinsic healing based on incorporation of an external healing agent; the other is intrinsic healing by the polymer itself. For incorporation of an external healing agent, the agent can be liquid [29,30] or solid [31,32]. While the solid healing agent such as thermoplastic particles can be directly dispersed in a polymer matrix [31–34], the liquid healing agent needs to be stored and released on demand through various approaches, such as using microcapsules [35,36], hollow fibers [37,38], and a biomimetic microvascular network [39,40]. For intrinsic healing systems, they include polymers with a thermally reversible covalent bond (TRCB) [41], ionomer [42], hydrogen bond [43], and supramolecule [4]. This is a fast growing area with an ample supply of literature. Several typical extrinsic self-healing systems are schematically shown in Figures 6.2 to 6.5.

While all of the above self-healing schemes have shown a certain success, there is still a long way to go before even the simplest biological healing mechanism can be replicated within these synthetic materials. In order to compare the various self-healing schemes on a fair basis, there is a need for a set of evaluation criteria. We believe that for a self-healing scheme, the following are desired features. (1) The healing should be autonomous. This suggests that minimal or no human intervention should be involved. The healing with a liquid healing agent satisfies this criterion because the capsules or hollow fibers will be autonomously broken by propagating cracks, which leads to the release of the contained monomers. When the released healing agent flows into the crack and contacts with the dispersed catalyst, in situ polymerization occurs and patches the crack. (2) The healing should be repeatable. Healing schemes based on physical change or reversible chemistry satisfy this requirement. For example, healing by a solid healing agent such as thermoplastic particles is through melting of thermoplastic particles, flowing into a crack by capillary force, wetting of the crack surface, diffusing into the fractured polymer matrix driven by the concentration gradient, bonding of thermoplastic molecules and polymer matrix molecules by physical entanglement, and cooling to form a solid wedge. Obviously, this healing process includes physical change only and thus is repeatable. (3) The healing should be efficient. A higher healing efficiency is a highly desired feature for self-healing systems. For intrinsic healing systems, because the broken bond can be fully recovered, the healing efficiency is usually high. For extrinsic healing systems using an external healing agent,

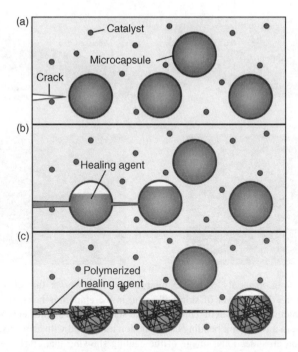

Figure 6.2 A microencapsulated healing agent is embedded in a structural composite matrix containing a catalyst capable of polymerizing the healing agent: (a) cracks form in the matrix wherever damage occurs; (b) the crack ruptures the microcapsules, releasing the healing agent into the crack plane through capillary action; (c) the healing agent contacts the catalyst, triggering polymerization that bonds the crack faces closed. *Source:* [35] Reproduced with permission from Nature Publishing Group

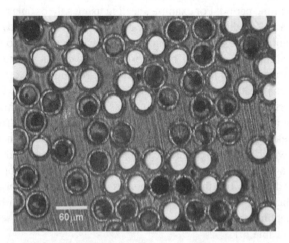

Figure 6.3 Self-healing by hollow fibers. The working principle is similar to the microcapsule approach but is more suitable for working with fiber reinforced polymer composites. *Source:* [37] Reproduced with permission from Elsevier

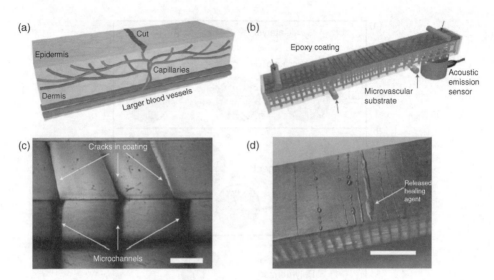

Figure 6.4 Self-healing materials with 3-D microvascular networks: (a) schematic diagram of a capillary network in the dermis layer of skin with a cut in the epidermis layer; (b) schematic diagram of the self-healing structure composed of a microvascular substrate and a brittle epoxy coating containing an embedded catalyst in a four-point bending configuration monitored with an acoustic-emission sensor; (c) high magnification cross-sectional image of the coating showing that cracks, which initiate at the surface, propagate towards the microchannel openings at the interface (scale bar = 0.5 mm); (d) optical image of a self-healing structure after cracks are formed in the coating (with 2.5 wt% catalyst), revealing the presence of excess healing fluid on the coating surface (scale bar = 5 mm). *Source:* [39] Reproduced with permission from Nature Publishing Group

this depends on the bonding strength of the healing agent. Healing efficiency lower or higher than 100% is possible. (4) The healing should be timely. It is desirable that the healing is completed in a short time period. For example, healing using a thermoplastic healing agent can be completed within 1 hour, depending on the heating and cooling rates. (5) The healing should be molecular. For almost all existing healing systems, healing is on a molecular-length scale, either by re-establishment of chemical bonds or molecular physical entanglements. (6) The healing should be structural. For load bearing structure applications, the crack is usually widely opened, such as cracks created by impact loading. Therefore, the key is how to heal cracks with a large width. Unfortunately, except for the microvascular healing system, almost all existing systems are unable to heal widely opened cracks. For example, for extrinsic self-healing, a large amount of healing agent is needed in order to fill the wide-opened crack with large damaged volume, which needs large microcapsules or thick hollow fibers. However, a large amount of healing agent will not only affect the physical/mechanical properties of the matrix but also create new defects when the healing agent is released from the large-sized containers. For intrinsic schemes, the fractured molecules need an external push to bring them in close proximity before healing occurs. The pros and cons of various existing self-healing schemes are summarized in Table 6.1.

From Table 6.1, no single self-healing scheme can satisfy all the requirements. What is particularly challenging is healing structurally or healing cracks with a wide opening. Mother Nature has developed a self-healing capability for almost all living things on various length

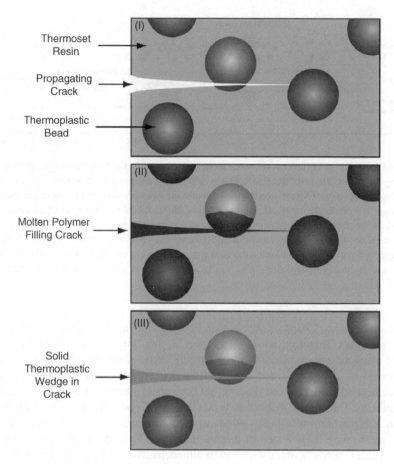

Figure 6.5 Self-healing by embedded thermoplastic particles in a thermosetting polymer matrix. *Source:* [31] Reproduced with permission from SAGE

Table 6.1 Summary of existing self-healing systems

Criteria	Autonomous	Repeatable	Efficient	Timely	Molecular	Structural
Microcapsule	Excellent	Poor	Good	Excellent	Excellent	Poor
Hollow fiber	Excellent	Poor	Good	Excellent	Excellent	Poor
Microvascular	Excellent	Good	Good	Excellent	Excellent	Good
Thermoplastic	Poor	Excellent	Good	Good	Excellent	Poor
Ionomer	Good	Excellent	Good	Good	Excellent	Good
TRCB	Poor	Excellent	Good	Good	Excellent	Poor
Hydrogen bond	Poor	Excellent	Good	Good	Excellent	Poor
Supramolecule	Poor	Excellent	Good	Good	Excellent	Poor

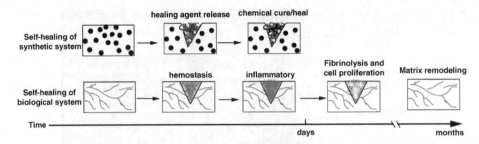

Figure 6.6 Synthetic and biological routes to healing. In synthetic materials (top), damage healing proceeds by an immediate response that actuates (triggers) the healing mechanism (e.g., the rupture of embedded microcapsules). Once triggered, the second stage involves transport of chemical species to the site of damage at a relatively rapid rate. During the final stage of healing, chemical repair takes place and can extend for several hours or days. In biological systems (bottom), wound healing follows three sequential steps, the first of which is an immediate inflammatory response, including blood clotting. In the second stage, cell proliferation and matrix deposition occur and can extend for several days. The long-term response is matrix remodeling, which sometimes extends for several months (adapted from Reference [14])

scales. As discussed in Chapter 2, the human body is amazing in healing wounds or injuries from cells to organs. Therefore, we can learn from Mother Nature to gain new insight into self-healing of synthetic polymers. Actually, a careful study of the review articles or books in References [1] to [28] reveals that almost all the existing schemes are bio-inspired. While biomimetic healing is our goal and progress has been made over the years, no existing system is truly biomimetic. One simple distinction is that biological systems need more than one step or more than one mechanism to heal wounds in a coordinated manner. The systems given in Table 6.1 are healing by one step or one mechanism. Therefore, a new generation of self-healing schemes needs a multistep and multimechanism strategy. Figure 6.6 shows a comparison between the self-healing process of a biological system and a synthetic system suggested in Reference [14]. It is noted that while Figure 6.6 shows a reduced time for the repair of the synthetic system, the man-made system is limited to narrow cracks.

Here, we would like to emphasize that we need to revisit the word 'self' in self-healing. From the language itself, 'self' means without external intervention or external help. Based on this rigorous definition, almost no existing scheme can be called 'self-healing' because more or less they need external help or intervention, such as heating in a thermoplastic based healing system. One may argue that the liquid healing agent contained in microcapsules or hollow fibers can perform 'self-healing' or autonomous healing. However, it is also limited to narrowly opened cracks. For wide-opened cracks, they need external help to narrow the crack first. Therefore, self-healing used in this book should be understood in the sense that minimal human intervention is needed during healing. For example, with a thermoplastic healing agent, the healing process is self-driven in that the flowing, wetting, diffusing, and entangling are without human intervention. Only external heating is needed. Also, while the current book is focused on restoring mechanical strength or load carrying capacity of polymer composite materials and structures, other physical, mechanical, and functional properties such as conductivity, color, toughness, etc., also need restoring or self-healing. It is expected that the healing efficiency will not be the same for different parameters to be recovered.

In this chapter, a biomimetic self-healing system based on a shape memory polymer matrix will be discussed and validated for healing a wide-opened structural-length scale crack (millimeter scale). The scheme is then extended to load carrying composite structures by repeated healing of a sandwich structure, a grid stiffened composite structure, and a 3-D woven fabric reinforced composite structure damaged by low velocity impact loading.

6.1   SMP Matrix Based Biomimetic Self-Healing Scheme

As discussed above, one immediate difference between biological and the existing synthetic healing mechanisms is that biological systems involve multiple-step healing solutions. For example, human skin healing processes rely on fast forming patches to seal and protect damaged skin before the slow regeneration of the final repair tissue takes place (Figure 6.6). Therefore, a multistep healing mechanism may be a new research direction. We have proposed a sequential two-step scheme by mimicking the biological healing process of, for example, human skin: close-then-heal (CTH). In CTH, the structural scale crack will be first narrowed or closed by a certain mechanism before the existing self-healing mechanisms such as micro-capsule or thermoplastic particles take effect. Thermally activated shape memory polymers (SMPs), if properly programmed or educated, can recover their original shape upon heating above their glass transition temperature (T_g), due to the autonomous, conformational entropy driven shape recovery mechanism. Because healing is to restore the original structural capacity and functionality while shape memory is to restore the original shape, we believe that the shape recovery functionality can be utilized to achieve the crack-closing purposes in CTH [44]. We design the system to work in such a way that the thermoset shape memory polymer (SMP) first fully closes or significantly narrows the crack opening by confined shape recovery (constrained expansion) (step 1) and then the thermoplastic particles melt and glue the two sides of the crack at molecular-length scale (step 2). Using an SMP based syntactic foam as an example, Li and Uppu [44] schematically demonstrated the working principles of the CTH scheme, as shown in Figure 6.7.

We envision that this CTH scheme will have the following special features:

1. Unique two-step healing mechanisms. As schematically shown in Figure 6.7, we start with an SMP based syntactic foam dispersed with a small amount of thermoplastic particles (Figure 6.7(a)). After that, we reduce the volume by constrained compression programming (e.g., 1-D compression programming in Figure 6.7(b)). After programming, a temporary shape or working shape is formed (Figure 6.7(c)). In Figure 6.7(d), a crack is created in the foam, for instance by impact. Figure 6.7(e) is the first-step healing (closing/narrowing the crack). The material is confined externally and the temperature is raised above T_{gs} (glass transition temperature of the SMP based syntactic foam). Due to shape memory function-ality, the material tends to recover its original shape with a larger volume as shown in Fig. 6.7(a). Due to the external confinement, however, the expansion is not allowed at the outer boundaries. As a result, the material is pushed toward the internal open space (crack) and tends to close/narrow the crack. Depending on the energy supplied during program-ming, that is, the amount of volume reduced, the crack can be fully closed or partially closed. Due to the cross-link nature of the thermoset SMP, molecular level continuity may not be established after this step even though the crack is closed at micro-length scale. It is noted that two aspects are critical for the success of this CTH scheme. One is volume

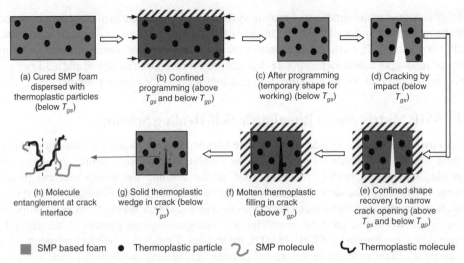

(a) Cured SMP foam dispersed with thermoplastic particles (below T_{gs})

(b) Confined programming (above T_{gs} and below T_{gp})

(c) After programming (temporary shape for working) (below T_{gs})

(d) Cracking by impact (below T_{gs})

(h) Molecule entanglement at crack interface

(g) Solid thermoplastic wedge in crack (below T_{gs})

(f) Molten thermoplastic filling in crack (above T_{gp})

(e) Confined shape recovery to narrow crack opening (above T_{gs} and below T_{gp})

▦ SMP based foam   ● Thermoplastic particle   ∿ SMP molecule   ∿ Thermoplastic molecule

T_{gs}:Glass transition temperature of the SMP based foam
T_{gp}: Glass transition temperature for amorphous thermoplastic particle or melting temperature for semi-crystalline thermoplastic particle

Figure 6.7 Schematic of the CTH scheme of the proposed smart foam: (a) to (e) complete closing and (f) to (h) ensure healing ($T_{gp} > T_{gs}$). *Source:* [44] Reproduced with permission, from Elsevier

reduction during programming and the other is external confinement during shape recovery. Without volume reduction by compression programming, constrained expansion cannot be achieved. It may be argued that tension programming can also close the crack by free shape recovery, which is true but does not work for actual structures because free shape recovery will change the geometry of the structures [45]. Therefore, we must run compression programming. Without external confinement, on the other hand, a free volume expansion may only partially close the crack or not close the crack at all. Therefore, compression programming and confined expansion are the two key criteria. One may argue that external confinement is impractical in real world structures. Indeed, external confinement can be coupled with architectural design of composite structures, such as sandwich structures where the fiber reinforced polymer skin provides out-of-plane constraint to the SMP core [46], grid stiffened SMP where the fibre reinforced polymer grid skeleton provides in-plane confinement [47], and 3-D woven fabric reinforced SMP where the 3-D fabric pre-form provides 3-D confinement [48]. Actually, if only the locally cracked volume is heated up for healing, the surrounding "cold" volume also provides confinement [49]. Therefore, external confinement is not difficult to realize. It is also noted that although a large strain can be recovered during the shape recovery process, the stress created is very small because the temperature is above T_{gs} and the foam is rubbery. This ensures that under the confined shape recovery, the adverse effect of the residual stress on the structure should be small. In Figure 6.7(f), the second-step healing starts. We first further raise the temperature above T_{gp} (the glass transition (for amorphous thermoplastic) or melting (for semicrystalline thermoplastic) temperature of the thermoplastic particles). The molten thermoplastic flows into the narrowed crack by capillary force and wets the two sides of the cracked surface. Because of the high mobility of the molten thermoplastic and concentration gradient from the thermoplastic healing agent to the SMP matrix, the thermoplastic molecules diffuse into the neighbouring thermoset SMP molecules and establish molecular level continuity. It is noted that, in order to speed up the diffusion, the temperature needs to be further increased up to the so-called bonding temperature, which is usually 15–20 °C higher than the melting temperature of the thermoplastic particles [33,34]. The material is then cooled down and the second healing step is completed (Figure 6.7(g)). Figure 6.7(h) schematically shows the entanglement of thermoplastic and thermoset SMP molecules at the crack interface after healing. The establishment of physical cross-links (or even some chemical cross-links through free radical reaction) completes the healing at molecular-length scale.

2. Autonomous, repeatable, efficient, controllable, and molecular-length scale healing. (a) The shape recovery process of the SMP is driven by conformational entropy through Brownian motion of the polymer segments and thus is autonomous. The only human intervention is to heat the material. If the foam becomes a conductor such as through incorporation of carbon nanotubes beyond the percolation threshold, the heating can be achieved by electricity or by a magnetic field if magnetic particles are embedded in the SMP matrix. (b) Because the SMP can experience many cycles of shape fixity and shape recovery and the thermoplastic particles can be melted and solidified many times, it is believed that the healing will be repeatable. (c) In the existing systems of healing thermosets by thermoplastics, the thermoplastics need to fully fill the cracks with a relatively wide opening. This inevitably needs a large amount of thermoplastic particles. In a study by Zako and Takano [50], up to 40% by volume of epoxy thermoplastic particles were used in order to fully recover the degraded stiffness of a glass fiber reinforced epoxy composite. In another study, it was found that up to 20% by weight of the thermoplastic was needed in order to achieve a healing efficiency of 70% in a thermoset resin with thermoplastic additives [51]. Due to the relatively wide opening of the crack, the thermoplastic within the crack behaves like the bulk, suggesting lower strength, lower stiffness, and lower healing efficiency. In the proposed two-step CTH scheme, the crack will be first closed or significantly narrowed by the confined expansion of the SMP matrix. Consequently, the thermoplastic within the crack behaves like a thin film, rather than the bulk, leading to higher strength, higher stiffness, and higher healing efficiency with only a small amount of thermoplastic particles, for instance 3–6% by volume [33,34]. (d) The shape recovery process can be controlled by varying the programming parameters such as 1-D, 2-D, 3-D, stress controlled, strain controlled, etc. Therefore, the healing can be controlled. (e) The healing is at the molecular level. Although the confined expansion may fully close the crack in the SMP matrix, the healing is at micro-length scale because the interaction or interdiffusion between the molecules at the crack interface is limited and the entanglement may be lacking. With the molten thermoplastic, its molecular mobility is very high and it can establish entanglement with the thermoset SMP matrix at the two sides of the crack. Thus molecular level healing can be achieved.

It is noted that in a recent review paper by Hu *et al.* [8], self-healing using the shape memory effect (SME) has been divided into two categories. One is unconfined shape recovery and the other is confined shape recovery. Rodriguez, Luo, and Mather [52] studied a blend system consisting of a cross-linked poly(ε-caprolactone) network (n-PCL) with linear poly(ε-caprolactone) (l-PCL) interpenetrating the network, which exhibited a combination of shape memory response from the network component and a self-healing capacity from the linear component. They called the self-healing using unconfined shape recovery as shape memory assisted self-healing (SMASH). Along the same lines as SMASH, Xiao, Xie, and Cheng [53] also observed healing using the so called reverse plasticity. We believe that these two approaches, CTH using confined shape recovery versus SMASH using unconfined shape recovery, have some fundamental differences although both work pretty well. The key for the success of the CTH approach is compression programming to reduce volume. Based on the definition of SMPs, the plastically deformed SMPs will restore their original shape (healing) upon external stimuli, as long as they are allowed free shape recovery. Therefore, it seems that SMPs do not need programming. This may be true on condition that (a) the shape recovery ratio is 100% and (b) the boundary of the SMP specimen is free (allows the specimen to go back to its original shape).

These fundamental requirements can be further explained as follows. If the shape recovery ratio is not 100%, the deformed shape cannot be fully recovered. Hence, the crack cannot be fully closed (healed). If the boundary of the specimen is not free, the free shape recovery is not fully allowed. Therefore, the shape cannot be fully recovered. In SMASH, although programming is not specified, it actually uses external load caused deformation as programming; that is, programming is coupled with external loading. In order to understand better why compression programming is essential in the CTH scheme for healing a wide-opened crack, let us examine Figure 6.8. Figure 6.8(a) shows an SMP rod with a microcrack at the center of the rod. One end of the rod is fixed and the other end is free and subjected to a tensile load, P. Figure 6.8(b) shows that the rod is tensile deformed and, after removal of the tensile load, there is a certain pseudo-plastic deformation with the crack further opened. Clearly, the tensile plastic deformation means that the rod is tensile programmed. After removal of the tensile load, the other end of the rod is also fixed. If we now heat the fixed–fixed rod in Figure 6.8(b), we will

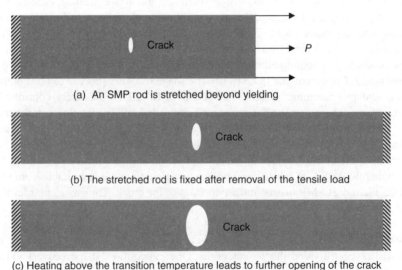

(a) An SMP rod is stretched beyond yielding

(b) The stretched rod is fixed after removal of the tensile load

(c) Heating above the transition temperature leads to further opening of the crack

Figure 6.8 Schematic of further crack opening during shape recovery (crack closing) process of tension programmed SMP rod specimen with constrained boundary condition

not close the crack. Instead, the crack may be further opened (see Figure 6.8(c)). The reason is that, when we heat the rod above the transition temperature, shape recovery will occur, that is, shrinkage of the rod. Due to the fixed boundary condition, however, this shrinkage is not allowed. As a result, a tensile stress (recovery stress) will occur in the rod. This tensile stress may cause further propagation of the crack. Additionally, the material surrounding the crack has been tensile programmed during the crack propagation process (actually programmed more than the materials away from the crack due to the tensile stress concentration at the crack tip). Owing to the free inner surface of the crack, the surrounding tensile programmed material tends to shrink, which may also cause further opening of the crack. Of course, if at least one end of the rod is free during the heating process, the crack will most likely be closed if the shape recovery ratio is 100%. Unfortunately, for real world structures, one structural component is usually constrained by its neighbours, that is, the boundary is not free, and thus the crack may not be closed. On the other hand, if the specimen is compression programmed before use and the amount of programming is higher than the tension deformation due to the external tension loading, the stored compression deformation may be sufficient to close the crack when the specimen is heated. This is because the pre-compression will counterbalance the tension deformation during shape recovery. If the pre-compression (programming) is higher than the tension by the external loading, constrained expansion will push the material towards the inner open space, that is, the crack. From this simple illustration, it is obvious that in real world load bearing structure applications, compression programming is the key for success.

The CTH approach is able to control the crack width, which can be closed by controlling the compression pre-strain level. A simple equation has been established by Li *et al.* [54] to correlate the crack width to be closed and the pre-strain level during compression programming. Cracks with different opening widths can be closed based on the level of compression programming (of course it is limited by the maximum allowed compression pre-strain level). In the unconstrained shape recovery approach such as the SMASH approach, it has not been demonstrated that it can close a wide-opened crack with constrained boundary. Also, it cannot control the crack width that can be closed because it depends on the external load to perform programming and relies on the shape recovery ratio and boundary condition of specimens to perform crack closing.

As shown in Figure 6.9, assuming that a piece of SMP is sandwiched in between two walls of a U-shaped rigid frame with full constraints at the two sides but free at the bottom, at the

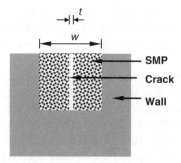

Figure 6.9   Schematic of an SMP with a crack to be closed by constrained shape recovery

temperature close to or above the glass transition temperature of the SMP, we can estimate the pre-strain level required to close the crack due to constrained shape recovery of the SMP:

$$\varepsilon = \frac{t}{Rw} \tag{6.1}$$

where ε is the required pre-strain to close the crack, t is the crack width, w is the width of the SMP, and R is the shape recovery ratio of the SMP. As an example, if the crack width is $t = 1$ mm, the SMP width is $w = 20$ mm, and the shape recovery ratio of the SMP is $R = 50\%$, the pre-strain level (compressive strain during programming) required to close the crack is $\varepsilon = 10\%$.

The CTH approach also has some other differences as compared to SMASH. For example, the SMASH approach focuses on non-load-bearing applications such as self-healing bladders, inflated structure membranes, architectural building envelopes, and coating [52,54]. On the other hand, the CTH approach focuses on load bearing structures. It also considers the boundary conditions to be encountered in real structure applications, as discussed above.

As indicated by Hu *et al.* [8], a difference between the SMASH approach and the CTH approach is that the CTH approach needs external confinement, while the SMASH does not need external confinement. While it seems that external confinement is an added requirement, as discussed above, it has been achieved through architectural design of composite structures such as a grid stiffened structure [47], a 3-D woven fabric reinforced composite structure [48], and a sandwich structure [46]. Therefore, external confinement is not difficult to maintain and repeated healing can be achieved by directly heating the structure to above the bonding temperature of the embedded thermoplastic particles. It is noted that Li, Ajisafe, and Meng [49] prove that heating can be conducted locally, surrounding the cracked region. In such a way, the needed external confinement in the CTH concept can be naturally provided by the surrounding unheated "cold region."

Here, we would like to further explain why the healing can be repeatable with only one time programming. It comes from the fact that each confined shape recovery actually serves dual purposes: one for self-closing internal cracks and the other for completing a new round of compression programming. This is because the confined shape recovery experiences heating of the SMP matrix above its T_g, applying a certain compressive stress to the matrix due to external confinement, and cooling down below T_g while maintaining the pre-strain, which is the process typical for strain-controlled programming. In other words, strain-controlled programming is coupled with confined shape recovery. Therefore, although it seems that only one programming is needed in Figure 6.7, each shape recovery actually has one prior programming to supply the energy. The only difference is that the subsequent programmings are automatically performed by coupling with each confined shape recovery. Therefore, in confined shape recovery, one "nominal" programming leads to several cycles of shape recovery.

We would like to emphasize that for healing a wide-opened crack, some may argue that we do not need the shape memory effect to close or narrow the crack first; we can directly inject strong adhesive into the crack. This seems a legitimate argument but it does not work on a number of occasions. For example, for laminated composite structures under transverse impact loading, the cracks are at the back surface, which is usually invisible and inaccessible for real world structures. In Figure 6.10, E-glass fibre reinforced epoxy laminated composite beams with a stacking sequence of $[0/90/0_6]_s$ and dimensions of 177.8 mm × 50.8 mm × 4.0 mm were

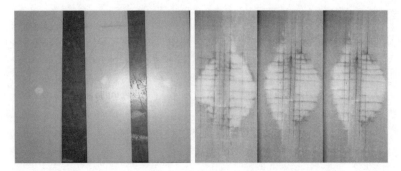

Figure 6.10 Low velocity impact damage in a laminated composite on the front surface (left) and the back surface (right)

impacted by a hammer with a weight of 35 N and velocity of 2 m/s. It is clear that the impacted surface (front surface) only shows a small indentation, which is visible, while significant cracking is found on the back surface, which is invisible. For sandwich structures, the crack may be in the core, which is both invisible and inaccessible for injecting. Also, it is well known that polymer is stronger as a thin film than as a bulk. Therefore, filling a wide-opened crack with a large amount of adhesive is not a wise choice. Consequently, closing or narrowing the crack first by constrained expansion and healing the defect with a thin film of thermoplastic, that is, the CTH scheme proposed in this book, is a better way of healing a wide-opened crack.

In the following, we would like to present a proof-of-concept study to validate this CTH scheme for healing wide-opened structural-length scale cracks [33,34].

6.1.1 Raw Materials, Specimen Preparation, and Testing

The materials used in this study included: *Veriflex*® polystyrene shape memory polymer (PSMP) from Corner Stone Research Group Inc. (glass transition temperature 62.0 °C, tensile strength 23 MPa, and modulus of elasticity 1.24 GPa at room temperature as provided by the manufacturer) and a thermoplastic polymer identified as copolyester (CP) from Abifor Inc., Switzerland (particle size $\leq 80\,\mu m$, density 1.3 g/cm^3, glass transition temperature determined by DSC as 17 °C and 70 °C, melting range 114–124 °C, and bonding temperature range 125–150 °C).

The *Veriflex*® PSMP is a two-part resin system. Part A is composed of styrene, divinyl benzene and vinyl neodecanoate. Part B is composed of benzoyl peroxide. The chemical structure for each component has been shown in Figure 3.2. The thermoplastic copolyester (CP) is composed of isopthalic acid, terepthalic acid and butane-1,4-diol. The chemical structure for each component is shown in Figure 6.11.

The experimentation process started with preparation of the composite. The shape memory resin was preheated for two hours at 75 °C just before the thickening and curing process began. This was conducted in order to prevent the denser CP particles from settling to the bottom of the resin. After that, the CP particles (3%, 6%, and 9% by volume) were dispersed in a beaker containing the preheated resin. The mixture was mixed to uniformity and poured into a steel mold with dimensions of 300 mm × 300 mm × 12.5 mm. The mold was sealed and the material was cured in an oven as follows: 75 °C for 12 hours, 90 °C for 3 hours, and 112 °C for 3 hours.

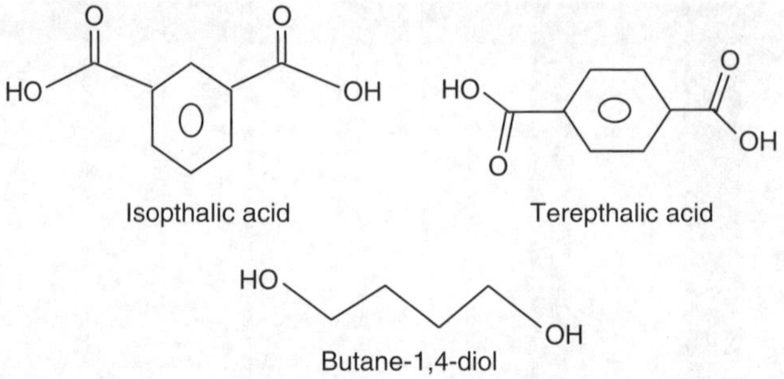

Isopthalic acid   Terepthalic acid

Butane-1,4-diol

Figure 6.11 Chemical structure of each component in CP. *Source:* [34] Reproduced with permission from Elsevier

Once the curing procedure was complete, the setup was cooled down and demolded to obtain the copolyester modified polystyrene shape memory polymer (CP-PSMP) composite.

Single-edge notched bend (SENB) specimens with dimensions of 120 mm × 25 mm × 12.5 mm were fabricated by machining the cured composite slab and programmed in compression in the length (longitudinal) direction in an adjustable metal frame on a CARVER Model 2697 compression molding fixture. The classical way of programming was used. The beam specimen was compressed at 80 °C to the designed pre-strain in the longitudinal direction. The strain was held while cooling down to room temperature. After that the load was removed and the compression programming was completed. For specimens with a CP content of 3% and 6%, the pre-strain level was 6.7%. For specimens with 9% CP content, the pre-strain levels used were 2.7%, 4.7%, 6.7%, 8.7%, and 10%, in order to evaluate the effect of the pre-strain level on the healing efficiency. After programming, the SENB specimens with notch dimensions of 1 mm wide and 12.5 mm deep were fabricated according to the ASTM D 5045 standard. One reason for using the SENB specimen is that the notch in the specimen served as structural-length scale damage (where two crack surfaces are not in contact). As such, the ability of the CTH scheme to close and heal the crack (notch) can be studied. Another reason is to determine the fracture toughness of the composite. Figure 6.12(a) is a schematic explanation of the specimen preparation, programming, and notching process.

In order to determine the effect of the CP content on fracture toughness of the composite, the SENB specimens were tested in three-point bending on a universal MTS 810 testing system. The span length was 100 mm and the loading rate was 10 mm/min following the ASTM D 5045 standard. The specimens were tested and fractured completely into two halves. This test was also used to determine the peak bending loads of the virgin unnotched control specimens and healed specimens in order to determine damage healing ability.

Following the two-step CTH self-healing scheme proposed by Li and Uppu [44], fractured specimens were placed in the adjustable rectangular steel frame as shown in Figure 6.12(b) at room temperature. Once the specimen was fitted into the frame, the frame was placed on the compression molding fixture, which was preheated to 150 °C. It is noted that the steel frame provided confinement in the length and width directions during heating (in-plane confinement). The top surface of the framed specimen was close to the top heating plate but not in direct contact. This facilitated uniform heating to the specimen without applying confinement in the thickness direction. Therefore, only 2-D in-plane confinement was used during healing.

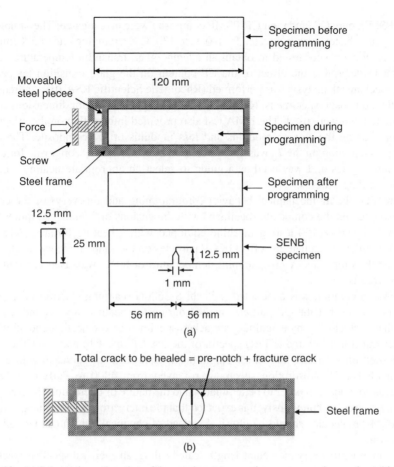

Figure 6.12 (a) Schematic explanation of the specimen preparation, programming, and notching process and (b) schematic of a fractured specimen in the steel frame ready for the two-step healing (top view). *Source:* [34] Reproduced with permission from Elsevier

The specimens were kept within the preheated fixture for 20 minutes. During this time period, the temperature within the specimen was gradually increased, first passed the T_g of the PSMP, which caused shape recovery or closing of the fractured surface and the pre-notch, and then caused melting of the CP, finally coming to the bonding temperature of the CP (between 125 °C and 150 °C) so that the CP molecules diffused and entangled with the PSMP molecules. After 20 minutes, the heating plates were tuned off and the specimen was cooled down to room temperature. This completed the two-step CTH self-healing scheme. Subsequent healing after fracture was conducted in a similar manner.

6.1.2 Characterizations of the Composite Materials

A differential scanning calorimetry (TA Instruments, Q100) test was performed in order to investigate the compatibility between the PSMP and CP. The glass transition temperatures of the pure PSMP, pure CP, and their CP-PSMP composite were investigated. The sample weight was 6.5 mg and the test was conducted from 0 °C to 130 °C at a ramping rate of 5 °C/min. Three effective specimens were tested to obtain an average glass transition temperature value.

In order to determine the effect of the CP content on the glass transition temperature, a dynamic mechanical analysis was performed (Rheometic Scientific RSA III) at a frequency of 1 Hz on the composite specimens. Rectangular tension specimens with dimensions of 36 mm × 11.5 mm × 2 mm were used. The DMA test also provided information on the effect that the CP content has on the storage modulus and loss modulus of the composite. The effect of repeated programming on the T_g was also studied on a 6% CP-PSMP composite for up to five programming cycles and was used as a guide to select an appropriate healing temperature during the healing process.

Compression thermomechanical behavior (programming and recovery) of the composite was investigated on the composite specimens with dimensions of 25 mm × 25 mm × 12.5 mm using a MTS QTEST 150 testing machine equipped with a heating furnace (ATS heating chamber) in order to evaluate the effect of CP particles on the shape memory functionality of the CP-PSMP composite. A programming temperature of 80 °C was used according to the DMA test results.

An SEM observation was conducted (Hitachi S-3600N scanning electron microscope) in order to (1) verify that the CP particles melted during the healing process and (2) visually validate the crack closing by examining a cracked specimen before healing and after healing.

Fourier transform infrared (FTIR) spectra of the 3% CP, 6% CP, and 9% CP composites were recorded on a Bruker Tensor 27 single-beam instrument at 16 scans with a nominal resolution of 4 cm^{-1}. Absorption spectra were saved from 4000 to 700 cm^{-1}. The FTIR analysis was conducted in order to better understand the nature of interactions between the two polymers. As discussed previously, it is expected that the interaction between the CP molecules and the PSMP molecules should be physical instead of chemical, which can be validated by FTIR spectra.

Finally, in order to verify molecular-length scale healing, an energy dispersive spectroscopy (EDS) (15 kV, super-ultrathin window (SUTW) Saphire detector, amplification process time (AMPT) of 25.6 μs) analysis was conducted using a Hitachi 3600 N scanning electron microscope equipped with an EDAX genesis detector. The rationale was that if the CP molecules diffused into the PSMP matrix, the chemical composition near the interface will show a certain gradient. In this study, a specially prepared EDS specimen was used. To prepare the EDS specimen, a SENB specimen made of pure PSMP was fractured and a very thin layer of CP was placed in between the fractured surfaces. Next, the EDS specimen was healed as described in Section 6.1.1. An EDS analysis was conducted at and around the healed interface of the EDS specimen.

6.1.3 Results and Discussion

6.1.3.1 DSC Test Results

As discussed previously, the thermoplastic healing agent and the thermosetting SMP matrix need to have a certain miscibility in order to have a higher healing efficiency. If the two materials do not have any miscibility, it will be difficult for the thermoplastic molecules to diffuse into the thermosetting SMP matrix and establish physical entanglement. If the two materials are fully miscible, they will become a uniform mixture at a certain temperature, which

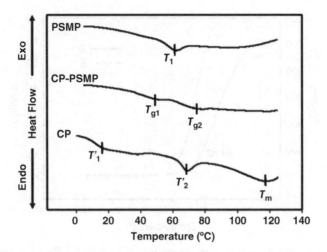

Figure 6.13 Typical DSC thermograms of CP, SMP, and CP-PSMP with 6% CP. *Source:* [34] Reproduced with permission from Elsevier

is clearly not desired. An ideal case is where the two materials have partial miscibility. In order to investigate the miscibility of the two polymers (PSMP and CP) within the composite, the single glass transition criterion [56] was adopted. Based on this criterion, PSMP and CP are compatible if the composite shows a single glass transition temperature that is in between the glass transition temperatures of the pure PSMP and CP. Figure 6.13 shows the DSC thermograms of PSMP, CP and CP-PSMP. The PSMP shows a single glass transition temperature ($T_1 = 62\,°C$). The CP, on the other hand, shows two glass transition temperatures ($T'_1 = 17\,°C$ and $T'_2 = 70\,°C$) and a melting temperature T_m ($118\,°C$), which is within the melting range of the CP (114–$124\,°C$ as provided by the manufacturer). Obviously, the CP used in this study is a copolymer as provided by the manufacturer. Between T_1 and T'_1, a single glass transition temperature T_{g1} ($50\,°C$) is observed for the CP-PSMP composite. This indicates some degrees of compatibility between the PSMP and one component of the CP copolymer. The CP-PSMP composite also shows a second glass transition temperature $T_{g2} = 72\,°C$. This should be an indication of the effect of the other component of the CP copolymer. Because the PSMP does not have the corresponding second glass transition temperature, the PSMP does not have compatibility with the second component of the CP copolymer. Therefore, it is concluded that the PSMP and CP only have partial compatibility. However, as will be shown in the DMA test results, the concentration of the second component in the CP copolymer may be very small. Therefore, the PSMP has a certain compatibility with the major component of the CP copolymer.

6.1.3.2 DMA Test Result

Figure 6.14 shows typical storage modulus (E')/loss modulus (E'')–temperature plots of the CP-PSMP composite. By taking the T_g as the temperature corresponding to the peak of the loss modulus, it is found that the CP-PSMP composite shows a single T_g at $54\,°C$. No other transition is observed above $54\,°C$. This suggests that the component causing T_{g2} ($72\,°C$),

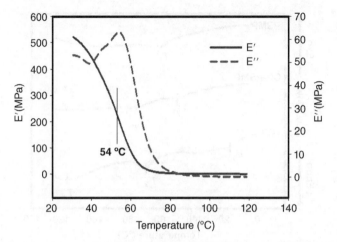

Figure 6.14 Storage modulus/loss modulus–temperature plots of the CP-PSMP composite with 6% CP

which is observed through the DSC test, is the second component in the CP copolymer and may only have a small concentration. In other words, the major component of the CP copolymer has a certain compatibility with the PSMP. It is noted that the T_g (54 °C) from the DMA test is 4 °C higher than T_{g1} (50 °C) from the DSC test. This trend has been reported before in another study [45]. Based on the T_g (54 °C), the programming temperature was taken as 80 °C, which was well above the T_g.

For comparison purposes, the DMA test results of the CP-PSMP composite with 3% and 9% CP are shown in Figure 6.15. From Figures 6.14 and 6.15, the glass transition temperature of the composite reduces as the CP content increases. This can be understood from the point of view of the rule-of-mixtures [52] because the major component of the CP copolymer has a lower glass transition temperature than that of the PSMP. Based on this result, a programming temperature of 80 °C was selected, which is well above the glass transition temperatures of the

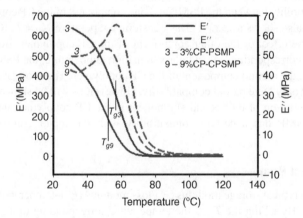

Figure 6.15 Typical plots of storage modulus (E') and loss modulus (E') with temperature for the CP-PSMP composite with 3% and 9% CP

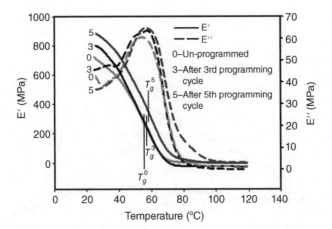

Figure 6.16 Typical plots of storage modulus and loss modulus with temperature for a 6% CP-PSMP composite specimen showing variations of T_g with programming cycle

composites, regardless of the CP content. From Figure 6.15, it can also be seen that the storage modulus of the composite with 3% CP is higher than that of the composite with 9% CP. This is consistent with the rule-of-mixtures as CP is softer than PSMP in the temperature range used.

For SMPs, thermomechanical cycles may change the physical/mechanical/shape memory properties. The reason is that each thermomechancial cycle may cause some ageing, damage, and unrecoverable plastic deformation. Figure 6.16 shows typical plots of the storage modulus and loss modulus with temperature for a 6% CP-PSMP composite specimen. The glass transition temperatures T_g^0, T_g^3, and T_g^5 correspond to the zero (unprogrammed), third and fifth thermomechancial cycles (one programming and one recovery cycle is called one thermo-mechancial cycle), respectively. From Figure 6.16, it is seen that the glass transition temperature of the composite increased slightly with the programming cycle. The T_g values were respectively $T_g^0 = 54\,°C$, $T_g^3 = 56\,°C$, and $T_g^5 = 57\,°C$. This result shows that for up to five programming cycles, $T_g^5 = 57\,°C$ is still well below the programming temperature of $80\,°C$. Using this result as a guide, we deduce that the increase in T_g for the 3% CP-PSMP and 9% CP-PSMP composites for the first five repeated programming cycles should also be very small. Thus, the programming temperature of $80\,°C$ was used for all the specimens in this study. It is noted that, in Figure 6.16, the storage modulus increases as the thermomechanical cycle increases. The reason may be that, with the increase in thermal cycles, the composite experienced longer lengths of time at a higher temperature, which facilitated diffusion of the CP into the PSMP matrix, leading to better physical entanglement between the CP and PSMP, and thus a higher stiffness. Another possible reason is that, with repeated thermal cycles, the PSMP molecules align more towards the loading direction, leading to an increase in stiffness in the loading direction.

6.1.3.3 Thermomechanical Behavior

Figure 6.17 shows a typical 3-D stress–strain–temperature profile of the CP-PSMP composite with 6% CP. Figure 6.17 depicts programming (ABCD), 1-D fully confined recovery (DEF)

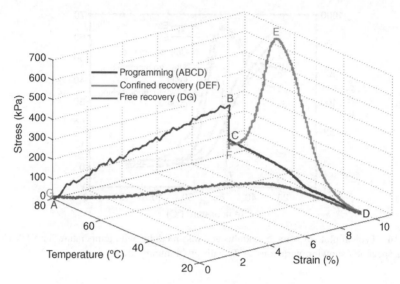

Figure 6.17 Typical thermomechanical behavior in terms of stress–strain–temperature of the CP-PSMP composite with 6% CP during programming (ABCD), confined stress recovery (DEF), and free strain recovery (DG). *Source:* [34] Reproduced with permission from Elsevier

and free recovery (DG) processes. Specimens were heated to 80 °C (at point A), compressed (strain-controlled mode) in the thickness direction to 10% pre-strain level (point B, loading rate of 1.3 mm/min), held at 80 °C for one hour to stabilize the stress (point C) and cooled down to room temperature (CD). While confined in the thickness direction, the specimens were heated back to 80 °C at an average rate of 0.5 °C/min in order to determine the recovery stress of the composite (DEF). The specimens were held at 80 °C for one hour to stabilize the stress (point F). Free shape recovery was also conducted. Programmed specimens were re-heated to 80 °C (0.18 °C/min) in an oven without applying any stress to determine the free shape recovery ability of the composite by measuring displacement in the thickness direction with change in temperature (DG). A linear variable differential transducer (Cooper Instruments LDT 200 series LVDT) was used to measure the displacement while the temperature was measured and recorded with a thermocouple instrument (Yokogawa Model DC-100). The average programming stress (at point C) was found to be (74 ± 2 kPa) while the average recovery stress (point F) was (45 ± 1 kPa). The shape fixity from the programming curve was found to be 98% while the shape recovery ratio based on the free shape recovery test was 98%, indicating good shape memory functionality of the composite. Again, here we show that even for an SMP based composite, the stress recovery ratio is lower than the strain recovery ratio, indicating good shape memory but poor stress memory.

Similar thermomechanical tests were conducted for the CP-PSMP composites with 3% and 9% CP, as shown in Figures 6.18 and 6.19, respectively. From Figures 6.17 to 6.19, it is seen that as the CP content increases, the peak recovery stress (point E) and stabilized recovery stress (point F) decreases. This is understandable because the PSMP is stiffer than the CP at the glassy temperature, leading to a higher peak recovery stress with a higher PSMP content. The reason is that the peak recovery stress is basically an accumulation of thermal stress due to

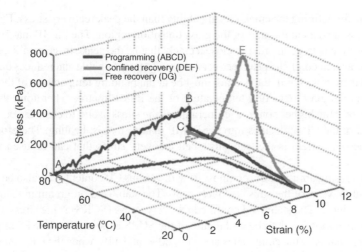

Figure 6.18 Typical thermomechanical behavior of the CP-PSMP composite with 3% CP. *Source:* [33] Reproduced with permission from IOP Publishing

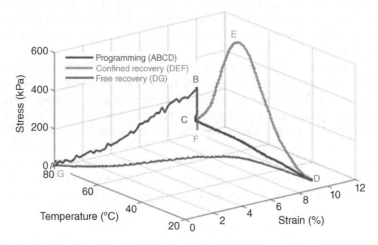

Figure 6.19 Typical thermomechanical behavior of the CP-PSMP composite with 9% CP. *Source:* [33] Reproduced with permission from IOP Publishing

thermal expansion of the composite specimen before the glass transition temperature is reached. For the stabilized recovery stress, it depends on the shape memory capability of the composite. Because the CP does not have a shape memory capability, the stabilized recovery stress reduces as the CP content increases. In the CTH approach, because the recovery stress is the driving force to push the cracked PSMP matrix in contact, a higher recovery stress is desired to close the wide-opened crack. Therefore, the ability for the composite to close cracks decreases as the CP content increases.

In Figures 6.17 to 6.19, we defined two recovery stresses – the peak recovery stress and the stabilized recovery stress. In the CTH approach, we believe that the stabilized recovery stress is more suitable for defining the crack-closing ability than the peak recovery stress. First, the peak recovery stress appears at the start of the glass transition region. The PSMP matrix is stiff and the stress required is large in order to close a crack; that is, the peak stress is unlikely to be able to close the crack. Second, the peak recovery stress is not sustainable and reduces quickly as the temperature increases. Third, the healing must be at the bonding temperature of the CP, which is higher than the glass transition temperature of the PSMP matrix. Therefore, the sustained recovery stress at the higher bonding temperature is the stress useful to close the crack and help the diffusion of the CP molecules into the PSMP matrix, that is, healing. Therefore, only the stabilized recovery stress is the true, sustained recovery stress that can be used in the CTH approach.

The crack-closing ability of the composites with various CP contents can also be understood from the point of view of the stress recovery ratio. The stabilized programming stress (SPS) is the stress at point C. The stabilized confined recovery stress (SCRS) is the stress at point F. If the stress recovery ratio is defined as the ratio of the SCRS at point F to the SPS at point C, the stress recovery ratios of the composites are 66%, 60%, and 44% when the CP contents are 3%, 6%, and 9%, respectively. Therefore, we again see that the crack closing capability reduces as the CP content increases. However, this does not mean that the less of the CP content the better of the healing efficiency. The reason is that the CTH needs two steps, close and then heal. The above conclusion is based on the first step. For the healing step, it is obvious that the CP content must at a certain level. Otherwise there is no sufficient CP to fill in the narrowed crack and diffuse into the fractured PSMP matrix, that is, no healing. Therefore, an optimized CP content exists in practice.

Figure 6.20 shows the evolution of recovery strain with time for the composite with CP contents of 3%, 6%, and 9%, respectively. The shape recovery (ability for the composite to recover the strain at point G) is larger than 98% for all three groups, indicating that the CP-PSMP composite retained its shape recovery ability and shape memory functionality with various CP contents. However, it can be seen that after the "lift-off" point L, 3% CP composite shows a steeper and sharper slope of recovery (higher recovery rate) as compared to the other two specimens. This indicates that the speed of recovery or shape recovery rate reduces with increasing CP content. For instance, when the time is equal to 3.5 hours, the 3% CP composite has already recovered 86% of the strain while the 6% and 9% specimens have only recovered

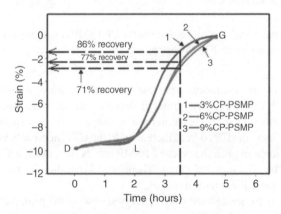

Figure 6.20 Typical recovery strain with time. *Source:* [33] Reproduced with permission from IOP Publishing

77% and 71% of the same property, respectively. Again, this is understandable because the CP does not have a shape memory ability so that the more the CP content is, the slower the shape recovery.

It is interesting to note that, while the CP content has a significant effect on the stress recovery, its effect on shape recovery is comparatively small. The reason is that shape recovery is basically a global or macroscopic behavior, while stress recovery is dependent more on the microstructure and internal parameters. For instance, the recovery stress depends on the stiffness at the recovery temperature, while the recovery strain does not. Therefore, the shape recovery ratio and the stress recovery ratio usually do not have the same value and the stress recovery ratio is usually lower than the strain recovery ratio. In this sense, the stress recovery ratio is a more rigorous indicator of shape memory functionality.

6.1.3.4 Damage Healing Ability Test Results

Figure 6.21 shows typical load deflection curves of (a) PSMP, (b) 3% CP-PSMP, (c) 6% CP-PSMP, and (d) 9% CP-PSMP. The peak bending loads (average of three specimens) of the originally un-notched, notched, and healed PSMP specimens were 1385 ± 4 N, 358 ± 2 N and 250 ± 3 N, respectively. These numbers were, respectively, 1260 ± 4 N, 321 ± 3 N, and 645 ± 2 N for 3% CP-PSMP composite, 1060 ± 3 N, 230 ± 2 N, and 693 ± 3N for 6% CP-PSMP

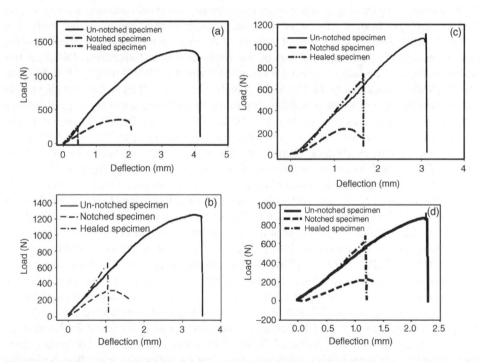

Figure 6.21 Typical load deflection curves of original, notched, and healed specimens after the first healing cycle: (a) PSMP. *Source:* [34] Reproduced with permission from Elsevier; (b) 3% CP-PSMP. *Source:* [33] Reproduced with permission from IOP Publishing; (c) 6% CP-PSMP. *Source:* [34] Reproduced with permission from Elsevier; and (d) 9% CP-PSMP. *Source:* [33] Reproduced with permission from IOP Publishing

composite, and $847 \pm 3N$, $220 \pm 2N$, and 631 ± 3 N for 9% CP-PSMP composite. From Figure 6.21, the following observations can be made. (1) The peak bending load decreases as the CP content increases. This is understandable because, first, as the CP has a lower strength than the PSMP matrix, the particulate composite should have a lower strength by the rule-of-mixture principle and, second, the stiffness of the CP particles is lower than that of the PSMP matrix at the test temperature. Hence the CP particles serve as stress concentration centers. Stress concentration unavoidably reduces the composite strength. (2) Notching significantly reduces the bending strength. (3) Using crack closing by confined shape recovery only (step 1 in the CTH approach), which is the case in pure PSMP, the peak bending load of the resulting "healed" specimen is lower than the peak bending load of the notched specimen, indicating that the pre-crack (notch) and newly created crack (complete fracture of the notched beam during the bending test) could not be healed by this step alone. The reason is that the constrained volume expansion can only narrow or close the crack, but not heal it at molecular-length scale.

Several approaches have been used to define the healing efficiency of self-healing materials such as the ratio of peak stress, peak strain, stiffness, or other properties of the healed specimen over the original undamaged control specimen [23]. It is noted that the healing efficiency should be determined using some physical constants of the virgin and healed specimens, such as fracture toughness [57,58]. However, because the notch was also healed in our study [33,34], the fracture toughness of the healed specimens could not be determined. Therefore, the definition used in this study was an arbitrary nonphysical definition of healing efficiency solely for the purpose of discussing trends in behavioral changes. To avoid confusion with the widely used terminology "healing efficiency," we used "damage healing ability" in our study. The damage healing ability is characterized in terms of the ratio (percentage) of the peak bending load of the healed specimen over that of the virgin un-notched specimen. Based on this definition, the damage healing ability for the pure PSMP, 3% CP-PSMP, 6% CP-PSMP, and 9% CP-PSMP is, respectively, 18.1%, 51.2%, 65.4%, and 74.4%. If the damage healing ability is defined as the ratio of the peak bending load of the healed specimen to the original notched specimens, these numbers become 69.8%, 200.9%, 301.3%, and 286.8%, respectively. The damage healing ability over 100% suggests that, using the notched specimens as the basis for comparison, both the newly created crack and the original notch are healed using the CTH scheme. The notched specimens before and after healing are shown in Figure 6.22.

Basically, the damage healing ability increases as the CP content increases. This is simply because the narrowed crack can be fully filled with a higher CP content. However, this does not mean that 9% CP is better than 6% or 3% CP. The reason is that the healing ability is calculated based on a different basis. If the healing ability is calculated on the same basis, for instance, the peak bending load of the unnotched PSMP specimens, the healing ability for PSMP, 3% CP-PSMP, 6% CP-PSMP, and 9% CP-PSMP is, respectively, 18.1%, 46.5%, 50.0%, and 45.6%. Obviously, 6% CP achieves the highest healing efficiency. Oversaturating the system with CP does not lead to a higher healing efficiency, nor result in a higher bending strength before damage. Considering that the CP-PSMP composite with 6% CP also has the highest healed strength, it is concluded that 6% CP should be close to the optimal CP content for this material system. Therefore, the advantage of the CTH approach is that a small amount of healing agent can achieve the maximum healing efficiency. This shows a distinct departure from healing by thermoplastic particles alone, where a higher thermoplastic particle content usually leads to a higher healing efficiency simply because it does not have the crack closing step and thus a higher content of the thermoplastic is needed to fill in the wide-opened crack.

Figure 6.22 A group of completely fractured 6% CP-PSMP notched specimens (left) before healing and (right) after healing, showing disappearance of the notch after healing

It is interesting to note that healing of the CP-PSMP composites is quite repeatable. From Figure 6.23, it can be seen that for five fracture/healing cycles, the proposed CTH healing mechanism as tested through CP-PSMP specimens is reasonably repeatable, for all three CP contents.

In order to investigate the effect of pre-strain level on the healing ability, five pre-strain levels, 2.7%, 4.7%, 6.7%, 8.7%, and 10%, were used on the 9% CP-PSMP composite. Figure 6.24 shows the change of the peak bending load of the CP-PSMP composite after the first healing for each pre-strain level. For convenience, healing efficiency or ability (peak load) at zero pre-strain is assumed to be zero because the specimen is completely fractured. It is seen that the peak bending load of the healed specimen increases with the pre-strain level before levelling off at 6.7% pre-strain. Further increasing the pre-strain (8.7% and 10%) does not increase the peak bending load in the healed specimen. This indicates that increasing the

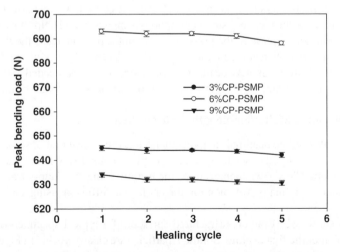

Figure 6.23 Average peak bending load with fracture–healing cycles for different CP contents. *Source:* [33] Reproduced with permission from IOP Publishing

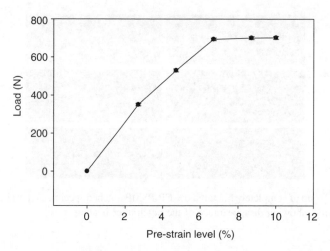

Figure 6.24 Effect of the pre-strain level on the peak bending load of the healed specimens with 9% CP. *Source:* [33] Reproduced with permission from IOP Publishing

pre-strain level further has no considerable effect on the damage healing ability. The reason is that the pre-strain serves two purposes: (a) it closes the crack through confined shape recovery and (b) it applies a compressive stress to help diffusion of the thermoplastic molecules. Based on the length of 120 mm and notch width of 1 mm of the SENB specimen, it can be calculated from Equation (6.1) that, if the shape recovery ratio is 100%, 0.83% pre-strain is sufficient to close the 1 mm wide notch (the fractured surfaces come to touch without any contact force). Therefore, the 10% pre-strain may not further serve the purpose of closing the crack. However, the recovery stress as shown in Figure 6.19 is very small. Because the "theoretical" pre-strain of 0.83% cannot apply any compressive stress at the crack surface, diffusion of the thermoplastic should be very slow, leading to a lower bonding strength and a lower damage healing ability. Thus, increasing the pre-strain level up to 6.7% sees a continuous increase in the damage healing ability because the recovery stress continuously increases and helps the diffusion of the CP molecules. Further increase in the pre-strain does not further increase the damage healing ability because the CP is oversaturated (the narrowed crack can only suck in a certain amount of molten CP and 6.7% pre-strain may be the maximum recovery stress needed to help diffuse the CP within the narrowed crack), leading to a levelling-off of the damage healing ability.

6.1.3.5 Validation of Molecular-Length Scale Healing

In order to validate that the healing is at molecular-length scale, we first observe the healing at micro-length scale using the scanning electron microscope (SEM) and the transmission electron microscope (TEM); we then examine the healing using the Fourier transform tnfrared (FTIR) technique and finally we monitor the molecule diffusion using energy dispersive spectroscopy (EDS).

Figure 6.25 shows SEM pictures (fractured surface) of a typical specimen (a) after initial fracture and (b) after the first healing. Solid CP particles are clearly visible in Figure 6.25(a). In Figure 6.25(b), solid particles are not seen but traces of melted, deformed, and debonded CP particles are identified. Also, the surface in Figure 6.25(b) is smoother compared to that in

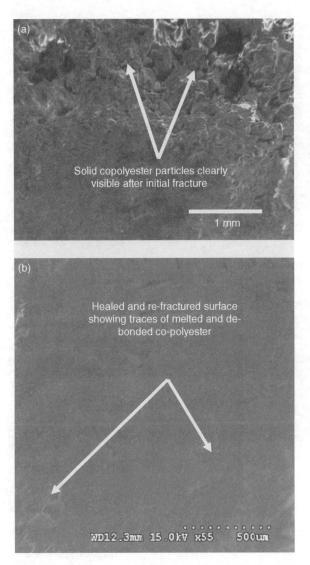

Figure 6.25 SEM images showing fractured surfaces of a typical 6% CP-PSMP composite (a) after initial fracture and (b) after the first healing cycle. *Source:* [34] Reproduced with permission from Elsevier

Figure 6.25(a). This is due to the compressive stress exerted on the surface during the crack-closing step (step 1 of the CTH approach), resulting from the shape memory effect by constrained shape expansion. This was necessary in order to keep both crack surfaces in intimate contact during the subsequent crack healing process (step 2 of the CTH scheme) via melting, sucking, wetting, diffusion and bonding of CP molecules across the crack interface.

Figure 6.26 shows the top surface view of SEM pictures of a typical specimen (a) after initial fracture and (b) after healing. In Figure 6.26(a), a crack originating from the end of the pre-notch is clearly visible. In Figure 6.26(b), the notch and the crack disappear after the two-step healing. The zoomed-in SEM picture shows good interfacial bonding.

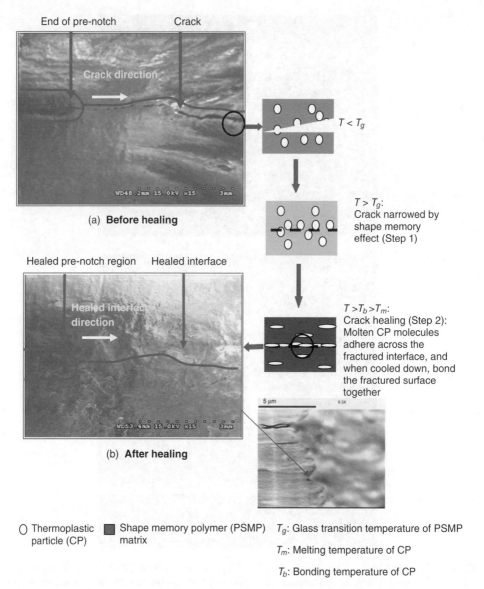

Figure 6.26 Top surface view SEM pictures of a typical specimen (a) after initial fracture and before healing and (b) after healing. *Source:* [34] Reproduced with permission from Elsevier

Figure 6.27 shows typical FTIR spectra of the PSMP, CP, and the particulate composite with 3%, 6%, and 9% CP contents. For the PSMP and CP, no significantly visible peaks are observed between 4000 cm^{-1} and 3000 cm^{-1} wavelengths. However, both PSMP and CP show peaks at 2800 cm^{-1}. From 2800 cm^{-1} down to 2000 cm^{-1}, CP shows no peak while PSMP shows a small peak at 2400 cm^{-1}. Both CP and PSMP show unique patterns from 1700 cm^{-1} to

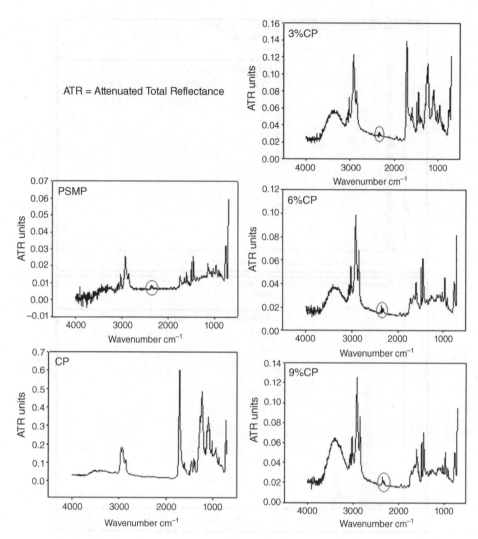

Figure 6.27 FTIR spectra of the pure CP, pure PSMP, and their composites. *Source:* [33] Reproduced with permission from IOP Publishing

$1000 \, \text{cm}^{-1}$. For the three particulate composites, a superposition of the individual CP and PSMP patterns is seen. For instance, the small peak at $2400 \, \text{cm}^{-1}$ on the PSMP spectrum (circled) is reflected on each of the CP-PSMP spectra. No new or significant peaks are identified on the spectra of the CP-PSMP composite, suggesting that no new functional groups or components are formed. This indicates that no chemical interactions or reactions occur between the PSMP and CP molecules. Thus, healing is a result of physical interactions between the two polymers, that is, diffusion of the CP molecules into the PSMP matrix, which will be validated by the EDS test.

Figure 6.28 shows EDS spectra of (a) pure PSMP and (b) pure CP with percentage counts of carbon and oxygen in the K energy level. The spectra were obtained to serve as baseline data.

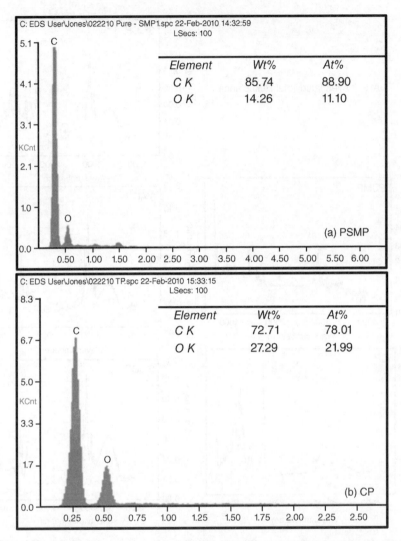

Figure 6.28 EDS spectra of (a) pure PSMP and (b) pure CP. *Source:* [34] Reproduced with permission from Elsevier

Top surface view SEM pictures showing the healed interface of the EDS specimen are presented in Figure 6.29 (left). The EDS analysis was performed at three points: directly at the interface, 0.1 mm away from the center of the interface, and 0.3 mm away from the center of the interface. EDS spectra corresponding to the analyzed points were recorded as shown in Figure 6.29 (right). Figure 6.30 shows the variation of carbon and oxygen counts in terms of weight percent with distance away from the interface. It is observed that the carbon count increases away from the interface (pure CP) while the oxygen count reduces in the same direction as the material approaches pure PSMP.

From Figures 6.28 (b) and 6.30, the oxygen and carbon counts at the interface are slightly different from the pure CP. The reason is that the counts obtained from the box immediately

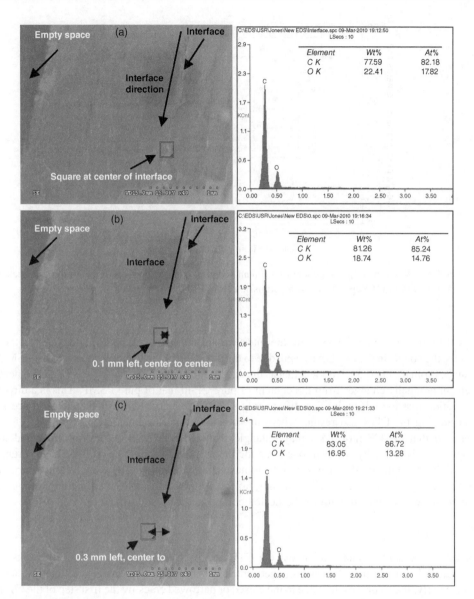

Figure 6.29 Top surface view SEM pictures (left) of the EDS specimen showing a healed interface, three analysis points ((a) at the center of the interface, (b) 0.1 mm left of interface and (c) 0.3 mm left of interface) and respective EDS spectra (right). *Source:* [34] Reproduced with permission from Elsevier

above the CP interlayer in the composite specimen were an average from an interaction volume, which includes the CP interlayer and may also include a small portion of the neighboring PSMP matrix. However, the interaction volume did not affect the composition in the box 0.1 mm and box 0.3 mm away from the interface. Based on Potts [59], the width of the interaction volume was about 3.4 μm. This suggests that the interaction volume for the

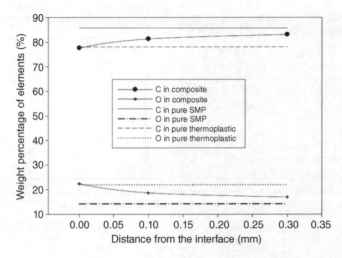

Figure 6.30   Variation of a component (oxygen and carbon) count with distance away from the healed interface. *Source:* [34] Reproduced with permission from Elsevier

box 0.1 mm and 0.3 mm away from the interface did not include the CP interlayer. In other words, the counts in the two boxes were due to the diffusion of the CP molecules into the PSMP matrix. From Figure 6.30, it is clear that as the distance from the interface increases, the element content gradually changes from the content of pure CP to that of pure PSMP. This is a direct indication that the CP molecules diffuse into the PSMP matrix during the healing process. Considering the FTIR test results where it was shown that the interaction is physical, it is believed that the CP molecules may entangle with the PSMP molecules, as illustrated in Figure 6.7. This is why we mentioned that the thermoplastic healing agent must have certain miscibility with the PSMP matrix in order to have a higher healing efficiency. Without certain compatibility, physical entanglement may not be formed and interfacial bonding strength, which determines the strength of the healed composite, may be low.

6.1.4 Summary

Based on the above systematic study, the CTH approach in healing wide-opened cracks is validated. It proves that the wide-opened crack can be healed repeatedly, efficiently, and molecularly. It also proves that, due to the closed or narrowed crack by the first step of the CTH approach, only a small amount of thermoplastic healing agent is needed to achieve optimal healing efficiency. Oversaturating the system with thermoplastic particles will not only reduce the overall mechanical properties of the composite but also reduces the healing efficiency. This is an obvious advantage as compared to self-healing using a thermoplastic healing agent alone, which needs a large amount of thermoplastic particles.

Next, we will present applications of the CTH scheme in several real-world composite structures. They are fiber reinforced syntactic foam cored sandwich structures, sandwich structures with a continuous fiber reinforced grid stiffened syntactic foam core, and 3-D woven fabric reinforced polymer composite structures. Of course, the core or the polymer composite is based on the same thermosetting PSMP as discussed above.

6.2 Self-Healing of a Sandwich Structure with PSMP Based Syntactic Foam core

The sandwich structure is a special class of structures that is fabricated by attaching thin but stiff skins to a light weight but thick core. Fiber reinforced polymer composite sandwich structures have been used in almost all man-made engineering structures, such as space shuttle, aircraft fuselage, ship hull, body of car, pressure vessel, piping, wind-turbine blade, bridge deck, building, offshore oil platform, waterfront fender, etc., due primarily to their high specific strength and stiffness, tailor-ability, and corrosion resistance. Transverse load induced impact damage has been a critical problem for composite sandwich structures. For example, for an armour-grade composite sandwich structure, the skin is responsible for eroding and breaking the projectile, carrying bending load, and protecting the core; the core is responsible for separating and fixing the skins, resisting transverse shear, carrying in-plane load, and providing other functionalities such as absorbing impact energy, shielding radiation, and insulating heat transfer. Therefore, the versatility of the sandwich construction comes from the core. Various types of core materials have been studied such as foam core (polymeric foam, metallic foam, ceramic foam, balsa wood, syntactic foam, etc.) [60,61], web core (truss, honeycomb, etc.) [62], 3-D integrated core [63], foam filled web core [63,64], laminated composite reinforced core [65], etc. While these core materials have been used with a certain amount of success, they are limited in one way or another. For example, the brittle syntactic foam core absorbs impact energy primarily through macro-length scale damage, significantly sacrificing residual strength [66–68], and the web cores lack bonding with the skin and also have empty spaces for easy perforation of projectiles [63,64].

6.2.1 Raw Materials and Syntactic Foam Fabrication

The syntactic foam was fabricated by dispersing glass microballoons and multiwalled carbon nanotubes into a shape memory polymer matrix. The shape memory polymer (polystyrene, CRG industries) has a T_g of 62 °C, tensile strength of 23 MPa, and modulus of elasticity of 1.24 GPa at room temperature. The foam was fabricated by dispersing 40% by volume of glass microballoons (Potters Industries Q-cel 6014, with bulk density of 0.08 g/cm^3, effective density of 0.14 g/cm^3, particle diameter range of 5–200 μm, average diameter of 85 μm, and crushing strength of 1.72 MPa), and 0.15% by volume of multiwalled carbon nanotubes (Cheap Tubes Inc., with density of 2.1 g/cm^3, diameter of 20–30 nm, and length of 20–30 μm) into the polymer matrix. Carbon nanotubes were used for a couple of reasons. One is that carbon nanotubes can enhance the strength, stiffness, and recovery ratio of the SMP matrix, which helps closure of the crack; the other is that carbon nanotubes have a potential to serve as a medium for heating and triggering shape recovery if its content is higher than the percolation threshold, such as by electricity. A woven roving fabric 7725 (Fiber Glast) was used as the top and bottom skin or face sheet reinforcement. The warp and fill yarns run at 0 and 90 degrees, respectively. Thus these cross-ply woven glass fabrics are strong in both directions.

A two-step procedure was used to prepare the syntactic foam. First, the carbon nanotubes were added to the resin matrix. The mixture was mixed with the assistance of an ultrasound mixer for 30 minutes at a frequency of 20 kHz (Sonics Vibracell VC 750W) and a three-roll

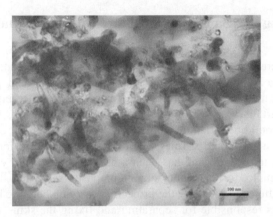

Figure 6.31   TEM picture of the smart foam. *Source:* [46] Reproduced with permission from Elsevier

mill for one pass (NETZSCH type 50). It was found that this combination was adequate to uniformly distribute the carbon nanotubes in the resin matrix; see a TEM picture in Figure 6.31. Second, microballoons and hardener were added to the carbon nanotube/resin mixture and mixed with a spatula for 15 minutes. It was then poured into an aluminium mold for curing. The process started with 24 hours of room temperature curing, followed by post-curing in an oven at 75 °C for 24 hours, 90 °C for 3 hours, and 100 °C for 3 hours. This curing cycle was chosen by a *trial and error* process because the curing cycle for the pure polymer recommended by the manufacturer cannot cure the foam. The thickness of the cured foam panel was 11.3 mm. It was then machined into 152.4 mm × 101.6 mm × 11.3 mm slabs for fabricating sandwich specimens. This particular dimension was selected in order to conduct compression after impact test using an anti-buckling test fixture.

6.2.2 Smart Foam Cored Sandwich Fabrication

The sandwich panels were fabricated using the vacuum assisted resin infusion molding (VARIM) system (Airtech). In order to keep chemical compatibility, the polymer used in the skin had the same shape memory polymer as the foam core. The whole setup is illustrated in Figure 6.32(a) and the resin infusion process is shown in Figure 6.32(b) and (c). The VARIM process involves the utilization of a distribution mesh to distribute the SMP uniformly through the glass fabric skins. A porous Teflon sheet was used in between the face sheet and the distribution mesh to avoid sticking the resin to the glass face sheet after curing. This helped the resin flow through the panel to penetrate the pores in the porous Teflon sheet and wet through the glass fabric skin uniformly. On top of the cured foam sandwich core the mesh was bonded to a spiral wrap tube, which was connected to a hose for polymer infusion. A similar arrangement of the Teflon sheet and distribution mesh was also adopted at the bottom face sheet. Meanwhile, a drain tube from the mesh was connected to an excess resin collecting container, which was connected to a vacuum pump. The whole system was put under vacuum using a vacuum bagging sheet. The vacuum pump pulls the resin through the

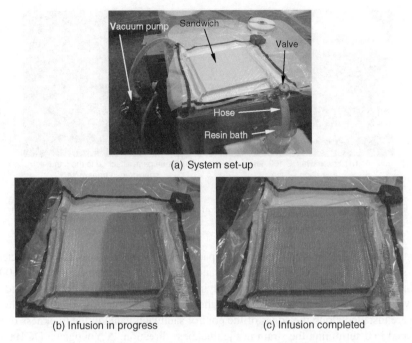

(a) System set-up

(b) Infusion in progress   (c) Infusion completed

Figure 6.32 Sandwich fabrication using VARIM system

skin and once the resin has passed through the entire skin, the system was closed using valves at both ends.

After maintaining the system under vacuum at room temperature for about 15 minutes, the setup was transferred into an oven for curing. The glass fabric sheet skins infused with the SMP resin were cured following a similar curing cycle as that of the SMP based syntactic foam core. A unique feature of this manufacturing process is that the curing of the sandwich face sheet was coupled with compression programming of the SMP foam core due to the vacuum used. After curing, the foam cored sandwich panels with a dimension of 152.4 mm × 101.6 mm × 12.7 mm were prepared for impact and self-healing tests.

6.2.3 Compression Programming

As indicated above, compression programming of the smart foam core was coupled with the curing of the sandwich panels. In this study, stress-controlled programming with a constant compressive stress of 0.05 MPa was used. The programming process started by heating the sandwich panel to 75 °C. After that, a pressure of 0.05 MPa was applied by the vacuum system. This temperature and pressure were maintained for 24 hours. Then the temperature was raised to 90 °C and maintained for 3 hours. The sandwich was then heated to 100 °C and the temperature was maintained for another 3 hours (step 1). After that the sandwich panel was cooled down to room temperature in about 6 hours while maintaining the stress level (step 2). After 0.5 hours at room temperature, the pressure was removed (step 3). This completed the three-step shape fixity process and the sandwich panels were ready for impact testing.

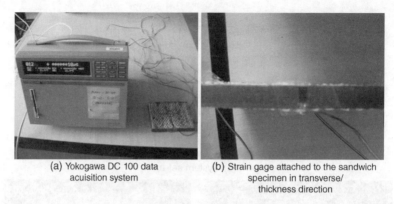

(a) Yokogawa DC 100 data
acuisition system

(b) Strain gage attached to the sandwich
specimen in transverse/
thickness direction

Figure 6.33 Yokogawa DC 100 data acquisition system and strain gage attachment

In order to examine the shape recovery capacity, the fourth step was also conducted on a control sandwich panel. The programmed sandwich panel was heated again to 100 °C to determine the shape recovery without applying any stress (free or unconstrained shape recovery). In order to quantitatively obtain the four-step thermal–mechanical cycle, strain gages were installed along the edges of the control sandwich panel in the thickness direction. This helped in determining the strain in the thickness direction. A Yokogawa DC100 device was used to monitor the changes of strain with time and temperature. A DAQ 32 plus software in conjunction with the data acquisition system was used to record the strain and temperature data. The temperature was monitored with the help of thermocouples connected to the same device. The Yokogawa DC 100 strain and temperature measurement setup along with the strain gage attached sandwich specimen is shown in Figure 6.33(a). Figure 6.33(b) highlights the strain gage attached to the smart foam cored sandwich structure in the thickness direction. The strain gages (Vishay Micromeasurements Inc.) were rated at a resistance of $350\,\Omega \pm 0.3\%$ and a gage factor of about $2.105 \pm 0.5\%$. The permissible strain limit on this gage was about 5% of strain.

The four-step thermal–mechanical cycle of the smart foam cored sandwich in the loading direction (transverse or thickness direction) in terms of stress, strain, and temperature is shown in Figure 6.34. From Figure 6.34, it is found that the shape fixity ratio $R_f = (529 \times 10^{-5}\,\text{mm/mm})/(534 \times 10^{-5}\,\text{mm/mm}) = 99\%$. The shape recovery ratio $R_r = (534 \times 10^{-5}\,\text{mm/mm} - 13 \times 10^{-5}\,\text{mm/mm})/(534 \times 10^{-5}\,\text{mm/mm} - 0 \times 10^{-5}\,\text{mm/mm}) = 97.6\%$. From the above calculation, it is clear that the shape fixity ratio is close to 100% and the shape recovery ratio is very high, suggesting good shape memory functionality of the smart foam. Also, an inflection point is found in the shape recovery step (step 4) around the glass transition temperature of the SMP (62 °C). The shape fixity results indicate that the smart sandwich structure was able to store a majority of the strain introduced into the system during the high temperature deformation process. A value of shape fixity close to 100% is also an indication of minimal springback of the specimen during the low temperature unloading process. This is very important because it maintains the dimensional stability of the specimen. Shape recovery values were also close to 100%, indicating good shape memory behavior of the smart foam cored sandwich. This result suggests that the shape memory functionality of the smart foam has a great potential to seal internal damage upon heating.

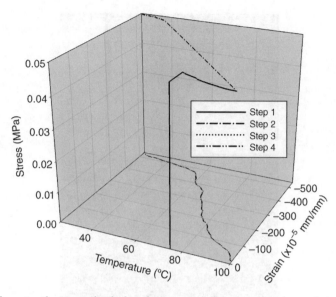

Figure 6.34 Four-step thermomechanical cycle (steps 1 to 3 are programming and step 4 is free shape recovery). *Source:* [46] Reproduced with permission from Elsevier

6.2.4 Low Velocity Impact Tests

Over the years, there has been mounting concern over the safety of fiber reinforced polymer composites subjected to low velocity impact (LVI). A low velocity impact on sandwich composites can cause various types of damage, such as interfacial deboning between the front face sheet and the core, indentation in the front face sheet, cracking of the back skin, cracking of the core, etc. These types of damage are very dangerous because they cannot be detected visually and lead to structural failure at loads well below design levels. Low velocity impact is not uncommon. Bird strike, hail impact, or even a drop of tool during routine inspection characterizes a low velocity impact event.

All the low velocity impact tests were performed with the help of an Instron DynaTup 8250 HV drop weight impact machine. The DynaTup machine shown in Figure 6.35 is equipped with an impulse data acquisition system. The fixture on which the specimen is mounted is housed within an environmental chamber at the bottom of the drop tower, which has a digital temperature controller. The velocity and impact energy can be varied by changing the hammer weight and dropping distance. The impactor is hemispherical in shape with an instrumented tup of capacity 15.56 kN and a diameter of 12.7 mm. The transient response of the specimen including velocity, displacement, load, and energy as a function of time can be recorded. The machine also possesses a velocity detector that measures the velocity of the tup. Pneumatic brakes are also provided that prevent multiple impacts on the specimen. The specimen can be either simply supported or clamped in the boundary by a pneumatic force up to 700 N. In this study, low velocity impact tests were performed on each sandwich panel at the same impact location (center of the panel) repeatedly at a velocity of 3 m/s and a hammer weight of 6.64 kg per ASTM D 2444. The impact tests were conducted at room temperature.

Figure 6.35   Low velocity impact machine

6.2.5 Characterization of Low Velocity Impact Response

In order to provide better understanding of impact/healing behavior, we would like to devote a section on characterization of low velocity impact response. At the instant when the projectile contacts the target, the impact process starts. Figure 6.36(a) shows a typical load/velocity versus displacement curve and Figure 6.36(b) shows a typical force/energy versus time curve using the DynaTup 8250 HV machine. It is noted that, based on the system design of the impact machine, the displacement and velocity are measured in terms of the projectile; the force is the contact force between the projectile and the target; and finally, the energy is a measurement of the work done by the projectile to the target.

During a low velocity impact, the stress wave has sufficient time to travel to the boundary and may be reflected back to the impact point, and may even be back and forth several times; thus the low velocity impact response is boundary controlled. In other words, the target responds to impact structurally or globally. At the same time, there is a local event. The contact between the projectile and the target is controlled by the local properties of the pair (target and projectile) and can be represented by the Hertz contact law. The global deflection of the target is driven by the local contact force from the projectile. Therefore, the local event is coupled with the global response. In order to understand the low velocity impact response better, let us try to decouple the two events.

If one is sitting on the target and observing the projectile, one can decouple the local indentation from the global deflection. In other words, one observes a typical indentation test. It

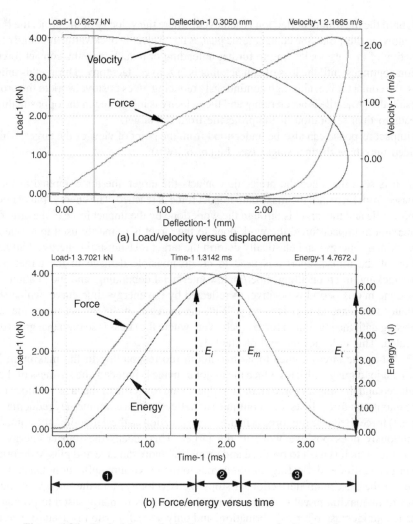

Figure 6.36 Typical low velocity impact response of a laminated composite

can be seen that, as the impact duration increases, starting from zero, the impact force (contact force) increases, until local damage or plastic deformation (permanent indentation) occurs. This shows the appearance of the maximum impact force in Figure 6.36(a). Following the peak force, the projectile is still moving along the same direction and further indenting the target because the velocity of the projectile corresponding to the peak impact force is not zero. Therefore, one can see that when the peak impact force is reached, the force almost plateaus (with slight decreases) as the deflection increases, indicating plastic flow of the material beneath the projectile, which leads to permanent indentation. The slight reduction in the contact force is also an indication of damage and reduction in load bearing capacity of the target immediately beneath the contact point. With a further increase in deflection, the reduction in the contact force becomes obvious, probably due to more damage in the materials involved in impact, until the maximum deflection is achieved, where the velocity of the projectile becomes zero. Following this point, the contact force pushes the projectile backwards (rebound of elastic deformation of the target), leading to further unloading or reduction in the contact force. This reduction continues until the loss of contact, that is, a zero contact force. During this unloading process, the contact force, although continuously reducing, does positive work to the projectile. As a result, the projectile is accelerating and the velocity is increasing in the opposite direction. This leads to final separation of the projectile from the target.

The impact response can also be understood from the point of view of the target, which can be divided into three distinct stages (see Figure 6.36(b)).

Stage 1. It is seen that once the projectile contacts the target, the force applied to the target increases from zero until the maximum impact force is reached. The energy transfer between the projectile and the target is through the work done by the impact force to the target. Up to the maximum impact force, the work done to the target is primarily used to accelerate the target (kinetic energy) and elastically deform the target (elastic strain energy). Only a small portion of the impact energy is used to create invisible damage in the target, such as microcracking, microbuckling, fiber/matrix interfacial debonding, and indentation. All the work done in this period is positive, as reflected by the energy–time curve, which shows a continuous increase in impact energy. With the increase of the impact force, the stress is increasing until the maximum force is achieved, where the stress is so high that gross damage is created. This indicates the end of Stage 1.

Stage 2. When the stress comes to its maximum (corresponding to the maximum impact force), major damage initiates. Once the major damage is induced, the stiffness of the target is reduced and the impact force starts to reduce. Due to inertia of the target, it does not stop at the maximum force. Instead, it continues to deflect, with a continuously reducing contact (impact) force. Because the force is in the same direction as the deflection, the contact force continuously does positive work to the target. This means the impact energy is still increasing, which is used to produce and propagate more damage and plastic deformation. The target continues deflecting until it comes to the maximum deflection. Once the target comes to the maximum deflection, it stops (zero velocity) and the energy of the target comes to its maximum value. In this stage, the energy is primarily used to propagate the damage and increase plastic deformation, and only a small portion is used to increase the strain energy. Since the target stops at the maximum deflection, the kinetic energy is zero and all the elastic energy is in the form of strain energy.

Stage 3. After that, the target rebounds and the projectile is pushed back by the target. In this stage, the direction of the impact force and the direction of the displacement are opposite (negative velocity). The impact force does negative work to the target or the target does positive work to the projectile. A portion of the strain energy transfers back to the projectile. As a result, the energy of the target is decreasing, and so are the impact force and deflection, until separation of the projectile from the target (zero impact force). Depending on the nature of the target, the deflection may be fully recovered (to zero) or cannot be fully recovered, leading to plastic deformation or even perforation.

At the end of Stage 3, that is, when the projectile is separated from the target, the target may still contain kinetic energy and strain energy (the velocity and the deflection may not be zero). Based on the work–energy principle, the integration of the load–deflection curve (the work done to the target) is equal to the absorbed energy due to damage and plastic deformation plus the remaining kinetic energy and strain energy:

$$W = E_t = E_a + E_r \tag{6.2}$$

where W is the work done to the target, which is equal to the area covered by the load–deflection curve (also equal to the impact energy recorded by the impact machine E_t); E_a is the absorbed energy due to damage and plastic deformation; and E_r is the residual elastic energy of the target (kinetic energy and strain energy). Obviously, the total absorbed energy E_t recorded by the impact machine is larger than the energy permanently absorbed by the target, E_a, that is, through damage and plastic deformation.

Generally, Stage 1 is called damage initiation. The energy corresponding to the maximum impact force is termed the initiation energy, E_i [69]. Stages 2 and 3 are termed damage propagation. The corresponding energy is called the propagation energy, E_p. In mathematical form, the propagation energy satisfies [69]

$$E_p = E_t - E_i \tag{6.3}$$

In some cases, for instance the structure has a very high strain energy at the end of Stage 2 (for high strength and high stiffness materials), it is found that $E_t < E_i$. This suggests that $E_p < 0$. Obviously, this is physically meaningless. In order to overcome this difficulty, we suggest that the energy increased at Stage 2 be used as the propagation energy:

$$E_p = E_m - E_i \tag{6.4}$$

where E_m is the maximum (peak) impact energy.

The rationality of this modification is as follows. (a) Stage 2 is the primary stage for damage creation and propagation. Therefore, the energy increment in Stage 2 is a reasonable estimation of the permanently absorbed energy. (b) Since it is always true that $E_m \geq E_i$, this ensures that $E_p \geq 0$. The difficulty of having negative propagation energy is overcome. (c) The results from both equations are consistant, that is, if the E_p is small from Equation (6.3), E_p from Equation (6.4) is also small, and vice versa. It is noted that E_p is by no means the permanently absorbed energy because it contains a small portion of elastic energy; for the same reason, E_i is not fully elastic because it contains a small portion of permanently absorbed energy. Because of the complexity of the problem, it is difficult to obtain the truly absorbed energy E_a and the above equations can only serve as rough estimations.

6.2.6 Crack Closing Efficiency in Terms of Impact Responses

After each impact, the specimen was brought to an oven for unconstrained shape recovery (healing) at a temperature of 100 °C for 3 hours. This impact–healing process was repeated for 7 cycles. For each impact/healing cycle at least five effective specimens were tested and the load and energy traces were obtained. The maximum impact force, maximum deflection, and impact duration were directly obtained from the load and energy traces. The maximum impact force, maximum deflection, impact duration, initiation energy, and propagation energy after each impact are summarized in Figure 6.37. It is noted that, among the various types of energies,

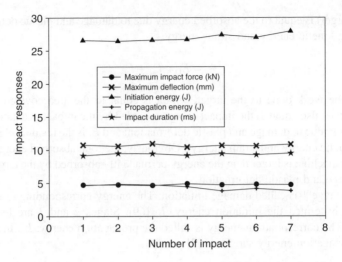

Figure 6.37 Impact response with impact cycles. *Source:* [46] Reproduced with permission from Elsevier

initiation energy and propagation energy are more relevant to evaluate the impact tolerance of the target. As discussed in 6.2.5, impact energy corresponding to the maximum impact force is defined as the initiation energy. Propagation energy is defined as the difference between the maximum impact energy and the initiation energy. These definitions have been used previously [70,71]. It has been suggested that the initiation energy is basically a measurement of the capacity for the target to transfer energy elastically; on the other hand, the propagation energy represents the energy absorbed by the target for creating and propagating gross damage and higher propagation energy usually suggests larger damage. From Figure 6.37, the impact response is statistically the same for each round of impact. This suggests that the damage induced by each round of impact has been effectively healed by the shape recovery process. The impact tolerance has been effectively recovered by self-healing. Actually, it seems that the impact tolerance after the seventh round of impact is slightly better than that after the first round of impact, as indicated by the slightly higher initiation energy and slightly lower propagation energy. This "abnormal" behavior suggests that repeated impact and healing cycles may have adjusted the microstructure of the foam, making it more beneficial for impact tolerance. Another possible reason is the post-curing effect caused by the healing process, which was conducted for 3 hours at 100 °C for each healing cycle.

6.2.7 Crack Closing Efficiency in Terms of Compression after Impact Test

Low velocity impacts might not create any damages visible to the naked eye on the sandwich structure. In practice, however, this type of damage can adversely affect the strength of the sandwich and hence result in low residual load carrying capacity. To have a better under-standing of crack closing capability in the sandwich structure, compression after impact (CAI) tests were performed on the impacted sandwich specimen. The testing was conducted using an MTS 810 machine and the fixture used was a "Boeing Compression after Impact Compression Test Fixture" according to the BSS7260 standard. The CAI tests were conducted

Figure 6.38 Compression after impact test setup

on sandwich specimens after each impact and each healing cycle. Each test contained five effective specimens. For comparison purposes, five effective specimens without impact were also tested using the same anti-buckling fixture as controls. The size of the specimen was 152.4 mm long, 101.6 mm wide, and 12.7 mm thick. A strain-controlled testing mode was used during the testing and the loading rate was 5.0 mm/min. Figure 6.38 shows an impacted sandwich specimen under the CAI test. Figure 6.39 shows the compressive stress-compressive strain behavior of the various specimens.

Figure 6.40 summarizes the variation of the residual yield strength with up to seven consecutive impact–healing cycles. Two observations can be made. (1) Each impact considerably decreases the CAI strength as compared to the control specimens. Also, the CAI strength gradually decreases as the impact cycle increases until the sixth impact cycle. For the seventh impact cycle, a significant rebound in CAI strength is observed. (2) Each crack closing cycle recovers a considerable portion of the compressive strength lost due to impact. However, the

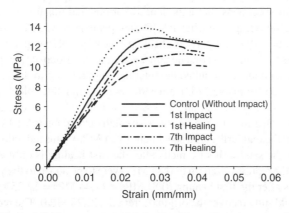

Figure 6.39 Typical CAI stress versus strain plots for various impact and healing cycles

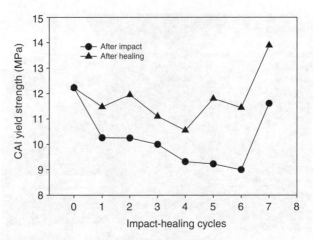

Figure 6.40 Change of CAI yield strength with impact–crack-closing cycles. *Source:* [46] Reproduced with permission from Elsevier

healed specimens still have a slightly lower strength than the control specimen. This tendency holds true until the sixth healing cycle. For the seventh crack closing cycle, its strength after healing is higher than the control specimens. These qualitative observations can also be detailed by quantitative analysis. The following quantitative estimation can be concluded from Figure 6.40. For instance, the CAI yield strength after the first impact is about 83.9% of the original yield strength without impact (control specimens, 12.23 MPa); after the first healing, the yield strength is about 93.8% of the original strength. This suggests that the self-healing has recovered a major portion of the lost strength. After the seventh impact, the CAI yield strength is about 94.9% of the original yield strength; after the seventh healing, the residual strength is about 113.6% of the original yield strength. This suggests that after seven rounds of impact–healing cycles, the sandwich specimens are actually gaining some strength. This result echoes the impact test results; that is, at the seventh impact–healing cycle, the sandwich panels are becoming better in resisting impact damage. It is believed that the healing process may have coupled with post-curing of the foam.

In order to understand the post-curing effect, the uniaxial compression test was conducted on a group of block foam specimens that experienced various additional post-curing time periods simulating the actual high temperature healing. The temperature was fixed at 100 °C and the period of post-curing was equal to the time period of healing. One group of five specimens was post-cured for 3 additional hours to simulate the first healing cycle; another group of five specimens was post-cured for 21 additional hours to simulate the total healing hours up to the seventh healing. The foam specimens were 25.0 mm long, 25.0 mm wide, and 12.5 mm thick blocks, which were tested flat-wise as per ASTM C 365 using an MTS 810 machine. The test results show that after 3 additional hours of post-curing, the average yield strength of the foam is 5.10 MPa, which is about 0.81 MPa lower than the regularly cured specimens (5.91 MPa). This reduction may explain why the CAI strength of the sandwich specimens after the first healing is only about 93.8% of the original strength. If the reduced strength of 0.81 MPa were to be added to the yield strength of the sandwich specimens after the first healing (0.81 MPa + 11.44 MPa = 12.25 MPa), it is clear that the first healing would fully recover the original strength (12.23 MPa). The reduction in the yield strength of the foam after 3 additional hours of post-curing may be due to the changed

Figure 6.41 Ultrapac ultrasonic C-scan equipment and a typical C-scan image

microstructure, which is unfavorable for the strength development. It is believed that the incorporation of carbon nanotubes, which were without any surface modifications, may affect the post-curing behavior of the foam. Of course, further study is needed to fully understand the mechanism. After 21 hours of additional post-curing, the yield strength of the foam becomes 7.60 MPa, which is about 7.60 MPa − 5.91 MPa = 1.69 MPa higher than the regularly cured foam without additional post-curing. This additional gain in strength is obviously due to the post-curing effect. If this additionally gained strength is removed from the sandwich panel after the seventh healing (13.90 MPa − 1.69 MPa = 12.21 MPa), it is clear that this corrected yield strength is close to the original strength (12.23 MPa) (control sandwich panels).

6.2.8 Ultrasonic and SEM Inspection

In order to examine impact damage and the crack-closing effect, ultrasonic inspection was performed on all specimens both after impact and after healing for each impact–healing cycle using a 1 MHz transducer for the foam cored sandwich. An UltraPac inspection machine from Physical Acoustics Laboratory was used in conjunction with UltraWin software to acquire the C-scan images and identify damages. Scanning electron microscope (SEM) observations of micro-length scale damage was conducted using a JEOL JSM-840A scanning electron microscope. The Ultrapac inspection machine is shown in Fig. 6.41.

Figure 6.42 shows the C-scan results of one sandwich panel before impact, after the first impact, after the first healing, until after the seventh impact and after the seventh healing. In these pulse-echo C-scan images, red color (gray area) represents an excess of 80% of the signal returning to the receiver, whereas blue color (dark area) indicates that 50–80% of the signal is being received. White color represents a complete attenuation of the ultrasound signal or 0% of the signal received. Therefore, the white spot at the center of the specimen indicates a certain type of damage. Two observations can be made: (1) the damage after each impact has been effectively healed, as evidenced by the disappearance of the white spot; (2) the microstructure has been changed after each impact–healing cycle, as indicated by the change in the color distribution. These observations are supported by the impact responses, which show that the healing has helped the sandwich panel not only recover but also slightly enhance its impact tolerance.

Figure 6.43 shows a comparison of a microcrack in the foam core immediately after impact and immediately after healing. It is clear that the microcrack length has been reduced and the microcrack opening has been narrowed after healing. It is noted that during healing the SEM sample, the shape recovery is stress-free, that is, unconfined free shape recovery. In the sandwich panel, the foam core directly under impact is partially confined by the skin and by the

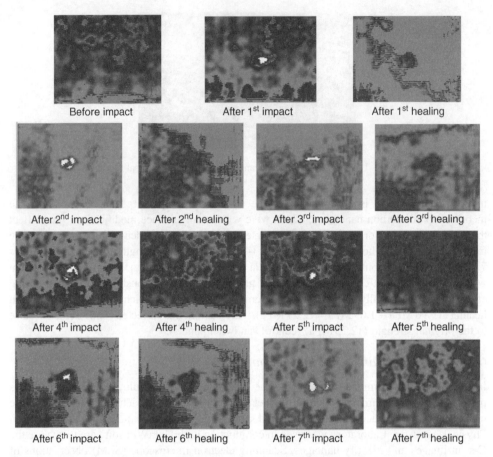

Figure 6.42 C-scan images of the sandwich panel after each impact and healing cycle. *Source:* [46] Reproduced with permission from Elsevier

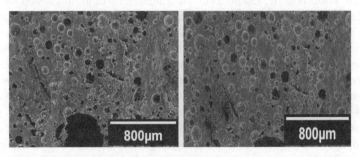

Figure 6.43 SEM pictures showing microcrack narrowing by free shape recovery. *Source:* [46] Reproduced with permission from Elsevier

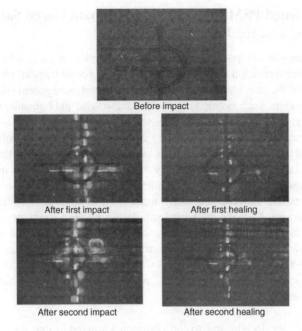

Before impact

After first impact   After first healing

After second impact   After second healing

Figure 6.44 Visual inspection of damage healing

surrounding materials. Therefore, the confinement may produce a certain stress, which may help in pushing the microcrack from two sides and thus may help close the crack more tightly. In other words, the microcrack in actual sandwich panels may have a better healing effect than that seen in Figure 6.43. This also proves our claim about the CTH scheme; that is, confined shape recovery is needed to close the crack effectively.

A visual inspection of the impacted sandwich specimen was also conducted after each impact–healing cycle. The images corresponding to the specimen before impact and after the first and second impact–healing cycles are shown in Figure 6.44. It is seen that after each impact cycle considerable damage occurred to the sandwich specimen. It is also seen that the damage has been effectively repaired or healed by each healing cycle. These visual inspection pictures echo the C-scan and SEM observations and suggest a high crack closing ability by partially constrained shape recovery.

6.2.9 Summary

Based on the comprehensive test program, it is seen that using the shape memory effect alone and with partial confinement by the sandwich skin, damage in the sandwich (skin and foam core) can be repaired. The lost CAI strength can be repeatedly recovered. It is noted that this study does not have thermoplastic particles for healing. Therefore, it only demonstrates the crack-closing capability by the shape memory effect. It also demonstrates that, with only partial confinement, the crack can be effectively narrowed. Of course, the CAI test itself is not very sensitive to cracks in the core. Therefore, it is expected that if other tests such as the bending test or tension test had been used, the recovered strength of the post-impact damaged sandwich specimens would have been lower.

6.3 Grid Stiffened PSMP Based Syntactic Foam Cored Sandwich for Mitigating and Healing Impact Damage

The impact response of composite panels is a very complex process, which involves stress wave transfer, damage creation and propagation, heat and sound transfer, phase change, etc. It has been found that the structural impact response is strongly dependent on several structural and material parameters, such as target materials, size, weight, and boundary conditions, which are designable, except the impact velocity, which is nondesignable [72–84]. Usually, the impact response can be divided into boundary controlled and wave controlled, which can be further categorized into four types based on the impact induced stress wave transfer in the target [76–78, 80] (see Table 6.2). As discussed by these previous studies, the quasi-static impact response usually results in less damage than other types of impact responses because the shear and flexural waves can be reflected many times by the boundary, which is sufficient to homogenize the stresses to make the peak load, deflection, and strain more or less in phase. Therefore, it is critical to control the response of a composite structure to impact in a quasi-static manner. This can be achieved by adjusting the material properties and geometry of the target. In the following, we will discuss the effect of each parameter on the impact response of composite sandwich structures, which will help us propose new composite sandwich structures.

1. *Material properties.* In addition to the common knowledge that the materials must be stronger, tougher, and lighter [79], it is envisioned that the materials must be less stiff during impact so that the impact duration can be extended and the elastic wave will have sufficient time to reach and be reflected by the boundary many times for a quasi-static response, as discussed by Olsson [76]. However, this requirement is contradictory to the structural requirement; that is, the material must be stiff to provide structural capacity. Therefore, an ideal material should be stiff under normal working conditions and flexible when impact occurs.

Table 6.2 Four types of structural impact responses and their characteristics

Impact response type	Wave propagation in target		Impactor characteristics
	Profile	Characteristics	
Hypervelocity		Response dominated by dilatational waves	Very short impact duration and very small mass
High velocity		Response by dilatational waves and transversally reflected several times	Short impact duration and small mass
Intermediate velocity		Response dominated by flexural and shear waves	Intermediate impact duration and intermediate mass
Low velocity		Quasi-static response	Long impact duration and large mass

2. *Weight effect.* Olsson [76] indicated that the type of impact response is governed by the ratio of the impactor mass versus target mass and not by impact velocity. There are two ways to obtain longer impact duration: (a) a larger mass ratio and (b) a lower plate stiffness. Therefore, a lighter weight and softer target will lead to a better impact response, even a quasi-static response to reduce the structural damage creation.

3. *Target size.* In order to ensure that the flexure stress wave and shear stress wave be transmitted and reflected many times by the boundary, and to homogenize the stresses sufficiently, a smaller sized target tends to result in a quasi-static response.

4. *Boundary conditions.* For a boundary-controlled impact response, a "stiff" boundary condition may lead the panel towards a high velocity impact response while a simply supported or "flexible" boundary condition may lead the panel towards a low velocity impact response.

In summary, in order to result in a quasi-static impact response, we need to design the target in such a way that (1) the target material must be stronger, tougher, stiffer, and lighter in normal service and become flexible during impact; (2) the panel or subpanel weight must be small; (3) the panel or subpanel size must be small; and (4) the boundary must be flexible.

The impact response of composite panels can also be understood from the point of view of energy balance mechanisms. As discovered in previous experimental and theoretical analyses [72–74,77], the impact energy dissipation can be generally divided into three major components:

1. *Energy transfer (reversible).* This portion of energy will be elastically dissipated in the form of elastic strain energy and kinetic energy due to elastic wave propagation, which is reversible by structural vibration and damping. Thus, it enhances impact tolerance in structural design. It is well known that the predominant longitudinal stress wave travels at the wave speed of disturbance and that the transverse wave is propagating simultaneously at a velocity lower than the disturbance wave propagation speed. With the propagation of elastic waves, kinetic energy will be transferred to the affected materials and the materials up to the wave front will flow back towards the impact point, inducing kinetic energy and elastic strain energy (see Figure 6.45) [81]. Therefore, an efficient elastic energy dissipation structure should have in-plane 2-D continuity such that it will dissipate the impact energy by transferring the stress waves in both the longitudinal and transverse directions.

2. *Energy absorption (irreversible).* It is clear that this portion of energy is dissipated by permanently damaging the target material and structure, which is the most lethal and undesired but inevitable energy dissipation mode. To minimize the adverse effect, an ideal structure is desired to have an energy absorption component. This component absorbs impact energy through "self-sacrificing" of material, causing damage, plastic deformation, and phase change. It is envisioned that this component should be "secondary" in terms of carrying the service load. It is mainly to absorb impact energy and to provide composite action to the primary load carrying component. Its damage will have a minimal effect on the load bearing capacity of the overall structure.

3. *Energy dissipation (irreversible).* Based on previous work about the energy based approach by Delfosse and Poursartip [73], the first two energy dissipation mechanisms possess over 90% of the impact energy, with the remaining impact energy being dissipated by the third mechanism – thermal energy. Depending on the nature of the impact (incident energy,

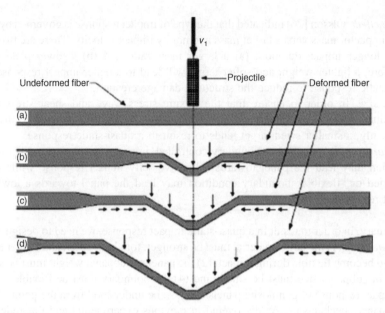

Figure 6.45 Configuration of a fibre before and after a transverse impact: (a) before impact; (b) to (d) after impact. *Source:* [81] Reproduced with permission from Elsevier

perforation or not, etc.), the temperature rise can be very high (76.6 °C [73]) or just a slight increase. Although the temperature rise may be small, an ideal structure must utilize the heat smartly and minimize its adverse effect.

In summary, an ideal composite structure should dissipate impact energy in the following ways: (1) it transmits elastic energy in terms of stress waves, which requires that the panel has a 2-D continuity for stress wave propagation; (2) there must have a "secondary component." Its purpose is not for carrying structural load; rather, it is for absorbing impact energy in terms of "self-sacrifice" or damage, preferably in the micro-length scale, so that the residual strength will not be considerably lost; and (3) the thermal energy needs to be utilized effectively.

Based on the above fundamental knowledge on the impact response of composite structures, we realize that the impact response can be designed or tailored towards a quasi-static response although we cannot control the incoming projectile. We also know that damage is one of the major mechanisms for impact energy dissipation. Therefore, in addition to minimizing the damage, we must ensure that the composite structures can self-heal their damage so that they can restore their original bearing capacity and withstand multiple impacts.

We believe that one such solution is a sandwich structure with a hybrid grid stiffened core – a continuous fiber reinforced polymer grid skeleton that is filled in with lightweight self-healing syntactic foam. A grid skeleton is an advanced grid stiffened (AGS) composite panel – a latticework of rigid, interconnecting beams in two, three, or four groups and directions. The state-of-the-art research shows that (1) an AGS panel is easily fabricated with filament winding, pultrusion, and tubes made from female molds [85–87]; (2) it can be analyzed using homogenization, smearing, and finite element modeling [88–97]; and (3) it is inherently resistant to delamination and crack propagation and the overall load carrying capacity can be

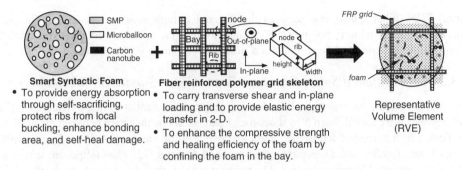

Figure 6.46 Unit cell schematic of the grid skeleton filled by smart syntactic foam

fully utilized because grid failure proceeds along the direction of the greatest strength [86,89]. Thus, this configuration proves to be an inherently strong and resilient arrangement for composite materials, without the material mismatch associated with laminated structures. It is envisioned that by filling the empty bays in the grid skeleton with a shape memory polymer (SMP) based self-healing smart syntactic foam (see Figure 6.46) the resulting composite sandwich would be an ideal structure for impact mitigation because (1) each cell or bay is a small panel with an elastic boundary, and thus tends to respond to impact quasi-statically; (2) the periodic grid skeleton, the primary load carrying component with 2-D continuity, will be responsible for transferring the impact energy elastically and providing the in-plane strength and transverse shear resistance; (3) the extremely lightweight self-healing syntactic foam in the bay, the secondary load carrying component, will be primarily responsible for absorbing impact energy through damage; (4) the grid skeleton and the foam will develop positive composite action, that is, the grid skeleton confines the smart foam to increase its strength and healing efficiency (confined shape recovery would significantly help in closing cracks), and the foam provides lateral support to resist rib local buckling and crippling. Furthermore, the foam will provide additional in-plane shear strength for bi-grids such as an orthogrid; (5) the core and skin will be fully bonded because the bay is fully filled, without the limitation of a web core; and (6) the impact damage will be self-healed so that the sandwich will be able to withstand multiple impacts. In summary, it is envisioned that the grid stiffened sandwich structure will be able to mitigate impact damage by responding to the impact quasi-statically and heal impact damage through confined expansion of the SMP based syntactic foam. Particularly, the grid skeleton and the SMP based syntactic foam perfectly match each other in that the foam possesses shape memory effect and the grid skeleton provides the most needed external confinement. With the sandwich skin, the confinement is actually in 3-D, which facilitates closing the crack in the core with arbitrary orientations.

In the following, we will present the test results of this new grid stiffened smart sandwich structure for closing (healing) low velocity impact induced damages.

6.3.1 Raw Materials

Glass fiber rovings (Saint Gobain) were used for dry weaving the ribs of the grid skeleton. The fiber rovings have a modulus of elasticity of about 70 GPa and tensile strength of 1700 MPa. They possess very low thermal conductivity and high corrosion resistance. A woven roving fabric 7725 (Fiber Glast) was used as the top and bottom skin or face sheet. The same polystyrene shape memory polymer (PSMP) and hollow glass spheres were used to prepare the smart syntactic foam. The foam consisted of 40% by volume of glass microballoons.

6.3.2 Grid Stiffened Smart Syntactic Foam Cored Sandwich Fabrication

The orthogrid skeleton was made by a dry weaving process similar to that in Reference [71]. Initially pins were nailed down on a sheet-rock board such that the space between the nails was 25.4 cm, which corresponded to the length and width of the bay area. After nailing down the pins, a layer of Teflon sheet was put on the board and then the 7725 glass fabric was laid on top of it as the bottom skin reinforcement. The fiber roving was then dry-wound around the pins in an orthogonal fashion. Next, the smart syntactic foam was prepared and poured into the bay area. The next step involved placing the top skin (7725 glass fabric) and sealing the whole system with a vacuum bag (Airtech). The vacuum system produced a compressive pressure on the top skin. Hence the top skin was pressed against the grid skeleton filled with the smart foam and eventually resulted in proper wetting of the grid skeleton and the skin by the foam. This ensured that the skin, the grid skeleton, and the foam in the bay became an integrated structure. After running the vacuum for about 15 minutes the system was transferred into an industrial oven for curing at 79.4 °C for 24 hours, followed by 107.2 °C for 3 hours and 121.1 °C for 6 hours to complete the total curing procedure. The system was then cooled down to room temperature and demolded. The slab was then machined into 152.4 mm × 101.6 mm × 12.7 mm specimens for conducting impact and compression after impact tests. The fiber volume fraction in the grid stiffened foam core is about 11%. The fiber weaving pattern and the finished grid stiffened sandwich panel can be seen in Figure 6.47 [47].

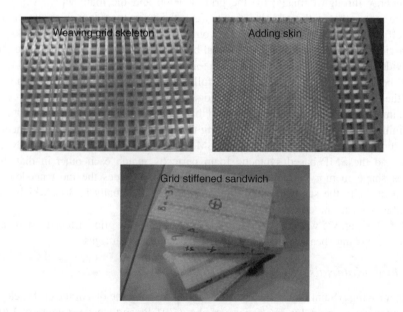

Figure 6.47 Grid weaving, fabrication, and finished grid stiffened sandwich structure. *Source:* [47] Reproduced with permission from IOP Publishing

6.3.3 Thermomechanical Programming

In order to make the sandwich specimens smart, they were subjected to a thermomechanical programming cycle. A compression molding fixture (CARVER Model 2697 compression molding fixture) with a top platen and a bottom platen was employed for the programming process. The top platen was fixed and the bottom platen was moved with the help of a lever. The top and bottom platens were able to be heated separately with different control knobs. The specimen was first inserted between the two platens. Then the platens were heated until the temperature reached 79.4 °C (above the T_g of the PSMP, which was 62 °C). Once the temperature in the specimen became uniform, it was compressed to the designed displacement (depending on the pre-strain levels) by moving the bottom platen up. The thermomechanical programming setup is shown in Figure 6.48. The displacement was measured by an LVDT (Cooper Instruments LDT 200 series) system, which had a stroke length of about 76.2 mm attached to the top platen and measured the movement of the bottom platen. The displacement was recorded on a Data Chart 2000 series data acquisition system. The recorded data were post-processed to an ASCI file for analyzing the data. Once the specimen was compressed at 79.4 °C to the designed pre-strain level, the heating was stopped and the platens were allowed to cool down to room temperature while maintaining the pre-strain constant (strain-controlled programming). Once room temperature was reached, the platens were released and the strain-controlled programming was completed. In this study, two pre-strain levels, 3% and 20%, were used to program the specimens, respectively. The purpose was to investigate how the pre-strain levels affect the healing efficiency. Also, these two strain levels resided on the linear elastic region and the plateau region of the foam under compression, respectively.

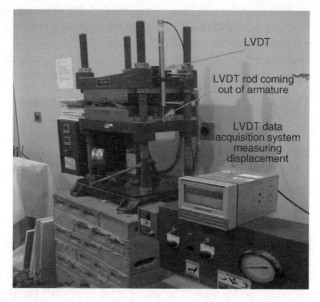

Figure 6.48 LVDT setup for thermomechanical programming and healing. *Source:* [47] Reproduced with permission from IOP Publishing

6.3.4 Low Velocity Impact Testing and Healing

Except for control specimens, low velocity impact tests were performed on each programmed specimen at the same impact location (center of the specimen) repeatedly using the Instron Dynatup 8250 HV drop tower machine. The tup nose is semispherical with a radius of 12.7 mm. The impact velocity was 4 m/s with a hammer weight of 6.64 kg, leading to an impact energy of 53.3 J. The test was conducted as per ASTM D 2444 standard at room temperature. For each impact, at least five effective specimens were tested and the load and energy curves were obtained. After each impact, the five specimens from each group were brought to the same compression molding equipment with both the top and bottom platens heated to the required temperature. Both of them were in contact with the specimen, thereby preventing free recovery in the thickness direction (see Figure 6.48). Because of the in-plane confinement by the grid skeleton and the transverse confinement by the platens, the foam in each bay was in 3-D confinement. The healing was conducted at a temperature of 121.1 °C for 3 h. The same process of impact and healing cycle was continued for seven rounds.

The impact energy is dissipated in terms of energy transfer (elastic strain energy and kinetic energy), energy absorption (through damage and plastic deformation), rising temperature, and sound. Usually, energy transfer and energy absorption constitutes the majority of energy dissipation. In some cases with perforation, energy absorption through temperature rise is considerable. As for the energy dissipation through sound, the amount of energy consumed is minimal. It is noticed that the energy transfer and energy absorption can be obtained through the energy curves in terms of initiation energy and propagation energy. The temperature rise in the specimen was obtained by an RAZ-IR infrared camera immediately before and after impact. The temperature profile immediately before and after impact of the smart sandwich specimen, which was measured by RAZ-IR Infrared Camera, is shown in Figure 6.49(a) and (b), respectively. It can be seen that, due to an impact at 53.3 J of energy, the temperature rise is about 12.4 °C, which surrounds a small area of the impact point. However, this is not enough to cause shape recovery of the PSMP foam because the temperature immediately after impact (39.2 °C) is still below the T_g of the foam (62 °C).

Figure 6.50 shows the effect of impact–healing cycles on the maximum impact load, initiation energy, and propagation energy. From Figure 6.50, the maximum impact force and

Figure 6.49 Infrared image showing temperature profile of the smart sandwich specimen (a) immediately before impact and (b) immediately after impact. *Source:* [47] Reproduced with permission from IOP Publishing

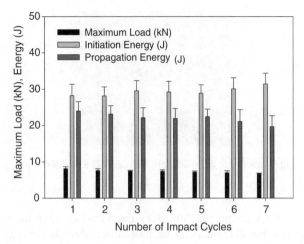

Figure 6.50 Effect of seven rounds of impact–healing cycles on the maximum impact load (kN), initiation energy (J), and propagation energy (J). *Source:* [47] Reproduced with permission from IOP Publishing

propagation energy continuously decrease and the initiation energy continuously increases as the impact–healing cycle increases. It can be visualized that the maximum impact force reduced from 8.11 kN for the first impact cycle to 6.71 kN for the seventh impact cycle. This reflects a decrease in the load bearing capacity by about 17%. Also, the initiation energy slightly increased and yet the propagation energy decreased slightly as the impact–healing cycle increased. The continuous reduction in propagation energy and increase in initiation energy suggests that more and more materials are involved in responding to impact as the impact–healing cycle increases. This implies that the amount of damaged materials under high impact energy become larger and larger and thus unrecoverable damage such as crushing of micro-balloons also becomes larger and larger. As a result, it is expected that the healing efficiency in terms of strength will become lower and lower.

6.3.5 Impact Response in Terms of Wave Propagation

A low velocity impact response or quasi-static impact response is a boundary-controlled impact response in which the flexural and shear waves generated due to impact have sufficient time to reach the boundary and be reflected back many times by the boundary. This type of impact results in less damage and hence is the most preferred impact response mode in composite structures due to the fact that the peak load, deflection, and strain are more or less in phase. In order to validate the claim, the wave propagation within the bay directly under the impact during the impact event was investigated. One strain gage was attached along one of the boundary ribs of the bay directly under the impact and the other diagonally on one of the nodes of the selected bay, see Figure 6.51. The strain gages were in turn connected to a strain gage conditioner (Vishay Micro-measurements). The strain gage conditioner was programmed such that 1 mV of voltage was equivalent to 1 με. The signal from the strain gage conditioner was amplified with the help of an oscilloscope and the oscilloscope was in turn connected to a laptop, which monitored both of the strain gage channels using a Wavestar software package. The data were saved and analyzed as an ASCI file.

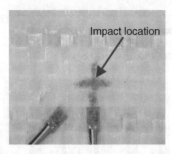

Figure 6.51 Strain gages attached to the sandwich. *Source:* [47] Reproduced with permission from IOP Publishing

Figure 6.52 shows the wave propagation along the boundary rib of the bay. The waveform from the strain gage attached diagonally on the node did not show any change during the impact event, possibly due to debonding of the strain gage from the sandwich under impact. Therefore, the data from the second strain channel are not shown here. From the impact data, the impact duration was about 4.7 milliseconds for this gauged specimen. The wave propagation during the impact can be easily identified by the voltage/strain peaks for the first 4.7 milliseconds. It is found that there are two crests and two troughs during this time period, which suggests that the impact wave reached and was reflected twice by the boundary of the bay directly under the impact. Therefore, it is concluded that the impact response of the bay directly under the impact is boundary controlled and is quasi-static. This validates the claim in this study.

6.3.6 Compression after Impact Test

The compression after impact (CAI) test was conducted to evaluate the residual strength after impact and the healing efficiency after confined shape recovery. Previous studies [70,71] have

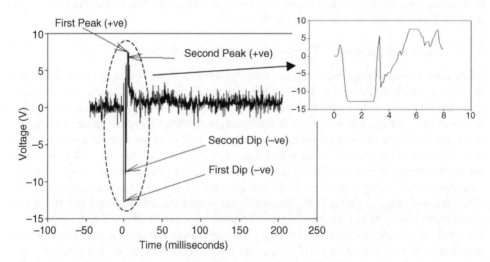

Figure 6.52 Wave propagation in the boundary rib of the bay directly under impact. *Source:* [47] Reproduced with permission from IOP Publishing

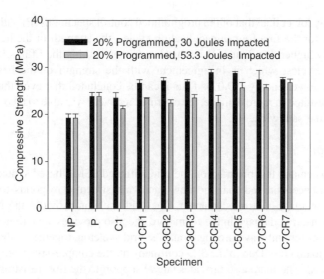

Figure 6.53 Effect of the impact energy on the CAI strength of the sandwich structure. *Source:* [47] Reproduced with permission from IOP Publishing

shown that the compressive strength of sandwich structures after a low velocity impact reduced considerably when compared with the unimpacted specimen. The remaining compressive strength of the specimen after impact is called the residual strength. Specimens without programming and programmed specimens without impact were also tested as controls. The testing was conducted using the MTS QTEST-150 machine and the fixture used was a "Boeing Compression after Impact Compression Test Fixture" in the BSS7260 standard. A strain-controlled testing was conducted at room temperature with a strain rate of 8.5×10^{-3}/min. For comparison purposes, the CAI strength for specimens impacted by 30 J of energy was also obtained.

Figure 6.53 shows the variation of residual compressive strength with impact–healing cycles at different impact energies (30 J and 53.3 J). The nomenclature used for each group of specimens is as follows: NP means control specimens without programming; P means control specimens after programming; C1 means compression after the first round impact; C1FR1 means compression after the first round impact and first round 2-D confined recovery (or "free recovery" (FR) without external confinement); C1CR1 means compression after the first round impact and first round 3-D confined recovery (or "confined recovery" (CR)); C3FR3 means compression after the third round impact and third round 2-D confined recovery; C3CR3 means compression after the third round impact and third round 3-D confined recovery; etc.

From Figure 6.53, the CAI strength of the specimens impacted by 53.3 J of energy is consistently lower than that impacted by 30.0 J of energy. However, it is interesting to note that as the impact–healing cycle increases, the difference between the two types of specimens becomes smaller. Actually, at the seventh round of the impact–healing cycle, the CAI strength of the two types of specimens is very close. This is because the CAI strength peaked at the fourth impact–healing cycle for 30.0 J of impact energy, while the CAI strength continuously increases up to the seventh cycle for the 53.3 J of impact energy, although at a decreasing rate. This is consistent with the change in initiation energy. While the strength of the specimens after healing is slightly lower than that of the programmed control specimens (P) without impact and without healing for the first four impact/healing cycles, the strength of the healed specimens becomes slightly higher than that of the control specimens during the fifth, sixth, and seventh impact–healing cycles, which is in agreement with the strength of the healed specimens impacted by the lower energy (30.0 J). Thus it can be concluded that even though the higher energy impact resulted in more damage, the confined recovery was able to recover and enhance the strength of the sandwich.

6.3.7 Summary

Based on the systematic test program on the grid stiffened PSMP based syntactic foam cored sandwich, it has been proved that the new sandwich structure responds to impact quasi-statically in a boundary-controlled manner. The grid skeleton provides the needed external confinement for healing damage in the foam core and also provides a continuous 2-D path for elastic energy transfer and wave propagation. The grid skeleton, together with the skin, works nicely with the foam core. Due to the 3-D confinement, the compressive strength of the foam has been enhanced. The foam also provides lateral support to the ribs, resisting local buckling and crippling of the ribs. Most importantly, due to the 3-D confinement, the smart foam can close or heal the impact repeatedly, efficiently, and almost autonomously (the only human intervention is by heating). The combination of the grid skeleton and PSMP based foam is ideal for impact mitigations.

6.4 Three-Dimensional Woven Fabric Reinforced PSMP Based Syntactic Foam Panel for Impact Tolerance and Damage Healing

Fiber reinforced polymer composite laminate has been the primary structural architecture for lightweight engineering structures. However, laminated composites are vulnerable to impact damage, even under a low velocity impact [98–102]. Low velocity impact is not uncommon. Even the drop of a tool during routine inspection characterizes a low velocity impact event. Under a low velocity impact, various matrix dominated or even fiber dominated failure modes occur, such as matrix cracking, delamination, matrix/fiber interfacial debonding, and fiber fracture. These types of failure modes lead the laminate towards structural failure at loads well below the design level. Therefore, improving the impact tolerance of laminated composites has been a popular research topic for many years. Among the three types of matrix dominated failure modes – delamination, matrix cracking, and matrix/fiber interfacial debonding – delamination is the most critical failure mode. Under in-plane compression, a delaminated laminate can experience reduction of the compressive buckling load well above 50% [102]. The key to improve delamination tolerance is to increase the transverse shear resistance of the laminate. Various approaches have been explored such as the Z-pin, that is, fiber reinforcement in the out-of-plane or thickness direction [103–105], 3-D woven fabric [106,107], etc. Among them, 3-D woven fabric is a recent development thanks to advancement in the textile industry.

Three-dimensional (3-D) fiber reinforced polymeric materials have been shown to have a good impact tolerance [108,109], making them attractive candidates in weight sensitive industries such as in aerospace, auto, and maritime. A number of studies have been conducted to understand the impact response of 3-D woven composites [108,109]. For example, Baucom, Zikry, and Rajendran [110] investigated the effects of fabric architecture on damage progression, perforation resistance, strength, and failure mechanisms in composite systems of comparable areal densities and fiber volume fractions, subjected to repeated impact. They found that 3-D systems survived more strikes before being perforated and absorbed more total energy compared to other systems. They reported transverse matrix cracking, fiber debonding from the matrix, fiber fracture, and fracture of Z-direction fiber tows as failure modes in the 3-D systems. Most recent developments in the area of 3-D woven fabric composites include: modeling the impact penetration of 3-D woven composite at the unit cell level [111], evaluating the transverse impact damage and energy absorption [112], studying the compressive responses and energy absorption [113], and investigating the damage mode by high strain rate loading [114].

As demonstrated in these previous studies, while 3-D woven fabric reinforcement has significantly improved the performance, including impact tolerance of lightweight composite structures, damage cannot be fully eliminated. Therefore, self-healing is still a desired feature for this type of composite structures. We believe that, like the grid stiffened composite structures discussed above, the CTH self-healing scheme perfectly matches with the 3-D fabric architecture; that is, the 3-D fabric provides the needed external confinement to the SMP matrix and the SMP matrix bonds the 3-D fabric and provides lateral support to the yarns to reduce local and global buckling. In the following, we will demonstrate the synergy between the SMP matrix and the 3-D woven fabric in an integrated composite panel and the ability to repeatedly heal low velocity impact damage.

6.4.1 Experimentation

6.4.1.1 Raw Materials

The only new raw material in this study was the 3-D woven fabric. In this study, 3-D E-glass woven fabric from Parabeam-Netherlands (thickness of 12.7 mm, aerial density of 1.6 kg/m^2, and compressive yield strength of 3.7 MPa) was used. The schematic fabric architecture and detailed fiber woven after impact damage are shown in Figure 6.54.

6.4.1.2 Fabrication of the Shape Memory Polymer-Based Healable Syntactic Foam

To fabricate the foam, copolyester (CP) thermoplastic particles (9% by volume) were dispersed in a beaker containing the PSMP matrix and mixed to uniformity. Next, glass microballoons (GMBs), 40% by volume, were dispersed into the CP-PSMP mixture by mechanical mixing to uniformly distribute the microballoons in the particulate matrix. Curing agent was then added and mixed manually to uniformity. This uncured foam was used to prepare the 3-D woven fabric reinforced composite panel. For addition testing, some of the mixture was then poured into a steel mold with dimensions of 300 mm × 300 mm × 12.5 mm. The mold was sealed and the material was cured in an oven as follows: 75 °C for 12 hours, 90 °C for 3 hours, and 120 °C for 3 hours. Once the curing procedure was completed, the setup was cooled down to room temperature and de-molded. The demolded panel was machined into various types of specimens for testing. Figure 6.55 shows the distribution of CP and GMB particles in the PSMP matrix.

6.4.1.3 Fabrication of 3-D Woven Fabric Reinforced Foam Panel

The 3-D woven fabric reinforced foam composite panel was fabricated in a steel mold using a gravity assisted method by impregnating the 3-D woven fabric with the uncured foam. The

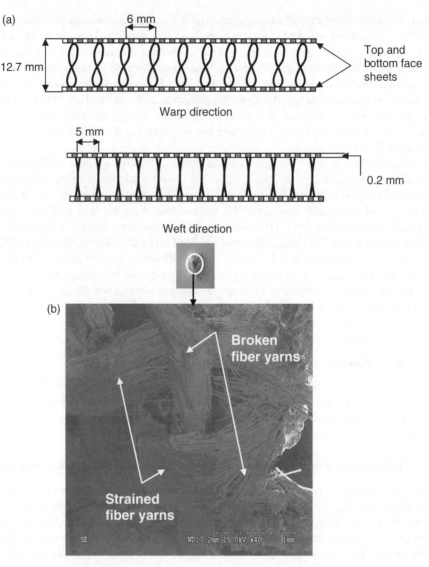

Figure 6.54 Three-dimensional woven fabric. *Source:* [48] Reproduced with permission from IOP Publishing

mold was made of a 254 mm × 254 mm × 12.7 mm detachable steel frame with the top and bottom plates having dimensions of 279 mm × 279 mm × 2.54 mm. First, one member of the frame was detached from the rest of the setup to create an open end. Next the open-ended frame was put on the bottom plate and placed on a flat table. Then the 3-D woven fabric with dimensions of 304.8 mm × 254 mm × 12.7 mm was placed within the frame such that 50 mm of the fabric extended out of the frame through the open end. Next, the top plate was carefully placed on top of the frame. Steel clips were then used to hold the plate and the frame assembly together. The setup was placed vertically on a flat table. Once the mold was ready, the uncured

Figure 6.55 Distribution of thermoplastic (CP) particles (irregular shaped) and glass microballoon (GMB) particles (spherical shaped) in the PSMP matrix

foam was poured gently into the mold from the open end (top) and allowed to descend to the bottom by flow under gravity. This continued until the 3-D woven fabric was completely filled with the foam. Finally, the setup was sealed and placed in an oven for curing. The material was cured as follows: 75 °C for 12 h, 90 °C for 3 h, and 120 °C for 3 h. Once the panel was cured, it was demolded. The panel was then machined into specimens with dimensions of 152.4 mm × 101.6 mm × 12.7 mm for impact testing and self-healing.

6.4.1.4 Dynamic Mechanical Analysis (DMA) of the Foam

The glass transition temperature and viscoelastic properties of the foam were determined using specimens with dimensions of 36 mm × 11.5 mm × 2 mm. The test was performed at a frequency of 1 Hz on a Rheometic Scientific RSA III instrument. The temperature was increased at a rate of 5 °C/min from room temperature to 120 °C. The setup of the test is shown in Fig. 6.56.

6.4.1.5 Thermomechanical Behavior of the Syntactic Foam

The thermomechanical behavior of the foam was investigated on specimens with dimensions of 25 mm × 25 mm × 12.5 mm using the MTS QTEST 150 testing machine equipped with a heating furnace (ATS heating chamber) in order to study the shape memory functionality of the foam in terms of shape fixity and shape recovery. The classical programming approach was used. The specimen was first inserted into the heating chamber at room temperature. The chamber was then heated up to 80 °C at a heating rate of 0.5 °C/min. It was then compressed to 10% pre-strain level and soaked at 80 °C for 1 hour while holding the strain constant. This allowed stress relaxation, which became a portion of the back stress. The chamber was then cooled down to room temperature while holding the pre-strain constant. Once the room temperature was reached, the load was removed, accompanied by a small springback. This completed the classical strain-controlled programming process. After that, the specimen was either subjected to free shape recovery without any external constraint or fully constrained stress recovery by heating the specimen to 80 °C at a heating rate of 0.5 °C/min.

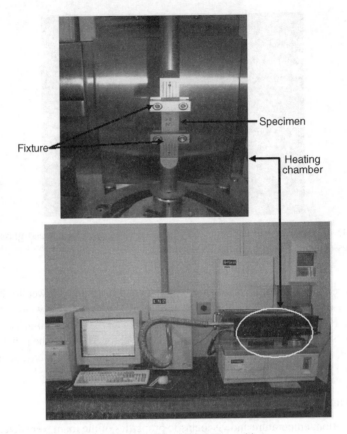

Figure 6.56 RSA III DMA machine

6.4.1.6 3-D Woven Fabric Reinforced Composite Panel Programming

Prior to low velocity impact testing, the 3-D woven fabric reinforced foam specimens were compression programmed in the transverse (thickness) direction. Because the composite panels were impacted transversely, which created indentation and damage, compression programming in the transverse direction allowed the panel to recover the indentation and heal internal damage. The programming was conducted in a strain-controlled mode at a pre-strain of 5% on a CARVER Model 2697 compression molding fixture. In order to program a specimen, the specimen was first placed within preheated platens (110 °C) of the fixture for 20 minutes. Then, with the help of a linear variable differential transducer (LVDT) system (Cooper Instruments LDT 200 series), the specimen was compressed to a 5% strain level along the transverse (thickness) direction and held at that temperature for another 15 min. After that, the heating switch on the fixture was turned off and the specimen was allowed to cool down to room temperature. Once at room temperature, the load was removed. With the removal of the load at room temperature, there was a small springback. Because of the large stiffness of the specimen at room temperature and the small programming stress, the springback can be neglected. This completed the typical strain-controlled three-step thermomechanical programming. Figure 6.57 shows a schematic of the programming setup.

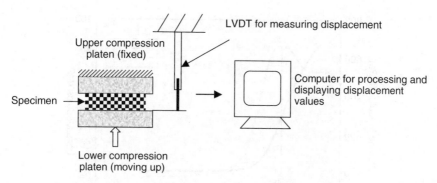

Figure 6.57 Schematic explanation of specimen programming

6.4.1.7 Impact Testing of 3-D Woven Fabric Reinforced Foam Panel

Repeated low velocity impact tests were conducted using an instrumented Instron Dynatup 8250 HV drop tower testing system. The tup nose is semispherical with a radius of 12.7 mm. The impact velocity was 3.5 m/s with a hammer weight of 6.64 kg, leading to an impact energy of 42 J. The test was conducted according to ASTM D 2444 standard at room temperature in order to determine the impact toughness and healing ability of the panel in terms of the number of impacts to perforation. The impact was conducted repeatedly at the center of the specimen until perforation. Two groups of specimens were compared. Group 1 (G1) specimens were healed after each impact until perforation. Group 2 (G2) specimens were impacted repeatedly without healing until perforation. At least three effective specimens were used for each impact–healing test.

6.4.1.8 Close-Then-Heal (CTH) Healing

In order to heal the G1 specimens after each impact, the specimen was placed in between the platens (preheated to 150 °C) of the fixture in Figure 6.57 for 20 minutes. At the beginning of the self-healing process, the top platen was about 0.7 mm above the top surface of the specimen because the maximum recovery in the thickness direction by the specimen is 12.7 mm (thickness of the specimen) × 5% (pre-strain) = 0.64 mm. Hence, the 0.7 mm gap was used to avoid touching the specimen and applying external constraint to the specimen by the platen. In other words, the needed confinement was provided by the 3-D woven fabric instead of the external confinement. The platens only provided uniform heating to the specimen. After 20 minutes of heating, the heat was turned off and the specimen was allowed to cool down to room temperature. This completed the CTH healing process.

6.4.2 Results and Discussion

6.4.2.1 DMA Test Results

Figure 6.58 shows typical plots of the storage modulus and loss modulus with the temperature for the foam. The glass transition temperature was taken as the temperature corresponding to the peak of the loss modulus curve. The glass transition temperature was found to be 49 °C. A

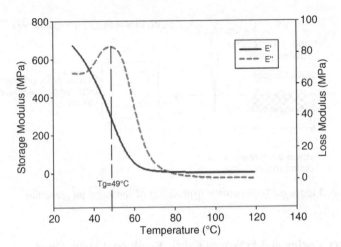

Figure 6.58 Typical plots of the storage modulus (E') and loss modulus (E'') with the temperature of the foam

couple of observations can be made. First, only one glass transition is seen, instead of the separated glass transition of the thermoplastic particles and the shape memory polymer matrix. This is an indication of partial miscibility between the two polymers, which is required for molecular level healing. Second, in previous studies, it has been proved that addition of glass microballoons increases the glass transition temperature of the foam due to the reduced molecule mobility of a thin layer shape memory polymer attached to the microballoons [45]. With the addition of 9% by volume of thermoplastic particles, it is seen that the glass transition temperature of the composite foam is lowered. This suggests that the thermoplastic particles have a higher activity than the glass particles, and leads the glass transition temperature of the foam towards the glass transition temperature of the thermoplastic particles. Based on this result, a programming temperature of 80 °C was used to study the shape memory behavior of the foam, which is well above the glass transition temperatures of the foam.

6.4.2.2 Thermo-mechanical Behavior of the Foam

Figure 6.59 shows a typical 3-D thermomechanical cycle plot of the foam. It consists of three-step programming (ABCD), fully constrained stress recovery (DEF), and free shape recovery (DG). The stabilized programming stress (SPS) is the stress at point C. The stabilized fully constrained recovery stress (SCRS) is the stress at point F. Stress recovery is defined as the ratio of the SCRS at point F to the SPS at point C. A high value of SCRS is an indication of good capability of closing cracks during confined shape recovery of the damaged composite. The stress recovery ratio of the composite is 60%. The shape fixity (ability of the composite to store its temporary shape in terms of strain, point D) and the shape recovery (ability of the composite to recover the strain from point D back to point G) were determined in terms of strain and were both greater than 98%. This indicates that the foam retained its shape memory functionality.

Again, here we notice the considerable difference between the stress recovery ratio and the strain (shape) recovery ratio. The reason is that shape recovery is a macroscopic measurement

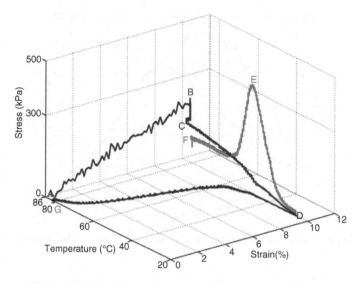

Figure 6.59 Typical 3-D thermomechanical cycle of the foam

of the shape memory ability, while stress recovery depends more on the microstructure or molecular structure change such as segmental sliding, chemical bond fracture, etc. Therefore, they have different recovery ratios. Because the CTH scheme depends more on the recovery stress to close the crack, it is seen that the crack-closing ability is limited by the lower recovery stress ratio. Overall, we must keep in mind that the shape recovery ratio is not the only way to evaluate memory ability. SMPs memorize some properties well, such as shape or strain, while they do not have the same memory for other properties such as stress, color, conductivity, etc. We can call this feature selective memory. Therefore, we should select the right indicator to evaluate memory ability based on the particular applications.

6.4.2.3 Impact Response and Healing Efficiency of 3-D Woven Fabric Reinforced Foam Panel

The impact response and healing efficiency can be evaluated by several parameters, for example, impact load, energy, and deflection with impact–healing cycles. Figure 6.60 shows a typical impact response at the first impact event of a G1 specimen and Figure 6.61 shows the same G1 specimen after 21 impact–healing cycles (at perforation). Figure 6.62 shows the variation of impact load with impact cycle for G1 (with healing) and G2 (without healing) specimens.

From Figure 6.61, the deflection at the end of the impact did not return to zero, suggesting that the hammer was not rebounded by the specimen; instead, the tup perforated the specimen. Comparing G1 with G2 in Figure 6.62, several observations can be made. (1) It is seen that the incorporation of thermoplastic particles slightly reduces the peak impact load (from 6 kN in Reference [48] to 5.3 kN in G2). However, the impact and perforation resistance of the foam is significantly increased (requiring up to 16 impacts to perforate the foam compared to only 5 impacts without thermoplastic particles in Reference [48]). Therefore, the soft thermoplastic particles help the foam absorb the impact energy, in addition to healing the foam molecularly.

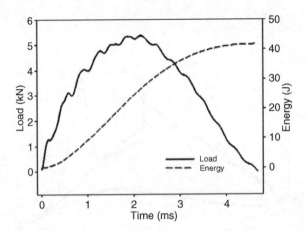

Figure 6.60 Typical impact load and impact energy change with time for the first impact event of a G1 specimen

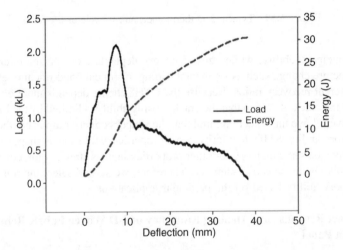

Figure 6.61 Typical load–deflection and energy-deflection curves of the G1 specimen at perforation (twenty-first impact)

(2) The close-then-heal (CTH) mechanism delays perforation in the G1 specimens by up to 5 impact cycles to a total of 21 impacts. For the foam without thermoplastic particles, that is, with crack closing but without molecular-length scale healing, perforation was delayed by only two impacts to a total of 7 impacts [48]. This indicates the importance by both steps in the CTH scheme: closing the crack can increase the impact resistance, while healing the crack molecularly further enhances the impact tolerance. (3) The first impact causes local densification around the impact region for the G2 specimen, as reflected by the sudden rise in peak load at the second impact cycle (Figure 6.62). (4) For the G2 specimens, it is seen that the load reduces very slowly for the first 12 impacts, before reducing more rapidly from the thirteenth impact to perforation at the sixteenth impact. This indicates that major unhealable damage was created in the 3-D fabric at the point of impact at the thirteenth impact cycle. Since the G1

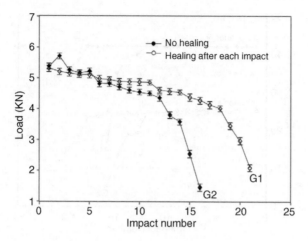

Figure 6.62 Variation of the impact load with impact number for both G1 and G2 specimens

specimens were healed after each impact, the creation of major damage in the material was delayed. For example, at the thirteenth impact, while the average peak impact load of the G2 specimen was dropped to about 3.6 kN, the peak load of the G1 specimen was still up at 4.6 kN. As can be expected, a similar trend exists for the change of deflection with impact numbers (see Figure 6.63).

As discussed in Reference [48], the impact tolerance and healing efficiency can also be understood from the point of view of the total impact energy, initiation energy, and propagation energy with impact–healing cycles. The change of impact energy and initiation energy with impact numbers for both groups of specimens is shown in Figure 6.64.

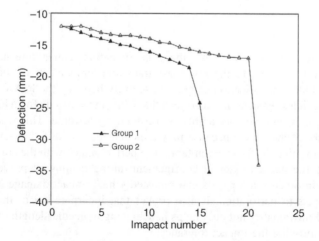

Figure 6.63 Variation of deflection with impact number

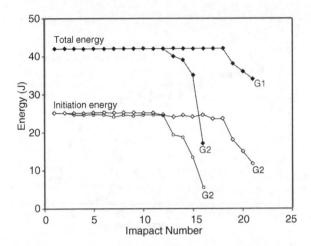

Figure 6.64 Variation of energy with impact number for both G1 and G2

It is seen that the total impact energy did not change much until the last few impacts for both G1 and G2. The reason is that toward the last few impacts, the tup begins to penetrate into the specimen and thus the tup shares a portion of the available energy. The initiation energy of the control specimens (G2) was fairly constant until the thirteenth impact and decreased sharply with increasing impact number, while the initiation energy of the G1 specimens was fairly constant until the eighteenth impact before decreasing steadily to perforation. Therefore, for G1, the healing delayed the onset of energy absorption by nonhealable damage until the eighteenth impact. This is a direct result of the CTH healing scheme adopted in this study. It also proves that the external confinement required by the CTH scheme can be naturally provided by the architectural design of composite structures. No special external confinement is needed.

6.4.3 Summary

From the test program, it is seen that the glass transition temperature has been shifted towards the lower side due to the low glass transition temperature of the thermoplastic particles. This is also an indication of good miscibility between the PSMP matrix and the thermoplastic particles selected. It is also seen that while the syntactic foam has a good shape memory ability, its stress memory ability is considerably lowered. This is an indication of selective memory of the PSMP because they are controlled by different levels of material structures. It shows that the 3-D woven fabric is a perfect match with the composite foam in terms providing the needed external confinement during healing. It proves that the 3-D woven fabric reinforced foam panels can repeatedly heal impact damage by delaying the number of impacts to perforation. It also proves that incorporation of the thermoplastic particles in the PSMP matrix not only helps in achieving molecular-length scale healing but also helps in improving the impact tolerance.

References

1. Meng, H. and Li, G. (2013) A review of stimuli-responsive shape memory polymer composites. *Polymer*, **54**, 2199–2221.
2. Gould, P. (2003) Self-help for ailing structures. *Materials Today*, **6**, 44–49.
3. Billiet, S., Hillewaere, X.K.D., Teixeira, R.F.A., and Du Prez, F.E. (2013) Chemistry of crosslinking processes for self-healing polymers. *Macromolecular Rapid Communications*, **34**, 290–309.
4. Herbst, F., Döhler, D., Michael, P., and Binder, W.H. (2013) Self-healing polymers via supramolecular forces. *Macromolecule Rapid Communications*, **34**, 203–220.
5. Aïssa, B., Therriault, D., Haddad, E., and Jamroz, W. (2012) Self-healing materials systems: overview of major approaches and recent developed technologies. *Advances in Materials Science and Engineering*, **2012**, Article ID 854203 (17 pages).
6. Zhang, M.Q. and Rong, M.Z. (2012) Theoretical consideration and modeling of self-healing polymers. *Journal of Polymer Science Part B: Polymer Physics*, **50**, 229–241.
7. Guimard, N.K., Oehlenschlaeger, K.K., Zhou, J., Hilf, S., Schmidt, F.G., and Barner-Kowollik, C. (2012) Current trends in the field of self-healing materials. *Macromolecular Chemistry and Physics*, **213**, 131–143.
8. Hu, J., Zhu, Y., Huang, H., and Lu, J. (2012) Recent advances in shape memory polymers: structure, mechanism, functionality, modeling and applications. *Progress in Polymer Science*, **37**, 1720–1763.
9. Zhang, M.Q. and Rong, M.Z. (2012) Design and synthesis of self-healing polymers. *Science China Chemistry*, **55**, 648–676.
10. Brochu, A.B.W., Craig, S.L., and Reichert, W.M. (2011) Self-healing biomaterials. *Journal of Biomedical Materials Research A*, **96**, 492–506.
11. Zhang, M.Q. and Rong, M.Z. (2011) *Self-Healing Polymers and Polymer Composites*, Wiley-VCH.
12. Hager, M.D., Greil, P., Leyens, C., van der Zwaag, S., and Schubert, U.S. (2010) Self-healing materials. *Advanced Materials*, **22**, 5424–5430.
13. Mauldin, T.C. and Kessler, M.R. (2010) Self-healing polymers and composites. *International Materials Reviews*, **55**, 317–346.
14. Blaiszik, B.J., Kramer, S.L.B., Olugebefola, S.C., Moore, J.S., Sottos, N.R., and White, S.R. Self-Healing Polymers and Composites. *Annual Review of Materials Research*, **40**, 179–211.
15. Murphy, E.B. and Wudl, F. (2010) The world of smart healable materials. *Progress in Polymer Science*, **35**, 223–251.
16. Meng, H. and Li, G. (2013) Reversible switching transitions of stimuli-responsive shape changing polymers. *Journal of Materials Chemistry A*, **1**, 7838–7865.
17. Hager, M.D., Greil, P., Leyens, C., van der Zwaag, S., and Schubert, U.S. (2010) Self-healing materials. *Advanced Materials*, **22**, 5424–5430.
18. Samadzadeha, M., Bouraa, S.H., Peikaria, M., Kasirihab, S.M., Ashrafic, A. (2010) A review on self-healing coatings based on micro/nanocapsules. *Progress in Organic Coatings*, **68**, 159–164.
19. Burattini, S., Greenland, B.W., Chappell, D., Colquhoun, H.M., and Hayes, W. (2010) Healable polymeric materials: a tutorial review. *Chemical Society Reviews*, **39**, 1973–1985.
20. Syrett, J.A., Becer, C.R., and Haddleton, D.M. (2010) Self-healing and self-mendable polymers. *Polymer Chemistry*, **1**, 978–987.
21. Amendola, V. and Meneghetti, M. (2009) Self-healing at the nanoscale. *Nanoscale*, **1**, 74–88.
22. Ghosh, S.K. (2009) *Self-Healing Materials: Fundamentals, Design Strategies, and Applications*, Wiley-VCH.
23. Wu, D.Y., Meure, S., and Solomon, D. (2008) Self-healing polymeric materials: a review of recent developments. *Progress in Polymer Science*, **33**, 479–522.
24. Yuan, Y.C., Yin, T., Rong, M.Z., Zhang, M.Q. (2008) Self healing in polymers and polymer composites. Concepts, realization and outlook: a review. *eXPRESS Polymer Letters*, **2**, 238–250.
25. Wool, R.P. (2008) Self-healing materials: a review. *Soft Matter*, **4**, 400–418.
26. Trask, R.S., Williams, H.R., and Bond, I.P. (2007) Self-healing polymer composites: mimicking nature to enhance performance. *Bioinspiration and Biomimetics*, **2**, 1–9.
27. van der Zwaag, S., Schmets, A.J.M., and Zaken, G. (eds) (2007) *Self-Healing Materials: An Alternative Approach to 20 Centuries of Materials Science*, Springer.
28. Balazs, A.C. (2007) Modeling self-healing materials. *Materials Today*, **10**, 18–23.
29. Dry, C.M. (1991) Alteration of matrix permeability and associated pore and crack structure by timed release of internal chemicals. *Ceramic Transactions*, **16**, 729–768.

30. Bleay, S.M., Loader, C.B., Hawyes, V.J., Humberstone, L., and Curtis, P.T. (2001) A smart repair system for polymer matrix composites. *Composites Part A: Applied Science and Manufacturing*, **32**, 1767–1776.

31. Zako, M. and Takano N. (1999) Intelligent material systems using epoxy particles to repair microcracks and delamination damage in GFRP. *Journal of Intelligent Material Systems and Structures*, **10**, 836–841.

32. Hayes, S.A., Jones, F.R., Marshiya, K., and Zhang, W. (2007) A self-healing thermosetting composite material. *Composites Part A: Applied Science and Manufacturing*, **38**, 1116–1120.

33. Nji, J. and Li, G. (2012) Damage healing ability of a shape memory polymer based particulate composite with small thermoplastic contents. *Smart Materials and Structures*, **21**, paper 025011.

34. Nji, J. and Li, G. (2010) A biomimic shape memory polymer based self-healing particulate composite. *Polymer*, **51**, 6021–6029.

35. White, S.R., Sottos, N.R., Geubelle, P.H., Moore, J.S., Kessler, M.R., Sriram, S.R., Brown, E.N., and Viswanathan, S. (2001) Autonomic healing of polymer. *Nature*, **409**, 794–797.

36. Lee, J., Zhang, M., Bhattacharyya, D., Yuan, Y.C., Jayaraman, K., and Mai, Y.W. (2012) Micromechanical behavior of self-healing epoxy and hardener-loaded microcapsules by nanoindentation. *Materials Letters*, **76**, 62–65.

37. Pang, J.W.C. and Bond, I.P. (2005) A hollow fibre reinforced polymer composite encompassing self-healing and enhanced damage visibility. *Composites Science and Technology*, **65**, 1791–1799.

38. Trask, R.S., Williams, G.J., and Bond, I.P. (2007) Bioinspired self-healing of advanced composite structures using hollow glass fibres. *Journal of the Royal Society Interface*, **4**, 363–371.

39. Toohey, K.S., Sottos, N.R., Lewis, J.A., Moore, J.S., and White, S.R. (2007) Self-healing materials with microvascular networks. *Nature Materials*, **6**, 581–585.

40. Williams, H.R., Trask, R.S., Knights, A.C., Williams, E.R., and Bond, I.P. (2008) Biomimetic reliability strategies for self-healing vascular networks in engineering materials. *Journal of the Royal Society Interface*, **5**, 735–747.

41. Chen, X., Dam, M.A., Ono, K., Mal, A., Shen, H., Nutt, S.R., Sheran, K., and Wudl, F. (2002) A thermally remendable cross-linked polymeric material. *Science*, **295**, 1698–1702.

42. Varley, R. and van der Zwaag, S. (2008) Towards an understanding of thermally activated self-healing of an ionomer system during ballistic penetration. *Acta Materialia*, **56**, 5737–5750.

43. Sijbesma, R.P., Beijer, F.H., Brunsveld, L., Folmer, B.J.B., Hirschberg, J.H.K.K., Lange, R.F.M., Lowe, J.K.L., and Meijer, E.W. (1997) Reversible polymers formed from self-complementary monomers using quadruple hydrogen bonding. *Science*, **278**, 1601–1604.

44. Li, G. and Uppu, N. (2010) Shape memory polymer based self-healing syntactic foam: 3-D confined thermomechanical characterization. *Composites Science and Technology*, **70**, 1419–1427.

45. Li, G. and Nettles, D. (2010) Thermomechanical characterization of a shape memory polymer based self-repairing syntactic foam. *Polymer*, **51**, 755–762.

46. Li, G. and John, M. (2008) A self-healing smart syntactic foam under multiple impacts. *Composites Science and Technology*, **68**, 3337–3343.

47. John, M. and Li, G. (2010) Self-healing of sandwich structures with grid stiffened shape memory polymer syntactic foam core. *Smart Materials and Structures*, **19**, paper 075013 (12 pages).

48. Nji, J. and Li, G. (2010) A self-healing 3D woven fabric reinforced shape memory polymer composite for impact mitigation. *Smart Materials and Structures*, **19**, paper 035007 (9 pages).

49. Li, G., Ajisafe, O., and Meng, H. (2013) Effect of strain hardening of shape memory polymer fibers on healing efficiency of thermosetting polymer composites. *Polymer*, **54**, 920–928.

50. Zako, M. and Takano, N. (1999) Intelligent material systems using epoxy particles to repair microcracks and delamination damage in GFRP. *Journal of Intelligent Material Systems and Structures*, **10**, 836–841.

51. Hayes, S.A., Jones, F.R., Marshiya, K., Zhang, W. (2007) A self-healing thermosetting composite material. *Composites Part A: Applied Science and Manufacturing*, **38**, 1116–1120.

52. Rodriguez, E.D., Luo, X., and Mather, P.T. (2011) Linear and crosslinked poly(ε-caprolactone) polymers for shape memory assisted self-healing (SMASH). *ACS Applied Materials and Interfaces*, **3**, 152–161.

53. Xiao, X.C., Xie, T., and Cheng, Y.T. (2010) Self-healable graphene polymer composites. *Journal of Materials Chemistry*, **20**, 3508–3514.

54. Li, G., King, A., Xu, T., and Huang, X. (2013) Behavior of thermoset shape memory polymer based syntactic foam sealant trained by hybrid two-stage programming. *ASCE Journal of Materials in Civil Engineering*, **25**, 393–402.

55. Luo, X.F. and Mather, P.T. (2013) Shape memory assisted self-healing coating. *ACS Macro Letters*, **2**, 152–156.

56. Petrovic-Djakov, D.M., Filipovic, J.M., Vrhovac, L.J.P., and Velickovic, J.S. (1993) DSC analysis of compatibility of poly(phenyl methacrylate) and some poly(xylenyl methacrylates). *Journal of Thermal Analysis*, **40**, 741–746.

57. Brown, E.N., Sottos, N.R., and White, S.R. (2002) Fracture testing of a self-healing polymer composite. *Experimental Mechanics*, **42**, 372–379.

58. Brown, E.N. (2011) Use of the tapered double-cantilever beam geometry for fracture toughness measurements and its application to the quantification of self-healing. *The Journal of Strain Analysis for Engineering Design*, **46**, 167–186.

59. Potts, P.J. (1987) *Handbook of Silicate Rock Analysis*, Springer Ltd, London, Hardcover, ISBN 0216917948.

60. Shutov, F.A. (1991) Syntactic polymer foams, in *Handbook of Polymer Foams and Foam Technology* (eds Klempner, D. and Frisch, K. C.), Hanser Publishers, pp. 355–374.

61. Griffith, G. (2002) Carbon foam: a next-generation structural material. *Industrial Heating*, **69**, 47–52.

62. Evans, A.G., Hutchinson, J.W., and Ashby, M.F. (1998) Multifunctionality of cellular metal systems. *Progress in Materials Science*, **43**, 171–221.

63. Hosur, M.V., Abdullah, M., and Jeelani, S. (2005) Manufacturing and low-velocity impact characterization of foam filled 3-D integrated core sandwich composites with hybrid face sheets. *Composite Structures*, **69**, 167–181.

64. Bardella, L. and Genna, F. (2001) On the elastic behavior of syntactic foams. *International Journal of Solids and Structures*, **38**, 7235–7260.

65. Hasebe, R.S. and Sun, C.T. (2000) Performance of sandwich structures with composite reinforced core. *Journal of Sandwich Structures and Materials*, **2**, 75–100.

66. Li, G. and Jones, N. (2007) Development of rubberized syntactic foam. *Composites Part A: Applied Science and Manufacturing*, **38**, 1483–1492.

67. Li, G. and John, M. (2008) A crumb rubber modified syntactic foam. *Materials Science and Engineering A*, **474**, 390–399.

68. Li, G. and Muthyala, V.D. (2008) A cement based syntactic foam. *Materials Science and Engineering A*, **478**, 77–86.

69. Agarwal, B.D., Broutman, L.J., and Chandrashekara, K. (2006) *Analysis and Performance of Fiber Composites*, 3rd edn, John Wiley & Sons, Inc., Hoboken, New Jersey.

70. Li, G. and Chakka, V.S. (2010) Isogrid stiffened syntactic foam cored sandwich structure under low velocity impact. *Composites Part A: Applied Science and Manufacturing*, **41**, 177–184.

71. Li, G. and Muthyala, V.D. (2008) Impact characterization of sandwich structures with an integrated orthogrid stiffened syntactic foam core. *Composites Science and Technology*, **68**, 2078–2084.

72. Jackson, W.C. and Poe, C.C. (1993) The use of impact force as a scale parameter for the impact response of composite laminates. *Journal of Composite Technology and Research*, **15**, 282–289.

73. Delfosse, D. and Poursartip, A. (1997) Energy-based approach to impact damage in CFRP laminates. *Composites Part A: Applied Science and Manufacturing*, **28**, 647–655.

74. Zhou, G. (1998) The use of experimentally-determined impact force as a damage measure in impact damage resistance and tolerance of composite structures. *Composite Structures*, **42**, 375–382.

75. Reddy, T.Y., Wen, H.M., Reid, S.R., and Soden, P.D. (1998) Penetration and perforation of composite sandwich panels by hemispherical and conical projectiles. *ASME Journal of Pressure Vessel Technology*, **120**, 186–194.

76. Olsson, R. (2000) Mass criterion for wave controlled impact response of composite plates. *Composites Part A: Applied Science and Manufacturing*, **31**, 879–887.

77. Abrate, S. (2001) Modeling of impact on composite structures. *Composite Structures*, **51**, 129–131.

78. Olsson, R. (2001) Analytical prediction of large mass impact damage in composite laminates. *Composites Part A: Applied Science and Manufacturing*, **32**, 1207–1215.

79. Cheeseman, B.A. and Bogetti, T.A. (2003) Ballistic impact into fabric and compliant composite laminates. *Composite Structures*, **61**, 161–173.

80. Olsson, R. (2003) Closed form prediction of peak load and delamination onset under small mass impact. *Composite Structures*, **59**, 341–349.

81. Naik, N.K. and Shrirao, P. (2004) Composite structures under ballistic impact. *Composite Structures*, **66**, 579–590.

82. Bart-Smith, H., Hutchinson, J.W., Fleck, N.A., and Evans, A.G. (2002) Influence of imperfections on the performance of metal foam core sandwich panels. *International Journal of Solids and Structures*, **39**, 4999–5012.

83. Wicks, N. and Hutchinson, J.W. (2004) Performance of sandwich plates with truss cores. *Mechanics of Materials*, **36**, 739–751.

84. Hou, W.H., Zhu, F., Lu, G.X., and Fang, D.N. (2010) Ballistic impact experiments of metallic sandwich panels with aluminum foam core. *International Journal of Impact Engineering*, **37**, 1045–1055.

85. Huybrechts, S.M., Meink, T.E., Wegner, P.M., and Ganley, G.M. (2002) Manufacturing theory for advanced grid stiffened structures. *Composites Part A: Applied Science and Manufacturing*, **33**, 155–161.

86. Li, G. and Maricherla, D. (2007) Advanced grid stiffened FRP tube encased concrete cylinders. *Journal of Composite Materials*, **41**, 1803–1824.

87. Han, D.Y. and Tsai, S.W. (2003) Interlocked composite grids design and manufacturing. *Journal of Composite Materials*, **37**, 287–316.

88. Kidane, S., Li, G., Helms, J.E., Pang, S.S., and Woldesenbet, E. (2003) Analytical buckling load analysis of grid stiffened composite cylinders. *Composites Part B: Engineering*, **33**, 1–9.

89. Huybrechts, S.M. and Tsai, S.W. (1996) Analysis and behavior of grid structures. *Composites Science and Technology*, **56**, 1001–1015.

90. Jaunky, N. and Knight Jr, N.F. (1996) Formulation of an improved smeared stiffener theory for buckling analysis of grid-stiffened composite panels. *Composites Part B: Engineering*, **27B**, 519–526.

91. Jaunky, N., Knight Jr, N.F. and Amburb, D.R. (1998) Optimal design of general stiffened composite circular cylinders for global buckling with strength constraints. *Composite Structures*, **41**, 243–252.

92. Hohe, J., Beschorner, C., and Becker, W. (1999) Effective elastic properties of hexagonal and quadrilateral grid structures. *Composite Structures*, **46**, 73–89.

93. Lennon, R.F. and Das, P.K. (2000) Torsional buckling behaviour of stiffened cylinders under combined loading. *Thin-Walled Structures*, **38**, 229–245.

94. Ambur, D.R., Jaunky, N., and Hilburger, M.W. (2004) Progressive failure studies of stiffened panels subjected to shear loading. *Composite Structures*, **65**, 129–142.

95. Rackliffe, M.E., Jensen, D.W., and Lucas, W.K. (2006) Local and global buckling of ultra-lightweight IsoTruss® structures. *Composites Science and Technology*, **66** (2), 283–288.

96. Jadhav, P. and Mantena, P.R. (2007) Parametric optimization of grid-stiffened composite panels for maximizing their performance under transverse loading. *Composite Structures*, **77** (3), 353–363.

97. Li, G. and Cheng, J.Q. (2007) A generalized analytical modeling of grid stiffened composite structures. *Journal of Composite Materials*, **41**, 2939–2969.

98. Ibekwe, S.I., Mensah, P., Li, G., Pang, S.S., and Stubblefield, M.A. (2007) Impact and post impact response of laminated beams at low temperatures. *Composite Structures*, **79** (1), 12–17.

99. Pang, S.S., Li, G., Helms, J.E., and Ibekwe, S.I. (2001) Influence of ultraviolet radiation on the low velocity impact response of laminated beams. *Composites Part B: Engineering*, **32** (6), 521–528.

100. Helms, J.E., Li, G., and Pang, S.S. (2001) Impact resistance of a composite laminate bonded to a metal substrate. *Journal of Composite Materials*, **35** (3), 237–252.

101. Li, G., Pang, S.S., Helms, J.E., and and Ibekwe, S.I. (2000) Low velocity impact response of GFRP laminates subjected to cycling moisture. *Polymer Composites*, **21** (5), 686–695.

102. Li, G., Pang, S.S., Zhao, Y., and Ibekwe, S.I. (2000) Local buckling analysis of composite laminate with large delaminations induced by low velocity impact. *Polymer Composites*, **20** (5), 634–642.

103. Bianchi, F. and Zhang, X. (2012) Predicting mode-II delamination suppression in z-pinned laminates. *Composites Science and Technology*, **72** (8), 924–932.

104. Grassi, M., Zhang, X., and Meo, M. (2002) Prediction of stiffness and stresses in z-fibre reinforced composite laminates. *Composites Part A: Applied Science and Manufacturing*, **33** (12), 1653–1664.

105. Vaidya, U.K., Kamath, M.V., Hosur, M.V., Mahfuz, H., and Jeelani, S. (1999) Low-velocity impact response of cross-ply laminated sandwich composites with hollow and foam-filled Z-pin reinforced core. *Journal of Composites Technology and Research*, **21** (2), 84–97.

106. Zic, I., Ansell, M.P., Newton, A., and Price, R.W. (1990) Mechanical-properties of composite panels reinforced with integrally woven 3-D fabrics. *Journal of the Textile Institute*, **81** (4), 461–479.

107. Hou, Y.Q., Hu, H.J., Sun, B.Z., and Gu, B.H. (2013) Strain rate effects on tensile failure of 3-D angle-interlock woven carbon fabric. *Materials and Design*, **46**, 857–866.

108. Mouritz, A.P., Bannister, M.K., Falzon, P.J., and Leong, K.H. (1999) Review of applications for advanced three-dimensional fibre textile composites. *Composites Part A: Applied Science and Manufacturing*, **30** (12), 1445–1461.

109. Bilisik, K. (2012) Multiaxis three-dimensional weaving for composites: a review. *Textile Research Journal*, **82** (7), 725–743.

110. Baucom, J.N., Zikry, M.A., and Rajendran, A.M. (2006) Low-velocity impact damage accumulation in S2-glass composite systems. *Composites Science and Technology*, **66**, 1229–1238.

111. Sun, B., Liu, Y. and Gu, B. (2009) A unit cell approach of finite element calculation of ballistic impact damage of 3-D orthogonal woven composite. *Composites Part B: Engineering*, **40**, 552–560.
112. Sun, B., Hu, D., and Gu, B. (2009) Transverse impact damage and energy absorption of 3-D multi-structured knitted composite. *Composites Part B: Engineering*, **40**, 572–583.
113. Huang, H.J. and Waas, A.M. (2009) Modeling and predicting the compression strength limiting mechanisms in Z-pinned textile composites. *Composites Part B: Engineering*, **40**, 530–539.
114. Walter, T.R., Subhash, G., Sankar, B.V., and Yen, C.F. (2009) Damage modes in 3D glass fiber epoxy woven composites under high rate of impact loading. *Composites Part B: Engineering*, **40**, 584–589.

7

Self-Healing with Embedded Shape Memory Polymer Fibers

In Chapter 6, we discussed self-healing of macrocracks in an SMP matrix through confined shape recovery (confined expansion) of the SMP matrix. However, in practice, conventional thermosetting polymers such as epoxy, vinyl ester, polyester, etc., which do not have shape memory capability, are usually used as matrix in load-bearing structures, such as in fiber reinforced polymer composite structures. Therefore, it is more desired to endow a conventional thermosetting polymer matrix with self-healing capability so that a large variety of engineering structures can benefit. In this chapter, we aim to bestow a conventional thermosetting polymer matrix with self-healing capability by using a small amount of SMP fibers. The SMP fibers can close macrocracks in the polymer matrix through constrained shape recovery (shrinkage) of the embedded cold-drawn SMP fibers and the embedded thermoplastic particles can achieve molecular level healing through the CTH scheme.

7.1 Bio-inspired Self-Healing Scheme Based on SMP Fibers

Again, this concept mimics the biological healing process such as human skin: sensing, actuating, closing, and healing (SACH); that is, when a cut occurs, the skin senses the pain and actuates blood bleeding; the doctor then sutures the skin together by surgery (close) before new cells gradually grow (heal). To mimic the self-healing of human skin, we propose to use a percolated carbon nanotube network as strain sensors and heaters so that the healing process can be triggered autonomously, or healing-on-demand; we will then use the autonomous shrinking of SMP fibers for stitching/closing macroscopic cracks and molten thermoplastic particles for healing molecularly. Using short SMP fibers as an example, Figure 7.1 illustrates how the system works using the close-then-heal (CTH) scheme.

Figure 7.1(a) shows a cured specimen with dispersed SMP short fibers, thermoplastic particles, and a percolated carbon nanotube network. For clarity, the carbon nanotube network and the thermosetting polymer, such as an epoxy matrix, are represented by one entity. Before mixing with the thermosetting polymer matrix, the SMP fibers have been trained by strain hardening through cold drawing. Figure 7.1(b) shows that a crack is created in the composite.

Self-Healing Composites: Shape Memory Polymer-Based Structures, First Edition. Guoqiang Li.
© 2015 John Wiley & Sons, Ltd. Published 2015 by John Wiley & Sons, Ltd.

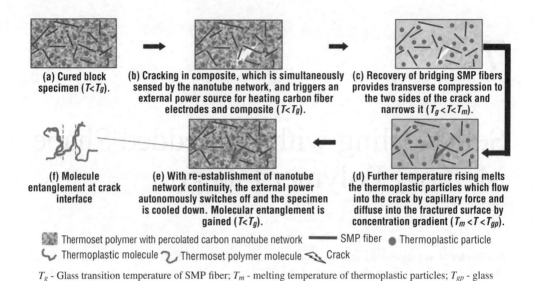

(a) Cured block specimen $(T<T_g)$.

(b) Cracking in composite, which is simultaneously sensed by the nanotube network, and triggers an external power source for heating carbon fiber electrodes and composite $(T<T_g)$.

(c) Recovery of bridging SMP fibers provides transverse compression to the two sides of the crack and narrows it $(T_g<T<T_m)$.

(f) Molecule entanglement at crack interface

(e) With re-establishment of nanotube network continuity, the external power autonomously switches off and the specimen is cooled down. Molecular entanglement is gained $(T<T_g)$.

(d) Further temperature rising melts the thermoplastic particles which flow into the crack by capillary force and diffuse into the fractured surface by concentration gradient $(T_m<T<T_{gp})$.

▨ Thermoset polymer with percolated carbon nanotube network ▬ SMP fiber ● Thermoplastic particle

↳ Thermoplastic molecule ∿ Thermoset polymer molecule ⌇ Crack

T_g - Glass transition temperature of SMP fiber; T_m - melting temperature of thermoplastic particles; T_{gp} - glass transition temperature of the thermoset polymer matrix; and T_c - curing temperature of the thermoset polymer matrix. Requirement: $T_{gp}>T_m>T_g>T_c$.

Figure 7.1 Schematic of the bio-inspired self-healing process of the proposed composite (2-D view)

Some SMP fibers may have been fractured or pulled out of the matrix. However, there are still some fibers bridging over the crack with varying embedded lengths and orientations due to the large ductility of SMP fibers. The cracking in Figure 7.1(b) is sensed by the embedded carbon nanotube network in terms of a fast increase in electrical resistance because of the damage to the carbon nanotube network by cracking. This signal is utilized to trigger an external power source, which heats the composite. If the carbon nanotube network is not sufficient to quickly heat the composite, some carbon fibers or metal wires can be embedded in the matrix, as electrodes. Once the temperature is above the glass transition temperature (amorphous) T_g or melting temperature (crystalline) T_m of the SMP fibers, the first healing step starts. In Figure 7.1(c), the fibers remember their original length and start to shrink. Due to the embedded length in the matrix, free recovery of the fibers is not allowed, leading to a compressive stress to the two sides of the crack. When properly designed, the compressive force may be sufficient to close the crack. Figure 7.1(d) shows that further heating leads to melting of the embedded thermoplastic particles, which will be sucked into the narrowed crack by a capillary force, adhered to the fractured surface due to chemical miscibility and diffused into the fractured surface by the concentration gradient and recovery stress. The re-establishment of continuity leads to a reduction in the electrical resistance, which will be utilized to switch off the external power, leading to cooling down to the ambient temperature and hardening of the thermoplastic wedge. A logic circuit can be designed to delay the switching off of the power source so that the molten thermoplastic has sufficient time to diffuse into the fractured matrix. Figure 7.1(e) shows that the molten thermoplastic wedge has been hardened after cooling down to room temperature and Figure 7.1(f) illustrates the physical entanglement of the thermoplastic molecules and matrix molecules. This completes one autonomous and molecular damage–healing cycle. As indicated by Li and Uppu [1], each confined shape recovery process also

Table 7.1 Comparisons of the two self-healing systems

Items for comparison	SMP matrix based self-healing system	SMP fiber based self-healing system
Matrix	SMP	Conventional thermosetting polymer
Fiber	No	Yes. SMP fiber in various forms such as 3-D woven, grid skeleton, unidirectional, short fiber, etc.
Programming	Volume reduction by compression	Lengthening by cold drawing
Mechanism for crack closing	Confined shape recovery by resisting free expansion of the SMP matrix (needs external confinement)	Constrained shape recovery by resisting free shrinkage of SMP fiber (does not need external confinement)
Cost	High. Large volume fraction of SMP as the continuous phase	Low. Small volume fraction of SMP fibers as the dispersed phase

represents a new programming cycle to the SMP fibers. Therefore, the damage–healing cycle is repeatable because the thermoplastic can also be repeatedly melted and hardened. It is noted that, similar to the SMP matrix based self-healing system discussed in Chapter 6, the biomimetic healing process is based on a physical phase change of materials, with no chemical reaction involved. This shows a distinct departure from other healing systems, which may require chemical reactions.

It is noted that while the sensing ability is introduced in Figure 7.1, we have not tested the sensing ability because our focus is on demonstrating the crack closing and healing process. However, we believe that the sensing function as proposed works because the carbon nanotube network as a damage sensing component has been well documented in the literature [1]. While a similar close-then-heal (CTH) scheme has been used in the SMP matrix based self-healing system in Chapter 6, there are some fundamental differences between the SMP matrix based self-healing and SMP fiber based self-healing, as will be discussed in this chapter. Table 7.1 summarizes the comparisons of the two self-healing systems.

7.2 SMP Fiber versus SMA (Shape Memory Alloy) Fiber

Neuser, Michaud, and White [2] and Kirkby *et al.* [3,4] proposed a very smart idea – a combination of microcapsule and shape memory alloy (SMA) wire for healing cracks in a conventional thermosetting polymer such as epoxy. It has been demonstrated that SMA wire helped the healing process and improved the healing efficiency. In their research, the cracks studied were in micro-length scale (about 150 μm), which were introduced in tapered double cantilever beam (TDCB) specimens (Figure 7.2(a)). Therefore, it is interesting to investigate whether SMA wires can close a wide-opened crack, which is the focus of our work. Based on the description in References [2] to [4], it is clear that the SMA wires have a very low interfacial bonding strength with the epoxy matrix, as evidenced by the clamps [3,4] and knots [2] used to facilitate transfer of the recovery force from the SMA wires to the epoxy matrix. Actually, the authors mentioned debonding of the SMA wire from the epoxy during the heating process [2]. We would like to use finite element analysis (FEA) to evaluate the ability for SMA wire to close a wide-opened crack. To simplify the FEA, we use a 2-D plane stress element and replace the three SMA wires by three pairs of concentrated line load along the thickness of the TDCB

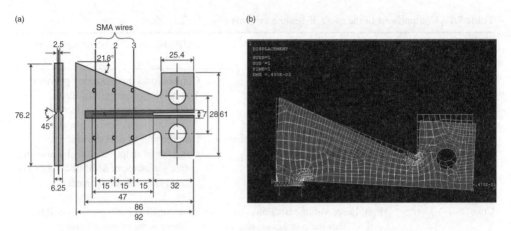

Figure 7.2 (a) TDCB specimen with three SMA wires. *Source:* [2] Reproduced with permission from Elsevier; (b) Finite element analysis of the TDCB specimen with a 1 mm gap

specimen. The line load is defined as the recovery force of each wire divided by the thickness of the TDCB specimen (6 mm). The width of the groove is 1 mm (mimicking a wide-opened crack) (see Figure 7.2(b)). The symmetry condition was used for the modeling. It is noted that the recovery load is not constant. It changes with the recovery strain. At this moment, we only know two control values from Reference [3]: a 240 MPa recovery stress at 0% recovery strain and a 0 MPa recovery stress at 2.3% recovery strain. If we assume a linear relationship between the recovery stress (σ) and strain (ε), we have $\sigma = (240\,\text{MPa}/2.3\%)(2.3\% - \varepsilon)$. An iterative process was used in FEA by using the ANSYS software package. The result is shown in Figure 7.2(b). The 1 mm gap becomes 0.92 mm, 0.70 mm, and 0.51 mm wide at the locations of the left wire, middle wire, and right wire in Figure 7.2(a), respectively, after activation of the SMA wires. In other words, the 1 mm gap cannot be closed. In summary, while the idea in References [2] to [4] is very innovative and works pretty well for microcracks, the challenge is still how to heal macrocracks.

We will now find out what is the "ideal" shape memory (SM) material for narrowing wide-opened cracks. For this reason, let us discuss an ideal model, as shown in Figure 7.3.

In this beam model (length $2L$), the two ends of the beam are clamped with a central crack of width $2W$. Now we want to narrow this crack using SM material. Let us insert an SM rod of length $2W$ with the same cross-section as the beam, and tightly glue it to the fractured two half beams. Assuming the SM material is pre-stretched so that it has a fully constrained recovery stress of σ_o (maximum recovery stress) and free shape recovery strain of ε_o (maximum

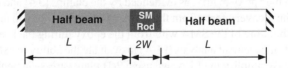

Figure 7.3 An ideal crack-closing model by the shape memory material

recovery strain). Now assuming that the recovery stress (σ) is linearly related to the recovery strain (ε):

$$\sigma = \sigma_o \left(1 - \frac{\varepsilon}{\varepsilon_o} \right)$$

(7.1)

where $0 \leq \varepsilon \leq \varepsilon_o$.

Further, it is assumed that the modulus of the beam is E and the temperature of the beam is uniform, and the SM rod is glued to the fractured beam by strong and insulating glue so that interfacial debonding and heat transfer can be avoided. Now the SM rod is activated by an external stimulus such as heat, light, magnetic field, electricity, etc., based on the particular SM material. Due to the SM effect, the SM rod tends to shrink, and applies the shape recovery stress (σ) to the fractured beam. Based on the mechanics of materials, the displacement of the beam at the fractured surface is (σ/E)L and the shrinkage of half of the SM rod is (εW) (we only need to consider half of the specimen because of symmetry). Based on the deformation compatibility (no interfacial debonding at the rod/beam interface), we have (σ/E)$L = (\varepsilon W)$, which leads to

$$\varepsilon = \frac{L\sigma_o \varepsilon_o}{E W \varepsilon_0 + L\sigma_0}$$

(7.2)

Based on this equilibrium recovery strain, the shrinkage (D) of the SM rod on each side of the rod is

$$D = \varepsilon W = \frac{L W \sigma_o \varepsilon_o}{E W \varepsilon_0 + L\sigma_o}$$

(7.3)

Let us now discuss the effect of the maximum recovery stress (σ_o) and maximum recovery strain (ε_o) of the SM rod on the crack closing ability. The SMA in Reference [3] is used as a reference SM material. Also, we fix the E at 500 MPa (modulus of EPON 828 at 80 °C [5]) and W as 1.0 mm, that is, a crack opening of 2.0 mm. We want to see how the maximum recovery stress, maximum recovery strain, and beam length affect the shrinkage of the SM rod, D. Figure 7.4 shows the comparison of six SM rods with the baseline SMA. From Figure 7.4, it is clear that all the SM rods lead to a larger closure of the gap than that of the SMA rod. Obviously, with the same maximum recovery strain, the ability to close the crack increases as the maximum recovery stress increases, and vice versa. Also, the crack closure efficiency increases as either the beam length L or $\sigma_o\varepsilon_o$ increases. It is interesting to note that ($0.5\sigma_o\varepsilon_o$) represents the energy density stored in the SM rod due to programming. This makes sense because the crack closing is a process that the SM rod does work to the beam, and the ability to do work (close the crack) depends on the energy stored in the rod. Also, it is noted that the largest strain in the beam is 0.6%, which is created due to shrinkage of the SM rod. This is much less than the ultimate tensile strain of EPON828 at 80 °C, suggesting that the beam will not be damaged in order to close the central wide-opened crack in the beam. Based on the above analysis, it is seen that an ideal SM fiber does not need to have a very high recovery stress; a reasonably lower recovery stress but higher recovery strain SM fiber also works. Actually, since the recovery stress becomes zero when the maximum recovery strain is

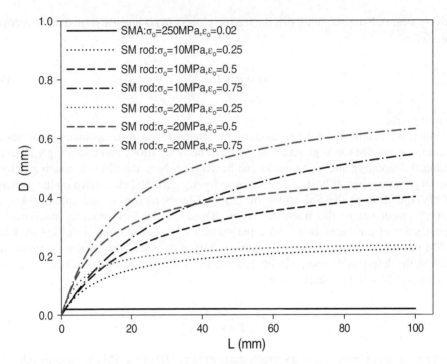

Figure 7.4 Comparisons of crack-closing ability of various shape memory materials

reached, increasing the recovery strain is more effective than increasing the recovery stress. Indeed, the stored energy density ($0.5\sigma_o\varepsilon_o$) is a better indicator of crack closing ability.

We believe that we have found such a fiber, that is, the shape memory polymer fiber. Various methods have been used to prepare the fibers for nonstructural applications such as in the textile industry [6–12]. In comparison with the SMPs in bulk forms, the smart fibers have a much higher mechanical strength, recovery stress, and recovery ratios because of the macromolecular pre-orientation of the soft segment and hard segment obtained during the spinning process [13]. The mechanical strength of the pre-oriented fibers can be five times more than that of SMP films [13]. The slight molecular pre-orientation in the smart fibers improves the recovery stress to about 2 MPa while that of the SMP bulk is less than 1 MPa [6,13]. In addition, a high recovery speed was observed on the pre-orientated smart fibers during the recovery process. Figure 7.5 shows test results of partially constrained recovery stress–recovery strain of semicrystalline shape memory polyurethane fibres (SMPFs). The fibers were either coated with a thin layer of epoxy coating to increase their shape fixity or without coating for comparison [14]. The fibers were cold-drawing programmed to 100% pre-strain before the partially constrained stress recovery test.

From Figure 7.5, it is estimated that the stored energy density is about 8.32 MJ/m^3 for the coated SMPF and 5.50 MJ/m^3 for the uncoated SMPF, which are higher than the SMA wire used in Reference [3], that is, $0.5 \times (240 \text{ MPa}) \times (2.3\%) = 2.76$ MJ/m^3. Therefore, it is expected that the SMPF will have better crack closing ability than the SMA wire.

In the following, we will report healing of thermosetting polymers by SMPFs in various configurations, including 1-D, 2-D, and 3-D.

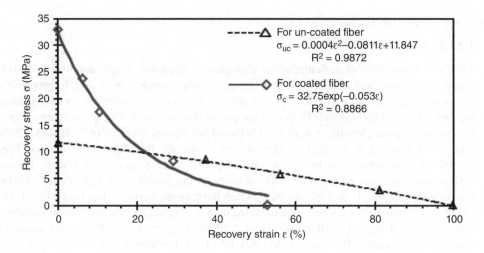

Figure 7.5 Recovery stress–recovery strain behaviour of coated and uncoated SMPFs. *Source:* [14] Reproduced with permission from Elsevier

7.3 Healing of Thermosetting Polymer by Embedded Unidirectional (1-D) Shape Memory Polyurethane Fiber (SMPF)

According to the biomimetic CTH mechanism, the ability for the SMPFs to close a wide-opened crack depends on the recovery force (more accurately, energy) applied to the two sides of the crack. Usually, as-spun SMPFs, which have very limited molecule alignment during the manufacturing process, have a very small recovery stress (less than 2 MPa) [15], which may not be sufficient to close wide-opened cracks. It has been found that strain hardening through cold-drawing programming can increase the recovery stress of thermoplastic SMPs [16–18]. It has also been found that cold-compression programming or 3-D confined programming of thermosetting SMPs can also increase the recovery stress [1,19]. Therefore, strain hardening through cold-drawing programming of thermoplastic SMPFs may enhance the healing efficiency of the proposed system. It is noted that, as discussed by Li and Shojaei [20], it is impractical to heat an entire structure with a large size in order to activate the constrained shape recovery of the embedded SMPFs and to heal a crack somewhere within the structure, for instance a crack in the fuselage of an aircraft. Ideally, heating should be localized and surround the cracked region, that is, "hot" in the cracked region and "cold" in the remainder of the structure. However, it is not clear if local heating can heal a wide-opened crack or not with embedded SMPFs. Therefore, in this section, we will demonstrate the capability for unidirectional SMPF to close macroscopic cracks and heal the narrowed cracks by the close-then-heal scheme. Particularly, we will investigate (1) the level of cold drawing on the healing efficiency of a thermosetting polymer composite embedded with thermoplastic SMPFs and thermoplastic particles, (2) the possibility of healing a wide-opened crack by localized heating surrounding the cracked region, and (3) the possibility of healing a wide-opened crack with fixed boundary conditions.

7.3.1 Experimentation

7.3.1.1 Raw Materials

EPON^{TM} Resin 828, an undiluted clear difunctional bisphenol A/epichlorohydrin derived liquid epoxy resin, and D.E.H. 24 curing agent by Dow Chemical were used to prepare the thermosetting polymer matrix. This is a low temperature curing polymer that can avoid shape recovery of the embedded SMPFs during the curing process. According to the manufacturer, the cured epoxy has a tensile strength of 69 MPa and modulus of elasticity of 2750 MPa. CAPA 6506 by Perstorp UK Limited, a high molecular weight linear polyester derived from caprolactone monomer, was used as the thermoplastic healing agent. It is a white powder with a molecular weight of 50 000. The density of the particle is 1.1 g/cm^3, the melting temperature is 58–60 °C, and 98% of the powders have a particle size of less than 0.6 mm. The SMPFs were polyurethane fibers manufactured by a melt spinning process. The polyurethane was synthesized using poly(butylene adipate) (PBA) (Sigma-aldrich, USA), 4′4-diphenyl-methane diisocyanate (MDI) (Sigma-aldrich, USA) and 1,4-butanediol (BDO) (Sigma-aldrich, USA) with PBA as the soft segment phase and (MDI + BDO) as the hard segment phase. Dibutyltin dilaurate was used as catalyst with a content of 0.02 wt%. The molecular weight of PBA is about $M_n = 650$. The average formula weight ratio was (MDI + BDO):PBA = 1021:300. Based on the formula weight ratio, the segmented SMPFs have about 77% by weight of hard segment domain and about 23% by weight of soft segment domain. Figure 7.6 shows a polarized optical microscope image of the SMPF.

7.3.1.2 Differential Scanning Calorimetry (DSC) Test

In order to determine the glass transition and melting temperature of the SMPFs, the thermal properties of the original SMPFs and the SMPFs after cold-drawing programming were investigated using a differential scanning calorimeter (Perkin-Elmer Diamond-4000) with nitrogen as the purge gas. Indium or zinc were used for calibration. Two thermal scans were conducted. The first scan started by cooling the SMPF sample, which was about 6.30 mg and length of about 2 mm, from ambient temperature to −50 °C; it was then heated to 240 °C, which completed the first scan. The sample was then quenched to −50 °C. Finally, the sample was scanned from −50 to 240 °C again, which completed the second scan. The first thermal scan was used to eliminate the thermal history during manufacturing and storage. For both scans, the

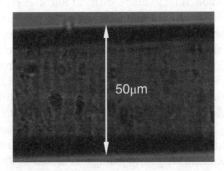

Figure 7.6   Polarized optical microscope image of the SMPF

Figure 7.7 DSC test machine

heating and cooling rate was the same, which was 10 °C/min. Figure 7.7 shows the DSC machine.

7.3.1.3 Tensile Test on SMPFs

In order to understand the stress–strain behavior of the SMPFs and strain-hardening during the cold-drawing process, MTS (Alliance RT/5, MTS Inc., USA) equipped with a 250 N load cell was used to test the fibers by tension at room temperature. The loading rate was 20 mm/min as per the ASTM D76/D76 M standard. The fiber was stretched until failure to determine the maximum tensile strength. Figure 7.8 shows the test setup.

7.3.1.4 Strain Hardening by Cold-Drawing Programming of the SMPFs

The SMPFs were cold-drawn to various pre-strain levels in order to increase their recovery stress. The cold-drawing process was coupled with the specimen preparation of the composite beams. The SMP filament from a spool was first wound by an improvised fiber winder into a fiber strand or bundle, with each strand consisting of 200 filaments (fibers). Steel nails (guiding pins) were driven into the two ends of a wood board based on the designed length and spacing between fiber strands for winding the fiber strands and providing pre-tension. The fiber strand was respectively stretched to 50% or 100% strain around the pins and continued winding and stretching until the designed volume fraction of the fibers was achieved. In this study, we selected the fiber volume fraction as 9.9%. The fiber strand without stretching (0% pre-strain) was also used as a control. We selected this fiber volume fraction because (1) it is similar to that used in grid stiffened composite panels, in which a good impact tolerance and load carrying

Figure 7.8 Fiber test setup

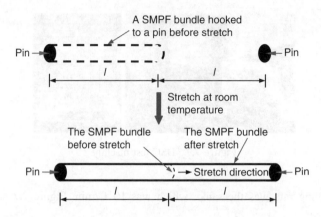

Figure 7.9 Schematic of an SMPF bundle being stretched to 100% pre-strain. *Source:* [21] Reproduced with permission from Elsevier

capacity has been exhibited [22–24]; (2) we wanted to demonstrate that even a low volume fraction of SMPFs can close a wide-opened crack to a certain level; and (3) we wanted to see the effect of strain hardening of SMPFs on the crack-closing efficiency. In other words, if we used too many SMPFs, the crack will be narrowed to the smallest level, regardless of the strain hardening effort, so that the effect of strain hardening on the crack narrowing capacity cannot be differentiated. A schematic of a fiber strand being stretched to 100% pre-strain is shown in Figure 7.9.

7.3.1.5 Beam Specimen Preparation and Crack Creation

Once the fiber strands were placed into the bottom wood board, four other pieces of wood boards were attached to the bottom board by nails to form a wooden mold. The mold had a length of 177.8 mm, width of 38.1 mm, and height of 5.0 mm. Therefore, the panel from this mold will yield three beam specimens with a length of 177.8 mm, width of 12.7 mm, and thickness (height) of 5.0 mm. Each beam specimen has four parallel SMPF strands in one layer. Because each strand has 200 fibers and the diameter of each filament is 0.05 mm, it is estimated that a total of 3200 fibers are needed to have a fiber volume fraction of 9.9%. From this estimation, there are four layers of fiber strands through the thickness for the strands without pre-tension. Figure 7.10 shows a schematic of the top view and side view of the beam specimen.

In order to investigate the healing efficiency in a controlled way, artificial cracks were needed. To create the artificial crack with an opening of 0.15 mm, an aluminum foil with thickness of 0.15 mm was cut and placed at the centre of the mold. The foil was machined into a comb shape so that it perfectly mated with the four fiber bundles for each beam specimen in Figure 7.10. A releasing agent was sprayed on to the foil so that it could be easily pulled out after curing, which creates a pre-crack in each beam specimen with SMPF strands bridging over the crack, as schematically shown in Figure 7.10.

Once the SMPF skeleton was prepared within the mold, the particulate polymer matrix was prepared. The thermoplastic particles (CAPA 6506) were first added to the EPON 828 resin while stirring by a mechanical mixer. The volume fraction of the thermoplastic particles was

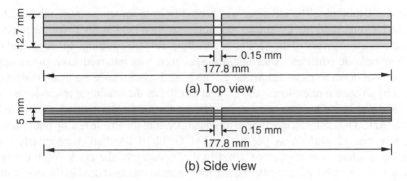

(a) Top view

(b) Side view

Figure 7.10 Schematic view of the beam specimen

7%, which was similar to the volume fraction of thermoplastic particles used previously in the SMP matrix based biomimetic self-healing system [25,26]. After about 10 minutes of stirring, the D.E.H. 24 hardener was added to the mixture at a volume ratio of 9 : 1 of resin:hardener and stirring continued for another 10 minutes. The mixture was then powered into the wood mold until the designed thickness of 5.0 mm was achieved. The mold was then left to cure at room temperature for 4 days. In the first 24 hours, an electric fan was used to quickly remove the heat generated during curing so that the SMPFs would not be recovered by the curing heat. After that, the panel was demolded and beam specimens with a length of 177.8 mm, width of 12.7 mm, and height of 5.0 mm were cut by a water jet machine for tensile testing.

7.3.1.6 Repeated Fracture/Healing Test on Beam Specimens

Prepared beam specimens with a pre-crack were brought to the MTS Q-TEST 150 machine for a healing test. The specimen was tightly clamped to the tensile test fixture, which was extended into a heating chamber. About 12.7 mm at each end of the specimens was clamped, which suggests that the gage length of the specimens was about 152.4 mm. Thermocouples were installed into the chamber to monitor the temperature. The Q-TEST 150 machine with the chamber is shown in Figure 7.11. In order to record the recovery stress while maintaining the fixed boundary condition, the tensile test mode was set up but the loading rate of the MTS

Figure 7.11 Q-TEST 150 machine with heating chamber

machine was 0 mm/min. The chamber was heated from room temperature (about 20 °C) at a rate of 2.67 °C/min until 80 °C. The temperature was then held at 80 °C for 20 minutes. This allowed the shape recovery of the SMPFs and melting, flowing, wetting, diffusion, and bonding of the thermoplastic particles. After that, the specimen was removed from the chamber and allowed to cool down to room temperature. The healed specimens were then nondestructively examined by an optical microscope and C-Scan to validate the healing at micro-length scale, as will be discussed later. Once the specimens were nondestructively inspected, they were brought back to the MTS Q-TEST 150 machine at room temperature for tensile testing until fracture. The loading rate was 12 mm/min as per the ASTM D638-10 standard. Immediately following fracture, the machine was stopped to avoid further opening of the crack. With the specimen clamped at the two ends by the tensile fixture, the machine was re-started in the tensile mode but with a zero loading rate. The healing process proceeded again by heating. The healing process followed the same procedure and used the same parameters as the first healing cycle. This fracture/healing cycle continued for seven cycles to evaluate the repeatability of the healing.

It is noted that the beam specimen was fixed at the two ends during the healing process. This is the most difficult condition for an in-service structure to heal, that is, the boundary is not allowed to move during healing. As discussed previously in Chapter 6, some SMP based healing studies by others used the free boundary condition for healing, which, as implied by the meaning of shape memory, works even without prior programming of the SMP. This is because the shape memory suggests that the material would restore the original uncracked configuration and the cracking process can be treated as ad hoc programming. Therefore, if the cracked specimen is allowed to recover freely, crack closing is almost guaranteed if the shape recovery ratio is about 100%. However, in practice, the structural component is almost always subjected to some type of global boundary constraints and local constraints by the material surrounding the crack. Therefore, a key criterion to evaluate the healing efficiency of self-healing schemes is to use a constrained boundary condition during the healing process.

7.3.1.7 Nondestructive Testing

In this study, both microscopic observation and C-Scan were conducted on the prepared specimens and healed specimens. An ultrasonic inspection was performed on the specimens before and after each healing cycle using a 5 MHz transducer for the specimen. An UltraPac inspection machine from the Physical Acoustics Laboratory was used in conjunction with UltraWin software to acquire the C-Scan images and identify damage and healing. The same transducer and setting were used for every specimen to view it before and after recovery for comparison. An Amscope optical microscope was also used to visualize the crack healing at macro-length scale and a JEOL JSM – 6390 scanning electron microscope (SEM) was used to visualize the crack healing at micro-length scale. Figure 7.12 shows the UltraPac C-Scan transducer.

7.3.2 Results and Discussion

7.3.2.1 DSC Test Results

Figure 7.13 shows the DSC test results of the as-spun SMPF and the same SMPF cold-drawn to 100%. From Figure 7.13, the DSC curve of the original SMPF shows a melt transition peak at about 167.9 °C, which can be ascribed to the melting transition of the hard segment phase. After 100% cold drawing, the DSC curve of the SMPF shows dual peaks at about 168.4 °C and

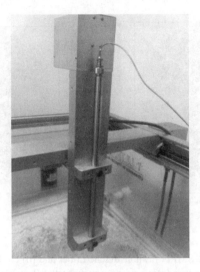

Figure 7.12 UltraPac C-Scan transducer

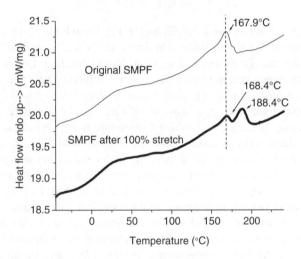

Figure 7.13 DSC results of the original SMPF and the SMPF after 100% cold drawing. *Source:* [21] Reproduced with permission from Elsevier

188.4 °C, respectively. The dual peaks indicate that two types of crystallites in the cold-drawing fibers are formed, that is, type I crystallites with a melting temperature at around 168.4 °C and type II crystallites with a melting temperature at around 188.4 °C. The type II crystallite corresponds to more perfect crystal structures or a larger crystal size, while type I crystallite corresponds to less perfect crystal structures or a smaller crystal size. Obviously, the original SMPF only has type I crystallites. Cold drawing partially transforms type I crystallites to type II crystallites, which indicates that a more stable hard segment phase is formed and a higher stiffness is obtained. The new type II crystallites are formed by stress induced crystallization (SIC). It is noted that the molecular weight of the soft segment is about $M_n = 650$, which is difficult to form a crystal

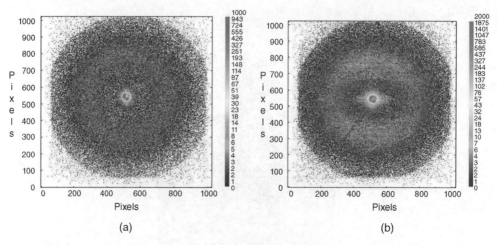

Figure 7.14 SAXS image for SMPF: (a) nonstretched fiber and (b) cold-stretched fiber up to 350% of strain leve. *Source:* [20] Reproduced with permission from the Royal Society

structure even under strain hardening conditions [16,17]. From Figure 7.13, the glass transition temperature is about 10 °C for both 100% cold-drawn SMPF and as-spun SMPF. This suggests that programming has less effect on the soft segment phase in this study. One reason may be due to the low shape fixity. When preparing the DSC sample, the holding time (stress relaxation time) during cold-drawing programming was almost zero, leading to very low shape fixity. In other words, the pre-stretch in the soft phase was not fully locked. Consequently, the effect of cold-drawing programming on the glass transition temperature is not very obvious. However, it is noted that, if the shape fixity is considerable, for instance allowing a longer time of stress relaxation during cold-drawing programming, the molecules in the soft phase domain will become aligned along the loading direction, that is, become more ordered, which will lead to an increase in the glass transition temperature for the cold-drawn SMPFs [14].

The effect of cold-drawing programming on the crystalline phase can be further evidenced by the small-angle X-ray scattering (SAXS) test. Figure 7.14 shows the image for SMPFs without cold-drawing programming and with cold-drawing programming [20]. The oriented diffraction scattering pattern after cold-drawing programming is an indication of the formation of stress-induced crystallization. It is noted that the SMPF used here is different from that used in Chapter 5 because they have different diameters, pre-strain levels, programming temperatures and cycles, and physical aging histories. This is why the SMPF used here is semi-crystalline, while the SMPF in Chapter 5 is amorphous.

7.3.2.2 Tensile Test Results of SMPFs

After cold-drawing programming with various pre-tensions, the SMPFs are tested again for tensile strength. Figure 7.15 shows the tensile test results of the SMPFs without pre-tension, with 50% pre-tension, and with 100% pre-tension, respectively. It is clear that pre-tension increases the stiffness (slope of the stress–strain curve) and strength (peak stress) of the SMPFs. Of course, like strain hardening by cold working on metals, the increase in strength and stiffness is as a result of reduction in ductility. Therefore, a balance or optimal strain hardening needs to be found that

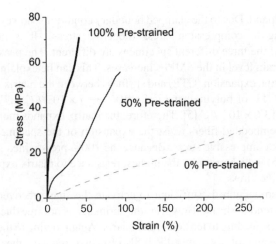

Figure 7.15 Tensile test results of SMPFs with various pre-tensions. *Source:* [21] Reproduced with permission from Elsevier

provides both the required recovery stress for healing and sufficient strength, stiffness, and ductility for carrying structural load. The reason for the enhancement in stiffness and tensile strength is due to the stress induced crystallization (SIC) in the hard segment domain, as shown in Figure 7.14, and molecule alignment of the soft phase domain along the loading direction.

7.3.2.3 First Healing Test Results

As discussed previously, the healing test is conducted with a clamped boundary, or zero recovery strain at the boundary. This is the worst case scenario in practice for self-healing, that is, healing under loading. The recovered stress and temperature with time for the first cycle of healing are shown in Figure 7.16 for specimens with different pre-tensions in SMPFs. From Figure 7.16, the recovery stress–time curves can be divided into three regions. In the first region, the specimen is heated until about 43 °C. As a result of thermal expansion, the beam

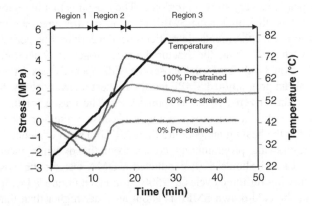

Figure 7.16 Recovery stress during the first healing test for pre-cracked beam specimens embedded with SMPFs having various pre-tensions. *Source:* [21] Reproduced with permission from Elsevier

specimen tends to expand. Due to the clamped boundary conditions, however, free expansion is not allowed, leading to compression stress (negative stress). It is noted that the peak compressive stress of the three different specimens are different. The peak compressive stress reduces as the pre-strain level in the SMPFs increases. This can be explained by the difference in coefficient of thermal expansion (CTE) and stiffness between the matrix and the SMPF. CTE tests show that the CTE of polyurethane is $3 \times 10^{-5} \sim 7 \times 10^{-5}/°C$ [27], while the CTE of EPON 828 is about $13.6 \times 10^{-5}/°C$ [5]. Therefore, the matrix expands more than the fibers or, in other words, the embedded fibers resist the expansion of the specimen, which leads to a reduction in the peak compressive stress. Because the 100% pre-strained fiber has the highest stiffness (Figure 7.15), which provides the highest resistance to matrix expansion, it yields the least peak compressive stress.

In region 2, the temperature is sufficient to activate the shape recovery of the embedded SMPFs. The SMPFs tend to shrink. However, the shrinkage is not free due to the constraint by the surrounding matrix, leading to tension of the matrix. Again, owing to the clamped boundary condition, the shrinkage of the embedded SMPFs first reduces the compressive stress continuously until zero. Further shrinking leads to a tendency of reduction in specimen length. This constrained shortening leads to the development of tensile stress, until a peak tensile stress is achieved, which is at about 60 °C. This is the end of region 2. It is noted that for the specimen with 0% pre-tensioned SMPFs, no tensile stress is developed due to the limited recovery ability of as-spun SMPFs. Taking the recovery stress as the difference between the peak tensile stress and the peak compressive stress, the specimens with 0% pre-strained fibers have the least recovery stress of 2.05 ± 0.06 MPa, the 50% pre-strained fibers have the recovery stress of 3.70 ± 0.03 MPa, and the 100% pre-strained fiber has the highest recovery stress of 4.86 ± 0.03 MPa. In region 3, with further heating and soaking at 80 °C for 20 minutes, the embedded thermoplastic particles melt and diffuse into the fractured matrix to complete molecular-length scale healing. Melting of the thermoplastic particles leads to softening of the matrix. Because of the softening of the matrix, the peak tensile stress reduces until stabilized, which shows that the specimen has a constant stiffness at the fixed healing temperature of 80 °C and the recovery stress of the SMPFs is stabilized. As discussed in Chapter 5, another reason for reduction in post-peak recovery stress is due to stress relaxation of the SMPF.

During the self-healing process, it is the stabilized recovery stress in region 3 that helps close the crack, not the peak recovery stress in region 2. The reason is that in region 2 the matrix is still very stiff and it is difficult for the SMPF to bring the fractured surfaces closer. Also, melting and diffusion of the thermoplastic particles need a certain time period, which requires sustained recovery stress to hold the fractured matrix together. Therefore, the stabilized recovery stress in region 3 is the stress that closes the wide-opened crack. In region 3, it is clear that the 100% pre-strain programmed SMPF has the highest recovery stress, followed by 50% pre-strain, and it is almost zero for the as-spun SMPF. The reason for the enhancement in recovery stress due to cold-drawing programming is because of the increase in stiffness of the cold-drawn SMPFs at the healing temperature (here 80 °C). Direct evidence of the increase in stiffness after cold-drawing programming comes from the dynamic mechanical analysis (DMA) test result. Li and Shojaei [20] conducted a DMA test for the as-spun SMPF and SMPF after cold drawing for three cycles to 250% strain (see Figure 7.17). It is clear that the storage modulus of the cold-drawn SMPF is about an order higher than that of the as-spun SMPF at the healing temperature of 80 °C and plateaus in a wide range of temperature, suggesting a significant increase in stiffness and thermal stability. Because the constrained

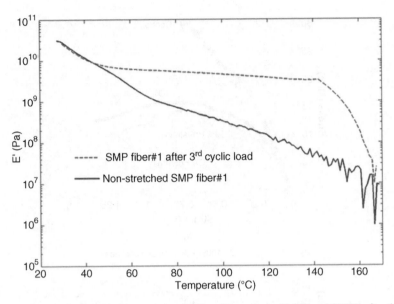

Figure 7.17 DMA test results of the storage modulus of an as-spun SMPF and SMPF after three-cycle cold drawing programmed to 250% with a diameter of 40 μm. *Source:* [20] Reproduced with permission from the Royal Society

recovery stress is in direct proportion to the stiffness of the SMPFs, the increase in stiffness at the healing temperature leads to the increase in the constrained recovery stress of the cold-drawn SMPFs.

7.3.2.4 Repeated Healing–Fracture Test Results

After the first tensile test (re-fracture of the healed specimen), the specimen was put back into the oven to heal, following the same procedure as the first healing. The specimens healed for the second time were tensile fractured again. These healing–fracture cycles proceeded for seven cycles. Figure 7.18(a) shows the tensile test results after the first healing and Figure 7.18(b) shows the tensile test results after the seventh healing. The control specimen in Figure 7.18(a) is the specimen without pre-cracking but embedded with nonprogrammed SMPFs (virgin specimen). From Figure 7.18(a), the beam specimen with the 100% pre-strained fibers has the highest residual tensile strength, while the specimen with 0% pre-strained fibers has the least. Defining the healing efficiency as the ratio of the peak residual tensile stress of the healed specimen over the tensile strength of the original uncracked specimens, the healing efficiency for the first healing is $53 \pm 3.7\%$, $44 \pm 1.0\%$, and $23 \pm 1.0\%$, for specimens with 100% pre-strained fibers, 50% pre-strained fibers, and 0% pre-strained fibers, respectively. Comparing Figure 7.18(b) with Figure 7.18(a), the healing is quite repeatable. There is little difference between the tensile test results after the seventh healing and those after the first healing. The reason is that both the SMPFs and thermoplastic particles experience only physical changes during the healing and fracture cycles. Because the healing temperature of 80 °C is lower than the melting temperature of the hard segment phase of the SMPFs, which is 168.4 °C, as shown

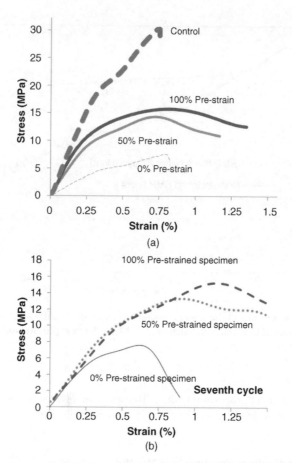

Figure 7.18 Tensile test results after (a) the first healing and (b) the seventh healing. *Source:* [21] Reproduced with permission from Elsevier

in Figure 7.13, the only change during each healing cycle is the physical phase change of the soft segment phase (glassy → rubbery → glassy). For the thermoplastic particles, the only change during each healing cycle is also a physical phase change (glassy → liquid → glassy). No chemical damage is involved in the healing and re-fracturing process. Therefore, the healing is repeatable.

It is noted that the residual tensile strength depends primarily on the strength of the thermoplastic particles and the thickness of the thermoplastic layer within the narrowed crack. The reason is that the molten thermoplastic forms a thin film in the narrowed crack under the recovery pressure by the SMPFs, similar to the adhesive layer in an adhesively bonded joint (here the fractured two half beams serve as the adherends). It has been well demonstrated that the tensile and shear resistance of adhesively bonded joints highly depends on the adhesive thickness. A thinner adhesive layer usually leads to higher peel and shear resistance [28–32]. This is one reason why a 100% pre-strained SMPF leads to a higher healing efficiency because

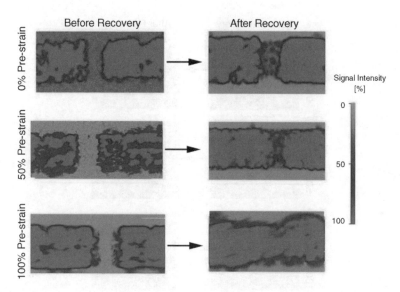

Figure 7.19   C-Scan images of thermosetting polymer beam specimens embedded with 9.9% by volume shape memory polymer fibers having pre-tensions of 0%, 50%, and 100%: (left) pre-crack before recovery and (right) crack closure after the first healing. *Source:* [21] Reproduced with permisson from Elsevier

it has a thinner thermoplastic adhesive layer, as will be demonstrated later using SEM. Also, the strength of the structure (the healed beam or joint) depends on the adhesive strength. Higher tensile strength thermoplastics can further increase the residual tensile strength and healing efficiency of the healed beams.

7.3.2.5   Nondestructive Evaluation

The C-Scan images of the beam specimens after fabrication and after the first healing are shown in Figure 7.19, the corresponding optical microscopic images are shown in Figure 7.20, and the SEM observations are shown in Figure 7.21. In these pulse-echo C-Scan images in Figure 7.19, red color (gray area) represents an excess of 80% of the signal returning to the receiver, whereas blue color (dark area) indicates that 50–80% of the signal is being received. The green color (light gray area) represents a complete alteration of the ultrasound signal or 0% of the signal received. This could mean a cracked region. It is clear from Figure 7.19 that a pre-crack has been created at the center of the specimens through the fabrication process. After the first healing, the pre-crack for the specimen with 0% pre-strained SMPFs is poorly healed as the color in the cracked region changes from green (light gray) to blue (dark); the specimen with 100% pre-strained SMPFs heals quite well as the color in the cracked region changes from green (light gray) to red (gray). This result is in agreement with the healing efficiency test results. The optical microscope test results and SEM results echo the C-Scan test results. It is seen in the SEM images that the crack in the specimen with 100% pre-strained SMPFs is closed to about 20 μm, while the crack in the specimen with 0% pre-strained fibers is closed to about 60 μm, demonstrating the increase in the crack-closure ability for SMPFs with strain hardening.

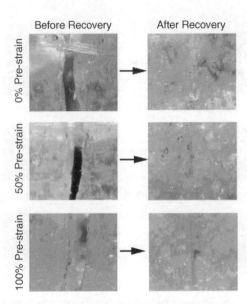

Figure 7.20 Optical images of the beam specimens after fabrication (left) before recovery and after first healing and (right) after recovery. *Source:* [21] Reproduced with permisson from Elsevier

Based on the parameters in this study, $L = 76.2$ mm, $W = 0.075$ mm, and the modulus of the EPON 828 epoxy at the healing temperature of 80 °C is about $E = 500$ MPa [5]. From Figure 7.16, the recovery stress for the 100% pre-tensioned SMPF is $\sigma_o = 4.86$ MPa and if the free shape recovery strain is assumed to be $\varepsilon_o = 100\%$, we found from Equation (7.3) that the value of $D = 0.068$ mm. For both sides of the crack, $2D = 0.136$ mm. Therefore, the original crack

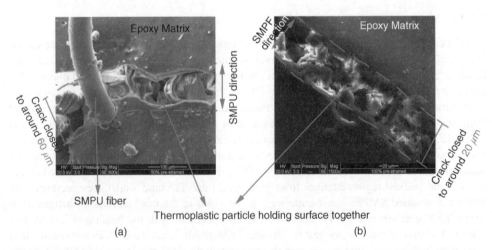

Figure 7.21 SEM observations of the healed crack: (a) SMPFs without pre-tension (crack narrowed to 60 μm) and (b) SMPFs with 100% pre-tension (crack narrowed to 20 μm). *Source:* [21] Reproduced with permisson from Elsevier

with a width of $2W = 0.15$ mm is narrowed to $(2W - 2D) = 0.14$ mm, or 14 µm, which is close to the SEM observation of around 20 µm in Figure 7.21 for the 100% cold-drawing programmed SMPF reinforced beam after healing. This experimentally validates the ideal model in Figure 7.3 and the derived Equations (7.1) to (7.3).

7.3.3 Summary

Based on the comprehensive manufacturing, testing, and characterization, it is seen that macroscopic cracks in a conventional thermosetting polymer matrix can be healed by embedded thermoplastic shape memory polymer fibers. It is found that the ability for the SMPFs to close a wide-opened crack depends on the recovery stress, or more accurately on the recovery energy. We emphasize recovery stress because the recovery stress of SMPFs is comparatively low. With the same fiber volume fraction, the higher the recovery stress, the higher the capability to close the crack, and the higher the healing efficiency. The increase in healing efficiency is not in direct proportion to the pre-strain level. While the healing efficiency almost doubles with 50% pre-strained fibers as compared to fibers without pre-strain, further pre-straining to 100% does not increase the efficiency at the same rate. This may be due to the saturated contribution of closing the crack on healing efficiency because it is the thickness of the thermoplastic thin film that predominates the recovered strength. The result in this study confirms that healing can be achieved by heating locally within a certain area surrounding the cracked region, which is very important for application in large structures because heating an entire structure is impractical. Most importantly, this study shows that the proposed CTH healing can be achieved under fully fixed boundary conditions, which is the most unfavorable condition for crack closing and molecular healing. It is also found that the mechanism for the increase in the constrained recovery stress and enhanced healing efficiency of the cold-drawn SMPFs is due to the increase in stiffness of the hard segment domain at the healing temperature. Cold drawing leads to molecular alignment and formation of some perfect crystals (stress induced crystallization) in the hard segment domain. Based on the biomimetic CTH scheme and the analogy of a healed beam specimen to an adhesively bonded joint, it is concluded that reducing the thermoplastic film thickness within the narrowed crack and increasing the strength of the thermoplastic thin film can further enhance the healing efficiency. Finally, the SEM observation of the crack width after healing validates the ideal healing model in Figure 7.3 and Equations (7.1) to (7.3) developed in this chapter.

7.4 Healing of Thermosetting Polymer by Embedded 2-D Shape Memory Polyurethane Fiber (SMPF)

In this section, the same SMPF used in the previous section will be used to reinforce the same epoxy matrix which is embedded with the same thermoplastic particles. However, instead of aligning the fiber in one direction, we align the fiber in two directions, forming an orthogrid skeleton. As discussed in References [22] to [24], a grid stiffened architecture endows the composite structure with a better impact tolerance and higher structural capacity as compared to its laminated composite counterpart with the same fiber volume fraction. As for the crack-closing capability, because of its 2-D nature, it can close an in-plane crack efficiently, which shows a distinct departure from the unidirectional fiber (1-D) reinforcement. A unidirectional SMPF cannot close an in-plane crack that is parallel with the fiber. The healing process of a crack in the bay area is schematically shown in Figure 7.22.

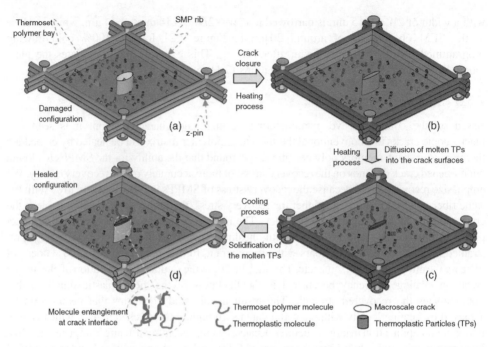

Figure 7.22 Schematic of the bio-inspired healing process of the proposed composite (3-D view). (a) A unit cell (bay) of an SMP grid (ribs and Z-pins) stiffened conventional thermoset polymer dispersed with thermoplastic particles. A macroscopic crack is introduced in the unit cell, which can be identified by visual or nondestructive inspections ($T < T_g$). (b) Crack-closure process through recovery of the SMP fiber ribs and Z-pins, when local heating is applied ($T > T_g$). (c) Further temperature rising melts the thermoplastic particles that flow into the crack by capillary force and diffuse into the fractured surface by the concentration gradient ($T > T_m$). (d) By cooling down to below the glass transition temperature, a solid wedge can be formed and molecular entanglement can be established ($T < T_g$). The magnified view shows the molecule entanglement at the crack interface. *Source:* [20] Reproduced with permission from the Royal Society

In order to validate the healing scheme in Figure 7.22, a finite element analysis was conducted on an SMP orthogrid stiffened thermoset polymer composite (see the schematic in Figure 7.23(a)). The central bay in Figure 7.23(a) contains a macroscale crack, which is an ellipsoidal hole with a major diameter of 20 mm, a minor diameter of 5 mm, and height of 5 mm, which suggested that the through-thickness crack has a maximum crack opening of 5 mm. In order to validate that crack closing can be conducted by heating around the cracked region, heating is conducted locally on the central bay and its surrounding four ribs. The following parameters were assumed: the SMP fiber reinforced rib has a modulus of 600 MPa and a maximum recovery stress of 10 MPa and the polymer composite in the bay has a modulus of 1000 MPa. The boundary of the grid stiffened panel was fixed. It is believed that, through localized heating, the elastic modulus of the thermosetting matrix is reduced. On the other hand, the programmed SMP fiber exhibits excellent thermal stability. As shown in Figure 7.17, the elastic modulus of the SMP fiber does not vary significantly within the temperature range

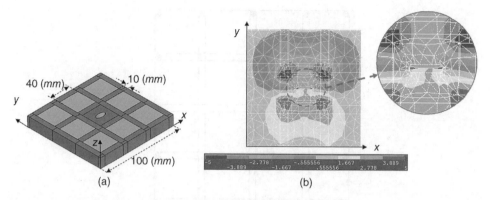

Figure 7.23 Closing of the macroscopic crack due to shape recovery of the SMP grid skeleton through localized heating: (a) an SMP orthogrid stiffened thermoset polymer composite with a macroscopic crack at the center of the central bay and (b) FEA results for the displacement field in the Y direction when stored stresses in SMP fibers are released by a local heating process. The magnified view shows closure of the macroscopic crack. *Source:* [20] Reproduced with permission from the Royal Society

investigated. Hence, it is expected that the macrocrack in the softened thermosetting matrix can be narrowed or closed by the shrinkage of the SMP fiber at the healing temperature.

Solid tetrahedral elements with 10 nodes are used. Once the localized recovery stresses overcome the stiffness of the bay the walls of the crack are collapsed together, as shown in the magnified view in Figure 7.23(b). The displacement field in the Y direction is also shown in Figure 7.23(b), which confirms that the walls of the crack march towards each other due to the applied localized compressive stresses from the SMP ribs.

7.4.1 Specimen Preparation

We started with preparing the SMPF grid skeleton. A dry weaving process was used. The SMPF wound around the guiding pins nailed to a wood board. The guiding pins were spaced 12.7 mm apart, which suggests that each bay is a square with a side length of 12.7 mm. The weaving pattern is schematically shown in Figure 7.24. The weaving continued until the SMPF volume fraction was 7%. After that, the Epon epoxy with 12% by volume of the CAPA thermoplastic particles was poured into the grid skeleton. The composite was vacuum bagged and cured at ambient temperature (22 °C) for 24 hours. Figure 7.25 shows the grid skeleton.

Once the grid stiffened composite panel was prepared, it was machined into beam specimens for damage and healing. The beam specimen has a length of 152.4 mm, width of 25.4 mm, and thickness of 12.7 mm. At the centre of the specimen, a transverse notch at both the bottom and top surfaces was created using a sharp blade. The depth of the notch was about 1 mm to avoid cutting the fibers. The notched beam specimen was then mounted on the MTS Q-TEST 150 machine to determine the tensile strength. The clamped length at each end was 12.7 mm. The specimen was stretched until fracture. After fracture, the specimen was further stretched to increase the crack opening. In this study, the crack opening was about 3 mm.

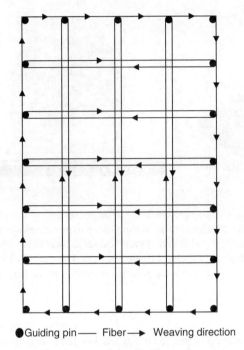

● Guiding pin —— Fiber —→ Weaving direction

Figure 7.24   Schematic of the dry weaving pattern

Figure 7.25   The grid skeleton through the dry weaving process

7.4.2 Self-Healing of the Grid Stiffened Composite

The cracked beam specimen was healed by directly putting the specimen in an oven. The temperature of the oven was set at 80 °C. It was found that within about 2 minutes, the crack with a 3 mm opening was closed, as shown in Figure 7.26(a), which was obtained using a high resolution CCD camera (Sony XCD-CR90) with a resolution of 3.7×3.7 μm/pixel.

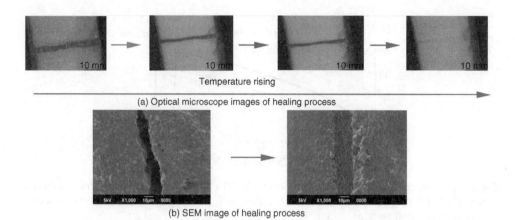

(a) Optical microscope images of healing process

(b) SEM image of healing process

Figure 7.26 (a) As the temperature rises, the crack is gradually narrowed/closed, as shown in the optical microscope images. (b) Further heating to the bonding temperature of the thermoplastic particles (80 °C) leads to diffusion of the molten thermoplastic molecules into the fractured matrix, establishing physical entanglement, as shown in the SEM images. *Source:* [33] Reproduced with permission from the Royal Society

The composite beam with the closed/narrowed crack was maintained in the oven for 15 more minutes and was then moved out of the oven for cooling. Figure 7.26(b) shows that the narrowed crack has been fully bonded by the molten thermoplastic, that is, healed.

TDCB (tapered double-cantilever beam) specimens were used to determine the healing efficiency and repeatability of the healing. The specific geometry of the TDCB can be seen in Figure 7.27(a). The healing efficiency was evaluated by a tensile test using the Q-TEST 150

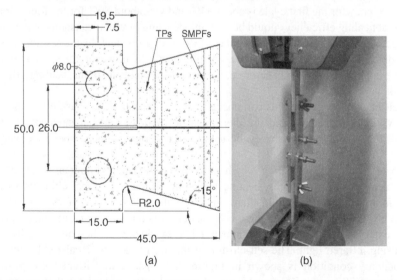

(a)　　　　　　　　　　(b)

Figure 7.27 (a) Schematic drawing of the TDCB specimen with dimensions (mm) and (b) the home-made holders for the test of the self-healing efficiency by tension. *Source:* [33] Reproduced with permission from the Royal Society

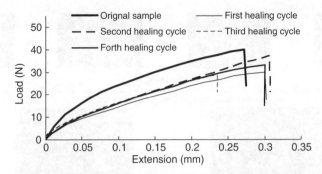

Figure 7.28 Maximum loads of the original and healed TDCB samples after different fracture–healing cycles. The healing efficiency of the first healing cycle is 94%. The healing process is repeatable. *Source:* [33] Reproduced with permission from the Royal Society

machine with lab-made specimen holders as shown in Figure 7.27(b). The protocol for evaluating the self-healing efficiency using the TDCB sample followed that of Reference [34]. Self-healing of the TDCB specimen was achieved simply by keeping the cracked TDCB samples in a preheated oven (Gruenberg Model L34HV104 from Lunaire Limited, USA) at 80 °C for 15 minutes. The oven was turned off after 15 minutes of soaking and the healing efficiency test on the healed specimen was conducted after 24 hours of resting at room temperature. After fracture, the specimen was healed again following the same procedure. The healing/fracture continued for four cycles. Typical tensile load-extension curves for the original and healed specimens are shown in Figure 7.28 for up to four healing–fracture cycles. The healing efficiency was defined as the ratio of the peak tensile load of the healed specimen over that of the original specimen without damage and without healing. Based on Figure 7.28, the healing efficiency for the first cycle is about 94% and was about 80% for the fourth fracture. It is noted that healing efficiency should be determined using physical constants of the virgin and healed specimens, such as fracture toughness [35]. Using the peak load does not consider the effect of deformability of the material. Based on Figure 7.28, however, the strain at fracture for each cycle is almost the same. Therefore, the peak load yields a similar healing efficiency as toughness.

The self-healing process of this composite system is repeatable because the shape recovery of the SMPFs is repeatable and the thermoplastic particles can be repeatedly melted and solidified. Figure 7.29 shows the cyclic tensile and thermal induced shape recovery of the SMPFs.

It is noted that the above healing is by heating the entire panel without any sustained load to the specimen. In large structures, heating the entire structure is impossible and usually the structure is subjected to external loading. Therefore, it is interesting to investigate whether the crack can be closed or not by heating locally under a fixed boundary condition (equivalent to under an external load). We have demonstrated crack closing of the composite by localized heating using infrared light. The self-closure of the macrocrack by localized heating under a fixed boundary condition is shown in Figure 7.30. The grid stiffened composite panel dimension was 300 mm × 100 mm × 5 mm. The crack was created by three-point bending using a Q-TEST 150 machine. The width of the crack was about 1 mm. From Figure 7.30(a), the crack closing was achieved by directly heating the specimen, which was clamped by the

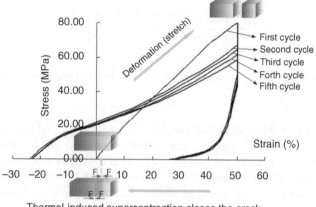

Thermal-induced supercontraction closes the crack
and exerts compress force on to the interface.

Figure 7.29 Cyclic tension and thermal-induced shape recovery of the SMPFs are repeatable. The shape recovery of the SMPFs closes macrocracks in the composite and exerts a sustained compressive force on the crack interfaces. *Source:* [33] Reproduced with permission from the Royal Society

grip of the machine. The localized heating was achieved using an infrared light lamp (EVA Medical Group LLC, USA). Figure 7.30(b) shows that the crack was closed. This result confirms that, similar to the 1-D SMPF reinforced beams, as discussed in the previous section, an SMPF grid stiffened beam can be healed by localized heating when the specimen is under external loading (here the fixed boundary condition).

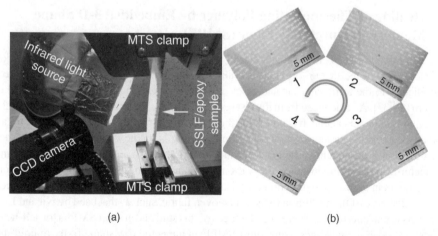

(a)   (b)

Figure 7.30 Self-closure of the macrocrack through localized infrared light heating under a fixed boundary condition. (a) Setup of the experiment for the localized heating using infrared light. The specimen was fixed on the MTS grips. (b) The crack-closing process of the macrocrack in the composite. The macrocrack was closed as a result of the constrained shrinkage of the embedded SMPF grid skeleton. *Source:* [33] Reproduced with permission from the Royal Society

Figure 7.31   Free body diagram shows the equivalence of the fixed boundary condition to an applied external force F

It is noted that the fixed boundary condition simulates an external load applied to the end of the beam when the healing starts. Figure 7.31 shows the free body diagram. In the diagram, the fixed boundary condition is replaced by the axial force F, shear force Q, and bending moment M. The recovery force by the SMPF is denoted as P. From Figure 7.31, when the SMPF recovers, it applies a tensile force P to the fractured surface. Based on static equilibrium, the clamped boundary applies the same force F to the boundary but in the opposite direction. Therefore, the fixed boundary condition simulates a force applied in the longitudinal direction that is equal to the shape recovery force. In other words, the fixed boundary condition simulates a cracked beam that is subjected to a longitudinal force which is equal to the recovery force. Because the recover force varies during the healing process, the equivalent external force also varies.

7.4.3 Summary

In this study, a thermosetting polymer composite embedded with 2-D SMPFs in an orthogrid pattern was prepared. A macroscopic crack as wide as 3 mm can be healed through the CTH scheme with a healing efficiency about 90% as determined by the tensile test. It is also found that the healing is repeatable. Self-healing through localized heating surrounding the cracked area with varying external loading (fixed boundary condition) was also conducted. It again shows the crack-closing ability by the SMPF grid skeleton under more realistic conditions.

7.5   Healing of Thermosetting Polymer by Embedded 3-D Shape Memory Polyurethane Fiber (SMPF)

There is no doubt that fibers perpendicular to the crack surface have the highest crack-closing efficiency. Therefore, in our previous studies, all the fibers were aligned in the direction perpendicular to the artificially created crack [20,21,33], with fibers either in a 1-D [21] or 2-D pattern [20,33]. As evidenced from the test results, the crack healing efficiency is quite high and the healing is quite repeatable. It is noted that these fiber patterns are especially effective for an in-plane crack. In actual structures, however, some loading is unknown or unpredictable, such as impact loading. Consequently, the crack location and orientation are also not known during the manufacturing process. For instance, the crack may be out-of-plane. For this type of random crack, aligning fibers in an in-plane direction only is not an ideal strategy. It is better to align fibers in three dimensions (3-D). In addition to the 3-D woven fabric, such as that used by Nji and Li [36], short fibers are an alternative. Therefore, the focus of this study is on short SMPFs for self-healing.

There are several challenges using short SMPFs. One is the low shape fixity of cold-drawn SMPFs. In a previous study, the shape fixity of cold-drawing programmed SMPU fibers (in this section, SMPF and SMPU fiber are used interchangeably) was 33% [37]. This number is even lower than cold-compression programmed thermosetting SMPs [19]. To achieve a higher shape fixity, one possible way is to coat the programmed SMPU fiber with fixing agents. The other challenge is the fabrication of a programmed short SMPU fiber reinforced composite. Without a hard shell, short SMPU fibers tend to cluster and entangle together. Again, coating with a fixing agent is necessary. Also, for a short fiber reinforced composite, the fiber length plays an important role in the mechanical properties [38–40]. However, it is not clear how the fiber length affects the self-healing efficiency.

Therefore, we aim to answer the following questions in this study: (1) a proper coating agent to enhance the shape fixity and dispersion ability of the strain hardened short SMPFs; (2) the ability for strain hardened SMPU fibers to repeatedly heal wide-opened cracks; and (3) the effect of fiber length on the healing efficiency.

7.5.1 Experiment

7.5.1.1 Preparation, Programming, and Coating of SMPFs

SMPF bundles were programmed by stretching to a strain of 100% at room temperature by the MTS machine (RT/5, MTS Inc., USA) equipped with a 250 N load cell. Each bundle consisted of 6 to 12 filaments. A half hour structural relaxation was followed. Then an acrylic conformal coating was sprayed on the stretched fibers, allowing 24 hour curing of the coating agent. As shown in Figure 7.32, the thickness of the conformal coating is much smaller than the diameter of the fiber; such a thin layer of coating could not hold the stretched fiber to improve its shape fixity. Hence, a two-step method was developed to improve the shape fixity of the stretched SMPFs. The stretched fiber was coated first with the acrylic conformal coating and then with epoxy. The reason why the epoxy coating only is not used is that the curing agent for epoxy is also an organic solvent for polyurethane fiber, which would be catastrophic for the stretched fiber. Therefore, the usage of the acrylic conformal coating is to protect the fiber against the organic solvent. Then the epoxy EPON828 was sprayed after 24 hour curing of the conformal coating. Once the two-step coating was completed, one week was allowed for curing the epoxy. Some programmed SMPFs did not have the coating layer. They were used as controls. It is noted that the SMPFs were coated as a bundle, not an individual filament (see Figure 7.32).

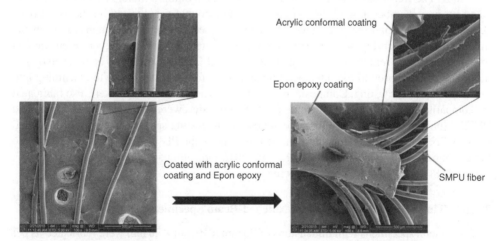

Figure 7.32 SEM observation of the pure SMPFs and coated programmed SMPFs with a 100% stretching ratio. *Source:* [14] Reproduced with permission from Elsevier

7.5.1.2 Programmed Short SMPF Preparation

The critical fiber length (L_c) of programmed short SMPFs is estimated by the shear-lag theory [41]:

$$L_c = \frac{\sigma_u d}{2\tau_m} \tag{7.4}$$

where σ_u is the ultimate tensile strength of the programmed SMPFs, d is the programmed fiber diameter, and τ_m is the shear yield strength of the matrix. Based on a tensile test result of the programmed SMPFs, the programmed SMPF at 100% pre-strain has an ultimate strength of 690 MPa. After programming, the diameter of the fiber is 0.028 mm. As provided by the manufacturer, the shear yield strength of the acrylic coating agent is 6.35 MPa. Therefore, the critical fiber length (L_c) is calculated based on Equation (7.4), which is 1.52 mm.

It is noted that the critical fiber length as determined by Equation (7.4) is the shortest fiber length required for effective shear force transfer. The fiber is supposed to be fully embedded in the matrix for a full shear stress transfer. Because of the 3-D dispersion of the short fibers with most of them inclined with respect to the loading direction, even for those fibers that bridge over the crack, the fiber length within one side of the fractured matrix may be smaller than that required for the shear-force transfer, although the other side might be sufficient. Therefore, the actual fiber length required for effective stress transfer may be much longer than the calculated critical fiber length. In this study, the coated programmed SMPFs were cut into short fibers with lengths of 4 mm, 7 mm, and 10 mm, respectively.

7.5.1.3 Preparation of the Composite

When preparing the composite, the resin and 10% by volume of thermoplastic particles were mixed first for 10 minutes. Second, 13% by weight of curing agent was mixed with the mixture for 2 minutes, then 5% by volume of short fibers were added to the mixture and mixed for 3 minutes. The well-mixed mixture was poured into an aluminum mold. After that, the mold was kept in a chamber at 4 °C for 24 hours. A thermocouple was located on the top surface center of the mold and the Yokogawa DC-100 was used to monitor the temperature of the mixture during the curing process. The first day of low temperature curing was to ensure that the curing heat was removed so that the embedded SMPU fibers would not be recovered. The mold was then exposed to room temperature and cured for three days prior to machining into beam specimens for testing. Composite beams without short SMPU fibers were also fabricated for comparison. For the sake of convenience, the composites were denoted as non-PUFRC and PUFRC (polyurethane fibre reinforced composite), suggesting specimens without SMPF and with SMPF, respectively. Depending on the fiber length, the PUFRC was further divided into three groups, 4-PUFRC, 7-PUFRC, and 10-PUFRC, indicating specimens with 4 mm SMPF, 7 mm SMPF, and 10 mm SMPF, respectively.

7.5.1.4 Three-Point Bending Test of Notched-Beam Specimens

Notched-beam specimens with dimensions of 90 mm × 10 mm × 20 mm were fabricated as per the ASTM D 5045 standard. The notch was machined by bandsaw. A schematic of the notched beam specimen is shown in Figure 7.33. We used single-edge notched-beam (SENB)

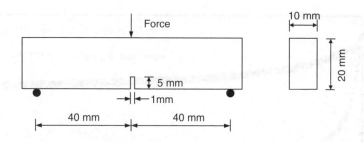

Figure 7.33 Schematic of the specimen and three-point bending test setup

specimens in order to create a structural scale crack. The three-point bending tests were carried out by the MTS (RT/5, MTS Inc., USA) system with a 250 N load cell at a loading rate of 10 mm/min. The beam specimens were fractured completely into two halves at room temperature. The process of the crack initiation and propagation was captured by a high resolution CCD camera (Sony XCD-CR90), which is equipped with a light source and a digital interface, which interlinks the computer and the camera. The CCD camera used in this test has a resolution of 3.7×3.7 µm/pixel. *Image-J* data acquisition software was used to store the image data from the camera during testing.

7.5.1.5 Close-Then-Heal (CTH) Self-Healing

It will be seen later that, while complete fracture of the notched beam might cause fiber pull-out for some fibers, those fibers that bridge over the crack were not fractured because of the high ductility of the SMPFs, which were the fibers leading to closure of the crack per the CTH scheme, suggesting that these beams were not separated into two halves. For the control beam without the programmed short SMPFs, the separated two half beams were put together manually, carefully aligning along the crack. The completely fractured beam specimen was then transferred into an oven. The oven was heated from room temperature (about 21 °C) to 80 °C at a rate of 5 °C/min. The specimen was held inside the chamber at 80 °C for 30 minutes. At this moment, the shape recovery effect of the programmed short SMPU fibers were activated, thus closing the crack automatically to micro-scale. Because of the melting and diffusion of the thermoplastic particles, a thermoplastic thin film formed at the interface, which served as an adhesive layer that glued the fractured matrix together. Cooling led to hardening of the thermoplastic film and healing of the fractured beam. This completed the first fracture–healing cycle. This fracture–healing continued for five cycles to evaluate the ability for repeatedly healing wide-opened cracks. During the fracture testing, the peak bending load was recorded, which was used to calculate the healing efficiency. At least three effective specimens were tested for each group of samples.

7.5.1.6 Nondestructive Testing and SEM Observations

The crack was measured by a UNICO zoom stereo microscope, equipped with an optional digital camera (AmScope MD35). Also, the healed crack was investigated using an optical microscope (VanGuard, USA) equipped with a camera (XLI, USA). The advantage of the nondestructive testing is to ensure that the damaged cracks and/or healed cracks have no interruption from some unknown factors.

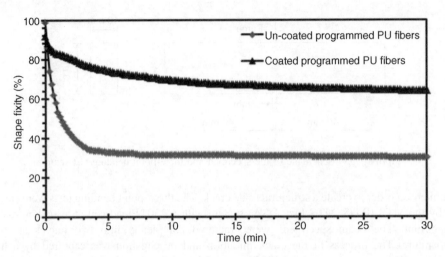

Figure 7.34 Shape fixity for uncoated and coated programmed SMPFs at room temperature (about 21 °C). *Source:* [14] Reproduced with permission from Elsevier

The fractured surfaces for specimens before healing and after healing were characterized using a scanning electron microscope (SEM) (Quanta 3D FEG field-emission electron microscope). This was used to investigate the crack interface, where an adhesive layer was formed by molten thermoplastic particles for bonding the two fractured surfaces.

7.5.2 Results and Discussion

7.5.2.1 Efficiency of Fiber Coating on Shape Fixity and Stress Recovery

The efficiency of the coating on fibers can be evaluated by the shape fixity ratio and stress recovery ratio as these are the parameters concerned with SMPFs without coating. As shown in Figure 7.34, the shape fixity of the programmed SMPFs has been significantly improved by the coating agent from 30% to 66%. It also shows that the shape fixity of an uncoated programmed SMPF relaxes fast for the first 3 minutes and becomes stabilized within 5 minutes. This is due to the fast structural relaxation at the early stage after the removal of the stretching load. However, in the case of a coated programmed SMPF, it keeps relaxing and is stabilized in about 30 minutes. This can be explained as follows. The removal of the stretching load leads to large springback and viscoelastic rebound in the first few minutes. However, due to the layer of coated fixing agent, free spingback of the fiber is not allowed, leading to slower structural relaxation of the fiber. Based on the shape fixity information, the coated programmed SMPU fibers were cut into short fibers and mixed with resin after 30 minutes of the removal of the programming stretching load.

The stress recovery behavior of the coated programmed and uncoated programmed SMPFs is shown in Figure 7.35. From Figure 7.35, the stabilized recovery stress for a coated programmed SMPF is 25.0 MPa, while in the case of an uncoated programmed SMPF, the recovery stress is 13.2 MPa. The difference between the recovery stresses for coated and uncoated programmed SMPFs is due to their different shape fixity ratios. The higher shape fixity ratio of the coated fiber leads to a larger conformational entropy change and better

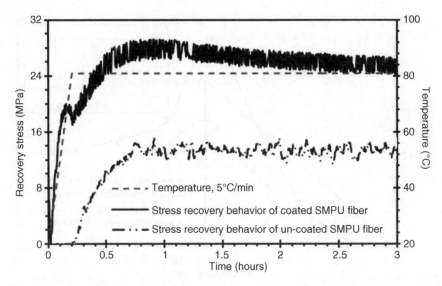

Figure 7.35 Stress recovery behavior of a coated and a uncoated programmed SMPF under temperatures from room temperature (about 21 °C) to 80 °C. *Source:* [14] Reproduced with permission from Elsevier

alignment of the molecules along the fiber direction [37,42]. Such a change would lead to a higher potential driving force for shape recovery under a thermal stimulus and thus a higher recovery stress.

7.5.2.2 Temperature Evolution During Curing of the Composite

One challenge of using SMPFs in a thermosetting polymer is how to avoid shape recovery of the SMPFs during the curing process as the curing of thermosetting polymers is usually exothermic. In this study, we put the mold in a chamber with a temperature of 4 °C in the first 24 hours of curing. Figure 7.36 shows the measured temperature of the composite in the first

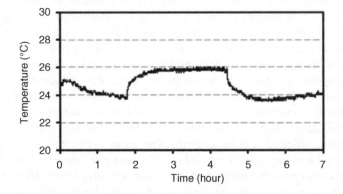

Figure 7.36 Evolution of the curing temperature during the curing process. *Source:* [14] Reproduced with permission from Elsevier

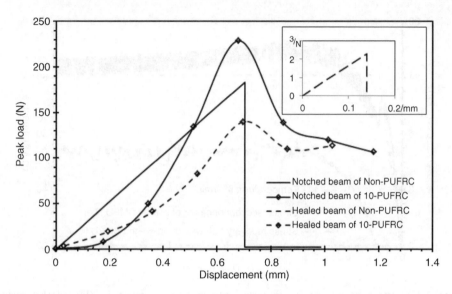

Figure 7.37 Typical load–deflection curves of notched and healed specimens after the first CTH cycle.
Source: [14] Reproduced with permission from Elsevier

7 hours of curing. From Figure 7.36, after about 2 hours of adding the curing agent, fast curing started, as evidenced by the temperature rising, which lasted for about two and a half hours, and eventually slowed down. The maximum recorded temperature was 26 °C, which is lower than the glass transition temperature of the cold-drawing programmed SMPFs, which is about 45 °C. This suggests that the SMPFs were not recovered in the matrix during the curing process.

7.5.2.3 Self-Healing Test Results

Typical load–deflection curves for four different groups of specimens (10-PUFRC notched beam, non-PUFRC notched beam, healed 10-PUFRC beam, and healed non-PUFRC beam) are shown in Figure 7.37. Several observations can be made. (1) Incorporation of SMPFs in the polymer matrix increased the peak bending load and the post-peak load bearing capacity, as shown by the comparison between the 10-PUFRC beam and the non-PUFRC beam. It is clear that the notched non-PUFRC beam fractured in a brittle manner, while the notched 10-PUFRC beam exhibited post-peak ductile failure, thanks to SMPF bridging over the crack. It is also noticed that the deflection corresponding to the peak load for the two types of beams are almost the same. The reason is that only 5% by volume of SMPFs was used. Therefore, the behaviour of the beam is controlled by the matrix before it fractures. The increase in peak load for the 10-PUFRC is due to the fiber reinforcement. (2) As the fractured non-PUFRC beam was manually aligned before healing, its healing efficiency can be neglected. The reason is that there is no sustained crack closure force applied to the beam to hold the two fractured half beams together. Therefore, the molten thermoplastic cannot be effectively diffused into the matrix. For the 10-PUFRC beam, the SMPFs provide a sustained compressive force to the two half beams, which not only helped in bringing the fractured half beams marching towards each other but also helped flowing, wetting, and diffusion of the molten thermoplastic into the fractured matrix. Therefore, it has a considerable healing efficiency. (3) The healed 10-PUFRC beam crests at a similar deflection, again due to the matrix dominated fracture before the peak load is achieved. Clearly, in order to increase the peak bending load further, we should either reduce the crack opening as a thinner adhesive layer usually has a higher peel or shear resistance [28–32], or increase the strength of the adhesive layer (adhesion and cohesion), or both. It is believed that better miscibility between the thermoplastic and thermosetting polymer matrix can enhance the adhesion strength and a stronger thermoplastic can increase the cohesion strength.

7.5.2.4 Effect of Fiber Length on Healing Efficiency

The healing efficiency is defined as the ratio of the peak bending load of the healed PUFRC beam over the peak bending load of the original non-PUFRC beam. The healing efficiencies for the first healing are about 52%, 56%, and 77% for the 4-PUFRC, 7-PUFRC, and 10-PUFRC, respectively. These numbers respectively become 42%, 52%, and 61% for the fifth healing cycle. Obviously, the healing efficiency reduces as the fracture–healing cycle increases. This is almost true for any self-healing system. Generally speaking, the healing is repeatable for a programmed short SMPF reinforced thermosetting particulate polymer composite. Also, as the fiber length increases, the healing efficiency also increases. It is believed that this increase will gradually approach that of continuous uniaxial fibers.

7.5.2.5 Nondestructive Inspection

Using the high resolution CCD camera, the crack opening during the three-point bending test and the crack narrowing due to springback of the SMPFs are obtained. Figure 7.38 shows the image immediately before removing the bending load and immediately after removing the bending load for the three types of SMPF reinforced specimens. A paper ruler has been attached to the specimen to facilitate measurement of the crack opening width. From Figure 7.38, the notch, the opened crack, and the SMPFs can be clearly identified. The bending crack width was about 1.8 mm immediately before removing the bending load. Because of the springback of the SMPFs, the crack was narrowed to 0.7 mm, 0.6 mm, and 0.32 mm for the 4-PUFRC, 7-PUFRC, and 10-PUFRC beams, respectively, after removal of the bending loads at room temperature. Therefore, the closure of the crack under the two-step CTH is initiated by the springback of the programmed short SMPFs. It is interesting to investigate why springback occurred as the springback of the SMPFs due to programming has been allowed before fabricating specimens. This is because the crack opening process during the bending test is also a cold-drawing process for fibers bridging over the crack. Longer fibers have a longer length embedded in the matrix and thus store a higher energy. Once the bending load is removed, the longer fibers spring back more and lead to a narrower crack.

The crack closing due to shape memory of the bridging SMPFs was obtained using an optical microscope. As shown in Figure 7.39, the cracks after the springback were further narrowed by the shape memory effect of the SMPFs. It is seen that the crack widths of 0.7 mm, 0.6 mm, and 0.36 mm in the 4-PUFRC, 7-PUFRC, and 10-PUFRC beams were narrowed to 0.037 mm, 0.031 mm, and 0.026 mm, respectively, after healing at 80 °C for 30 minutes. Clearly, the 10 mm long SMPFs lead to the largest crack closure among the three fiber lengths, which is in agreement with the healing efficiency test results.

The healed crack interface was also investigated using a polarized optical microscope. In Figure 7.40, the narrowed cracks are filled in with the healing agent (i.e., thermoplastic).

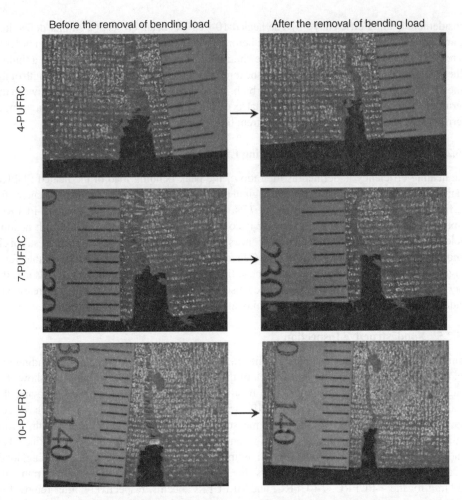

Figure 7.38 Crack narrowing due to springback of the SMPFs bridging over the crack immediately before and after removal of the bending load as captured by the CCD camera. *Source:* [14] Reproduced with permission from Elsevier

Discrete thermoplastic particles within the matrix are also identifiable. This result confirms that a thin film thermoplastic was formed within the narrowed crack. The film seems very dense. This suggests that the sustained recovery compressive force by the SMPFs facilitate the flow and dispersion of the molten thermoplastic within the narrowed crack. Furthermore, the sustained compressive force also helps the system drive the diffusion of the molten thermoplastic molecules into the thermosetting polymer matrix, that is, forced diffusion, which assists in producing a stronger bonding strength at the thermoplastic film/thermosetting polymer interface.

7.5.2.6 SEM Observation Results

In order to have a clear view of the distribution of the SMPFs and thermoplastic particles in the matrix, as well as the texture of the thermoplastic thin film within the narrowed crack after

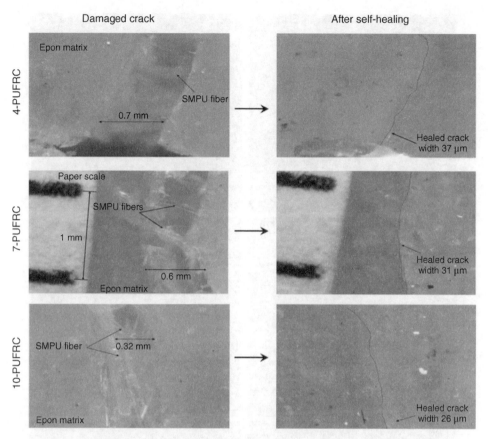

Figure 7.39 Effect of fiber length on crack closure during constrained shape recovery of short programmed SMPFs embedded in composite beams. *Source:* [14] Reproduced with permission from Elsevier

Figure 7.40 The healed crack interfacial investigation (the dots in dark color are thermoplastic particles; the lines in dark color are the healed crack interface). *Source:* [14] Reproduced with permission from Elsevier

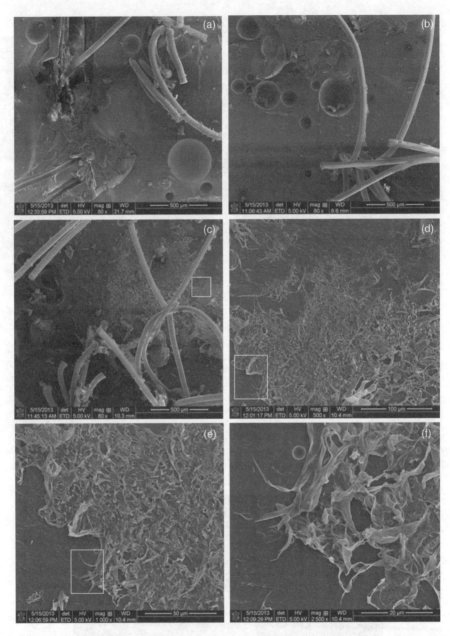

Figure 7.41   SEM fractography for a three-point bending damaged beam after the first fracture (a) and (b) and after the fifth fracture (c), (d), (e), and (f). *Source:* [14] Reproduced with permission from Elsevier

healing, a SEM was used. Figure 7.41(a) and (b) show the cracked surface after the first fracture. From Figure 7.41(a) and (b), the spherical objects observed are the thermoplastic particles. It is also seen that the coating layer on the fibers was broken during the crack opening process in the bending test. This is because the fibers are ductile while the coating epoxy is brittle. The large strain in the fiber during crack opening fractured the coating layer, but not the fibers. This is in agreement with the above discussion that the bridging fibers spring back once the bending load is removed, which narrows the crack to a certain level. In addition, the directions of fiber bundles on the crack surface are multiorientated, indicating 3-D dispersion of the coated short SMPF bundles. Therefore, the SMPFs can close a crack in any orientation and are not limited to the notched direction in this study. This is the major advantage of using a short fiber reinforcement.

In order to examine the thermoplastic thin film, we used a sequential observation by zooming in an area consecutively. Figure 7.41(c) to (f) show the successive zooming-in results of the thermoplastic thin film after the fifth fracture. Successive zooming-in means the white squared area in Figure 7.41(c) is enlarged in Figure 7.41(d), while the white squared area in Figure 7.41(d) is enlarged in Figure 7.41(e), and finally the white squared area in Figure 7.41(e) is enlarged in Figure 7.41(f). In other words, the microscope focused on the same area but in an increasing magnification. With increasing magnification, flowing traces and branching of the flow of the molten thermoplastic are clearly seen, an indication of a rough fractured surface. From Figure 7.41(c) to (f), self-similarity of the thermoplastic thin film can be identified. This further confirms that the fractured surface is not only rough but is also a fractal surface.

7.5.3 Summary

In summary, the three questions on 3-D short SMPFs for healing a wide-opened crack have been answered. It is found that the two-step coating method effectively improves the shape fixity ratio of the programmed short SMPFs from 30% to 66%. Also, the coated fiber has a good shape memory effect (i.e., higher recovery stress). The fabrication method proposed in this study is able to avoid the loss of the shape memory effect of the programmed short SMPFs by curing the composite at a low temperature in the first 24 hours of curing. It is also found that healing of the composite is repeatable. The healing efficiency increases as the fiber length increases for the three fiber lengths investigated due to the need of a certain embedded fiber length for shear-force transfer. The crack-opening process is also a cold-drawing programming process for fibers bridging over the crack, which leads to springback and some closure of the crack when the external load is removed. In other words, the crack closing process consists of two stages. One is due to springback of the bridging fibers immediately after removing the external load and the other is due to the shape memory effect of the bridging fibers. Finally, the fractured surface is very rough and the thermoplastic thin film is formed in a self-similarity pattern, or fractal manner.

References

1. Li, G. and Uppu, N. (2010) Shape memory polymer based self-healing syntactic foam: 3-D confined thermo-mechanical characterization. *Composites Science and Technology*, **70**, 1419–1427.
2. Neuser, S., Michaud, V., and White, S.R. (2012) Improving solvent-based self-healing materials through shape memory alloys. *Polymer*, **53**, 370–378.
3. Kirkby, E.L., Michaud, V.J., Månson, J.A.E., Sottos, N.R., and White, S.R. (2009) Performance of self-healing epoxy with microencapsulated healing agent and shape memory alloy wires. *Polymer*, **50**, 5533–5538.
4. Kirkby, E.L., Rule, J.D., Michaud, V.J., Sottos, N.R., White, S.R., Månson, J.A.E. (2008) Embedded shape-memory alloy wires for improved performance of self-healing polymers. *Advanced Functional Materials*, **18**, 2253–2260.

5. Miskioglu, L. and Burger, G.P. (1985) Material properties in thermal-stress analysis. *Research and Development (R&D) Journal of the South African Institution of Mechanical Engineering* (SAIMechE), 27–32.

6. Meng, Q.H. and Hu, J.L. (2009) A review of shape memory polymer composites and blends. *Composites Part A: Applied Science and Manufacturing*, **40**, 1661–1672.

7. Meng, Q.H., Hu, J.L., Shen, L., Hu, Y., and Han, J. (2009) A smart hollow filament with thermal sensitive internal diameter. *Journal of Applied Polymer Science*, **113**, 2440–2449.

8. Meng, Q.H. and Hu, J.L. (2008) Study on poly(ε-caprolactone)-based shape memory copolymer fiber prepared by bulk polymerization and melt spinning. *Polymers for Advanced Technologies*, **19**, 131–136.

9. Meng, Q.H. and Hu, J.L. (2008) The influence of heat treatment on properties of shape memory fibers: I. Crystallinity, hydrogen bonding and shape memory effect. *Journal of Applied Polymer Science*, **109**, 2616–2623.

10. Meng, Q.H. and Hu, J.L. (2008) Self-organizing alignment of carbon nanotube in shape memory segmented fiber prepared by polymerization and melt spinning. *Composites Part A: Applied Science and Manufacturing*, **39**, 314–321.

11. Meng, Q.H. and Hu, J.L. (2008) A temperature-regulating fiber made of PEG-based smart copolymer. *Solar Energy Materials and Solar Cells*, **9**, 1245–1252.

12. Meng, Q.H., Hu, J.L., and Yeung, L.Y. (2007) An electro-active shape memory fibre by incorporating multi-walled carbon nanotubes. *Smart Materials and Structures*, **16**, 830–836.

13. Hu, J., Meng, H., Li, G., and Ibekwe, S. (2012) An overview of stimuli-responsive polymers for smart textile applications. *Smart Materials and Structures*, **21** (5), paper 053001 (23 pages).

14. Li, G. and Zhang, P. (2013) A self-healing particulate composite reinforced with strain hardened short shape memory polymer fibers. *Polymer*, **54**, 5075–5086.

15. Meng, Q.H. and Hu, J.L. (2008) Study on poly(ε-caprolactone)-based shape memory copolymer fiber prepared by bulk polymerization and melt spinning. *Polymers for Advanced Technologies*, **19**, 131–136.

16. Ping, P., Wang, W., Chen, X., and Jing, X. (2005) Poly(ε-caprolactone) polyurethane and its shape-memory property. *Biomacromolecules*, **6**, 587–592.

17. Wang, W., Jin, Y., Ping, P., Chen, X., Jing, X., and Su, Z. (2010) Structure evolution in segmented poly(ester urethane) in shape-memory process. *Macromolecules*, **43**, 2942–2947.

18. Zotzmann, J., Behl, M., Feng, Y., and Lendlein, A. (2010) Copolymer networks based on poly(ω-pentadeca-lactone) and poly(ε-caprolactone) segments as a versatile triple-shape polymer system. *Advanced Functional Materials*, **20**, 3583–3594.

19. Li, G. and Xu, W. (2011) Thermomechanical behavior of thermoset shape memory polymer programmed by cold-compression: testing and constitutive modeling. *Journal of the Mechanics and Physics of Solids*, **59**, 1231–1250.

20. Li, G. and Shojaei, A. (2012) A viscoplastic theory of shape memory polymer fibers with application to self-healing materials. *Proceedings of the Royal Society A:Mathematical Physical and Engineering Sciences*, **468**, 2319–2346.

21. Li, G., Ajisafe, O., and Meng, H. (2013) Effect of strain hardening of shape memory polymer fibers on healing efficiency of thermosetting polymer composites. *Polymer*, **54**, 920–928.

22. John, M. and Li, G. (2010) Self-healing of sandwich structures with grid stiffened shape memory polymer syntactic foam core. *Smart Materials and Structures*, **19**, paper 075013 (12 pages).

23. Li, G. and Chakka, V.S. (2010) Isogrid stiffened syntactic foam cored sandwich structure under low velocity impact. *Composites Part A: Applied Science and Manufacturing*, **41**, 177–184.

24. Li, G. and Muthyala, V.D. (2008) Impact characterization of sandwich structures with an integrated orthogrid stiffened syntactic foam core. *Composites Science and Technology*, **68**, 2078–2084.

25. Nji, J. and Li, G. (2012) Damage healing ability of a shape memory polymer based particulate composite with small thermoplastic contents. *Smart Materials and Structures*, **21**, paper 025011 (10 pages).

26. Nji, J. and Li, G. (2010) A biomimic shape memory polymer based self-healing particulate composite. *Polymer*, **51**, 6021–6029.

27. Nieuwenhuyse, E. (2006) *Thermal Insulation Materials Made of Rigid Polyurethane Foam*, Report No. 1, Brussels, Belgium: Foundation of European Rigid Polyurethane Foam Associations.

28. Ji, G., Ouyang, Z., and Li, G. (2012) Local interface shear fracture of bonded steel joints with various bondline thicknesses. *Experimental Mechanics*, **52**, 481–491.

29. Ji, G., Ouyang, Z., and Li, G. (2012) On the interfacial constitutive laws of mixed mode fracture with various adhesive thicknesses. *Mechanics of Materials*, **47**, 24–32.

30. Ji, G., Ouyang, Z., and Li, G. (2011) Effects of bondline thickness on Mode-II interfacial laws of bonded laminated composite plate. *International Journal of Fracture*, **168**, 197–207.

31. Ji, G., Ouyang, Z., Li, G., Ibekwe, S.I., and Pang, S.S. (2010) Effects of adhesive thickness on global and local Mode-I interfacial fracture of bonded joints. *International Journal of Solids and Structures*, **47**, 2445–2458.

32. Li, G., Ji, G., and Ouyang, Z. (2012) Adhesively bonded healable composite joint. *International Journal of Adhesion and Adhesives*, **35**, 59–67.

33. Li, G., Meng, H., and Hu, J. (2012) Healable thermoset polymer composite embedded with stimuli-responsive fibers. *Journal of the Royal Society Interface*, **9**, 3279–3287.

34. Rule, J.D., Sottos, N.R., and White, S.R. (2007) Effect of microcapsule size on the performance of self-healing polymers. *Polymer*, **48**, 3520–3529.

35. Brown, E.N., Sottos, N.R., and White, S.R. (2002) Fracture testing of a self-healing polymer composite. *Experimental Mechanics*, **42**, 372–379.

36. Nji, J. and Li, G. (2010) A self-healing 3D woven fabric reinforced shape memory polymer composite for impact mitigation. *Smart Materials and Structures*, **19**, paper 035007 (9 pages).

37. Zhang, P. and Li, G. (2013) Structural relaxation behavior of strain hardened shape memory polymer fibers for self-healing applications. *Journal of Polymer Science Part B: Polymer Physics*, **51**, 966–977.

38. Cottrell, A.H. (1964) Strong solids. *Proceeding of the Royal Society*, **A282**, 2–9.

39. Takao, Y. and Taya, M. (1987) The effect of variable fiber aspect ratio on the stiffness and thermal expansion coefficients of a short fiber composite. *Journal of Composite Materials*, **21**, 140–156.

40. Mulligan, D.R., Ogin, S.L., Smith, P.A., Wells, G.M., and Worrall, C.M. (2003) Fibre-bundling in a short-fibre composite: 1. Review of literature and development of a method for controlling the degree of bundling. *Composites Science and Technology*, **63**, 715–725.

41. Agarwal, B.D., Broutman, L.J., and Chandrashekhara, K. (2006) *Analysis and Performance of Fiber Composites*, 3rd edn, John Wiley & Sons, Inc, Hoboken, New Jersey.

42. Nguyen, T.D., Qi, H.J., Castro, F., and Long, K.N. (2008) A thermoviscoelastic model for amorphous shape memory polymers: incorporating structural and stress relaxation. *Journal of Mechanics and Physics of Solids*, **56**, 2792–2814.

8

Modeling of Healing Process and Evaluation of Healing Efficiency

Modeling of the self-healing process and evaluation of healing efficiency have been topics of intensive research for years [1]. Indeed, they evolve with new self-healing methodologies. Modeling of self-healing first depends on the particular self-healing system. For this purpose, self-healing systems can be divided into two groups [2]. One is a coupled or active system in which crack propagation activates the healing process, that is, an autonomous system without human intervention. This may include some examples in the extrinsic self-healing systems incorporating a liquid healing agent such as containers using microcapsules [3], hollow fibers [4,5], microvascular system [6], etc. The other group is an uncoupled or passive system where the healing mechanism is not autonomous and requires some human intervention. This group includes some intrinsic self-healing systems that need to manually bring the fractured surfaces in contact, such as employing a thermally reversible covalent bond [7], using ionomers [8], etc.; this group also includes some extrinsic healing systems that need external heating such as incorporating thermoplastic particles [9,10] and implementing a biomimetic two-step close-then-heal (CTH) system [11–20]. In this chapter, we will focus on the self-healing system used in this book, that is, an uncoupled CTH healing system using embedded thermoplastic particles as the healing agent [11–20].

Self-healing occurs at an interface. An understanding of molecule interaction at the interface is essential to design and develop new self-healing schemes and to evaluate existing systems. Cracking of polymer starts with yielding, crazing, crack initiation, propagation, and coalescence, which is controlled by chain slippage, network stretch, interchain fracture, and chain scission. As a reverse process of cracking, healing depends on the physical and chemical nature of the polymer surface. Chemical and physical interactions and molecular connectivity achieved across the interfaces determine the healing efficiency [1]. Raghavan and Wool [21] identified 10 fundamental polymer–polymer interfaces, including a combination of liquid surface, virgin as-cast surface, new solid surface by fracture, and new treated solid surface by fracture. In the following, we will focus on the healing of a thermosetting polymer matrix by a molten thermoplastic under pressure, which is created by the shape memory effect, that is, a combination of a liquid and new solid surface under pressure.

Self-Healing Composites: Shape Memory Polymer-Based Structures, First Edition. Guoqiang Li.
© 2015 John Wiley & Sons, Ltd. Published 2015 by John Wiley & Sons, Ltd.

From the very beginning of the self-healing study, healing efficiency has been a primary parameter to evaluate the effectiveness of the self-healing system. Because most self-healing approaches target recovery of the mechanical properties, various mechanical test methods and mechanical parameters have been used to evaluate healing efficiency. For example, Raghavan and Wool [21] used compact tension (CT) specimens, Brown, Sottos, and White [22] and Brown [23] used tapered double cantilever beam (TDBC) specimens, Plaisted and Nemat-Nasser [24] and Plaisted, Amirkhizi, and Nemat-Nasser [25] used double cleavage drilled compression (DCDC) specimens, Li and John [11], John and Li [12], and Nji and Li [13] used anti-buckling compression after impact (CAI) specimens, Nji and Li [16,17] used single-notched beam (SNB) specimens, Li, Ajisafe, and Meng [19] used fully constrained tensile specimens, and Li, Ji, and Ouyang [26] used double cantilever beam (DCB) specimens, etc. For the parameters used to evaluate the healing efficiency, several parameters have been used, including the ratio of strength of the healed specimen over that of the virgin specimen [11–13,16,17,19], the ratio of toughness of the healed specimen over that of the original specimen [21–26], etc. Because toughness is a measurement of both strength and ductility, it is more appropriate to evaluate the healing efficiency using toughness rather than using strength.

The type of specimens also affects the evaluation results. Because cracking is intimately related to fracture toughness, most types of specimens are designed to have a stable and controllable crack propagation speed so that the repeated fracture/healing test can be conducted. Another issue is the type of healing toughness that is considered. For instance, Mode I fracture toughness represents the resistance to opening or peeling a crack; Mode II toughness represents the resistance to an in-plane shear crack; and Mixed Mode I&II fracture toughness represents the resistance to a mixed opening and in-plane shear crack. Currently, the majority of the test specimens measure the Mode I toughness. In this chapter, in addition to the Mode I toughness, we will introduce two types of specimens to evaluate the Mode II and Mixed Mode I&II toughness.

8.1 Modeling of Healing Process

8.1.1 Modeling of Healing Process Using Thermoplastic Healing Agent

In the CTH scheme, which is the focus of this book, the healing is achieved by molten thermoplastic particles. Once the embedded thermoplastic (TP) is heated to melting, several microscopic healing steps will proceed, including: (a) molten thermoplastic being sucked into the narrowed crack by capillary force, (b) wetting the fractured surfaces by the molten healing agent, (c) diffusion of the molten healing agent into the fractured matrix, and (d) molecule randomization and entanglement. The steps (b) to (d) are the same as those proposed by Wool and O'Connor [27], who focused on a thermoplastic–thermoplastic interaction. Figure 8.1 shows a schematic of the molecular-length scale healing process. In this schematic, we use the SMP fiber reinforced thermosetting polymer as an example. Also, we use TP fibers instead of TP particles. The discussion also applies to an SMP matrix based self-healing system with TP particles or fibers.

Previously, we have modeled the damage–healing process within the continuum damage mechanics framework [2,28–30], in which the healing process is represented by a healing variable. While phenomenological modeling in the continuum damage mechanics framework helps understanding of healing from the macroscopic point of view, it cannot understand the

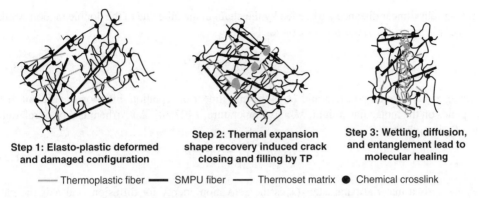

**Step 1: Elasto-plastic deformed
and damaged configuration**

**Step 2: Thermal expansion
shape recovery induced crack
closing and filling by TP**

**Step 3: Wetting, diffusion,
and entanglement lead to
molecular healing**

—— Thermoplastic fiber ▬▬ SMPU fiber —— Thermoset matrix ● Chemical crosslink

Figure 8.1   Schematic of the molecular healing process

healing process, which occurs at molecular-length scale. Therefore, in the following, we will focus on modeling the diffusion process under pressure, which is unique for this CTH healing scheme as other systems may not have the pressure from the recovery force.

Our approach of modeling the diffusion process is divided into the following steps. First, based on the volume fraction of the TP within the thermoset matrix, we can calculate the amount of TP available to flow into the crack using stereology and statistics [31]. The minimum amount of TP needed is determined in such a way that the volume of the available TP, i.e., the TP that can migrate into the crack, is equal to the volume of the crack. Otherwise, only a portion of the crack can be filled and only a portion of the crack surface can be wetted. Second, based on the surface area that is wetted, we will conduct a one-dimensional diffusion analysis.

Self-diffusion of random coiled chains in the bulk was studied by Edwards [32] and de Gennes [33], who developed a reptation model of a chain confined to a tube. As shown in Figure 8.2, a single chain of TP with a contour length of L is diffused in a tube, which represents the topological constraints on its motion imposed by the matrix and by other chains in the bulk. At the same time, the tube contour represents another random walk. At a given time t_r, the

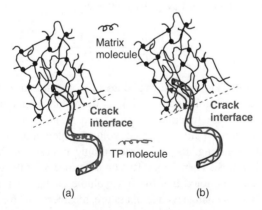

Matrix
molecule

Crack
interface

Crack
interface

TP molecule

(a)   (b)

Figure 8.2   TP molecule diffusion process via a reptation model for a chain confined to a tube: (a) TP and matrix chains in contact prior to diffusion and (b) TP molecules diffused into the matrix

average curvilinear distance χ travelled by the chain in the tube due to the double random walk process is given by Wool and O'Connor [27] as

$$\langle \chi^2 \rangle^{1/2} = \alpha \langle 2D_c t_r \rangle^{1/4} \tag{8.1}$$

where α is a material constant and D_c is the curvilinear or reptation diffusion coefficient that depends on the molecular weight, M, and temperature, T [27,34]. An Arrhenius type of relation for D_c is given by

$$D_c \approx \alpha_1 \frac{1}{M} T e^{-Q_o/RT} \tag{8.2}$$

where α_1 is a material parameter, Q_o is the activation energy for diffusion, and R is the gas constant. The activation energy for the diffusion depends on both diffusing and host phase molecular structures, which are available for a wide range of polymer pairs in the literature [27].

Now let us consider the effect of recovery stress by the shape memory polyurethane fibers (SMPFs) on the diffusion process. The recovery stress, σ, enhances the diffusion process by pushing the TP chains into the matrix, which helps overcome the diffusion barriers. In other words, it reduces the diffusion activation energy. Therefore, the diffusion coefficient D_c can be rewritten as

$$D_c \approx \alpha_1 \frac{1}{M} T e^{-(Q_o-W)/RT} \tag{8.3}$$

where W is the work done by the recovery stress on one mole of diffusing molecules.

In order to determine W, we can do the following. First, from the partially constrained stress recovery test, we can obtain $\sigma = f(\varepsilon)$, in which ε is the recovery strain [20]. The recovery strain can be measured from the narrowing of the crack. The crack narrowing is due to the diffusion of the molten TP into the host matrix. It is reasonable to assume that only the free volume within the matrix can be occupied by the diffusing TP chains. We can also assume that the TP diffusing front fully fills in the free volume available on its path forward. With these assumptions in mind, we can find the recovery strain of the SMPFs due to diffusion of the molten TP chains into the thermoset polymer matrix as

$$\varepsilon = \frac{(A\chi V_f)/A}{(M/\rho)/A} = \frac{\rho A \chi V_f}{M} \tag{8.4}$$

where ρ is the density of the TP, A is the wetted area by one mole of molten TP on the crack surface, χ is the average diffusion depth, and V_f is the volume fraction of the free volume available in the matrix, which is a function of temperature. As indicated by Hu [35], the volume in polymers consists of three parts, occupied volume, interstitial free volume, and hole free volume. Generally speaking, the hole free volume is easier to host guest molecules than the interstitial free volume. Therefore, diffusion is a process for the guest molecules to fill in the hole free volume. It is noted that the free volume changes with temperature. While the occupied volume does not change with temperature, the interstitial free volume may change linearly with temperature as it reflects the change in the interatomic distance. The portion of the free volume that takes the largest impact by rising temperature should be the hole free volume. Clearly, the polymer matrix has a higher hole free volume at the healing temperature than at the lower temperature, which provides more space for easier diffusion of the TP molecules. In other words, diffusion occurs more easily at a higher temperature.

Once the recovery strain is obtained, the work done by the recovery stress of the SMPFs on one mole of molten TP is

$$W = \int_0^{\varepsilon_\infty} f(\varepsilon) d\varepsilon \qquad (8.5)$$

where ε_∞ is the maximum recovery strain, which is determined by replacing χ with χ_∞ in Equation (8.4), where χ_∞ is the maximum depth of diffusion during the healing process. Because the diffusion depth χ depends on D_c, an iterative algorism will be used in the computation, that is, starting with assuming a χ value, from which we can calculate χ in Equation (8.1). The process proceeds until the difference between the calculated result and the assumed value is converged.

Now the healing can be linked to the number of newly generated constraints (physical cross-links) between diffused TP chains and the matrix molecules. Let us assume upon generation of each ith constraint that a stress of σ_i^d is recovered; the strength recovery rate due to diffusion of the TP chain into depth of χ_∞ is therefore

$$\dot{\sigma}_d = \int_0^{\chi_\infty} \sigma_i^d \, d\chi \qquad (8.6)$$

The macroscopic healing efficiency, k, can be measured through integrating the strength recovery rate, $\dot{\sigma}_d$, during the entire healing process, $0 < t \leq t_{max}$, as follows:

$$k = \left[\int_0^{t_{max}} \dot{\sigma}_d \, dt \right] / \sigma_f \qquad (8.7)$$

where σ_f is the fracture strength of the composite.

In addition to modeling, the molecular healing effect can also be experimentally validated. The following approaches can be used. (1) Microscopic observation, where the crack-closing effect can be examined using a high resolution CCD camera (e.g., Sony XCD-CR90) [20] and SEM (e.g., S-3600N variable pressure SEM) [18]. The fractured surface can be zoomed in with an SEM to obtain the area that has been wetted by the TP. This can provide information for the percentage of the area that allowed diffusion. (2) The depth of diffusion can be directly obtained using energy dispersive X-ray spectroscopy (EDS) (e.g., S-3600N) [26]. The idea is that with the diffusion of the TP into the thermoset matrix, the composition of certain signature elements will change as the distance from the crack interface increases (such as carbon or oxygen). The content of the elements with depth yields a composition profile, from which the depth and amount of TP diffused can be obtained [16,26].

8.1.2 Summary

In summary, the molecular healing process depends on many factors. In addition to the compatibility between the healing agent and the matrix, which determines the diffusion barriers, other factors have a significant effect on the healing process. Generally speaking, because the healing temperature affects the free volume, diffusion coefficient, and viscosity, a higher healing temperature leads to a higher healing efficiency. This is why it is stressed that the healing temperature is usually higher than the melting temperature of the thermoplastic healing agent [16]. Of course, the temperature must still be within a certain limit so as not to cause degradation of the healing agent and the matrix. Clearly, the healing time must be sufficient to allow diffusion, randomization, and entanglement of the healing molecules. A higher recovery stress and recovery strain of the embedded SMPU fibers help overcome the diffusion barrier and thus enhance healing efficiency. While the healing agent with a smaller molecular weight speeds up the diffusion process, it usually leads to lower strength. Therefore, a balance must be sought.

8.2 Evaluation of Healing Efficiency

As discussed above, while various parameters have been used to evaluate the healing efficiency, fracture toughness is more appropriate. In recent years, the cohesive zone model (CZM), which was first proposed by Barenblatt [36] and Dugdale [37], has been widely used to describe interfacial debonding or fracture, which can be used to determine the fracture toughness. The concept of CZM was initially motivated to deal with the nonphysical stress singularity near a crack tip in linear elastic fracture mechanics in terms of atomic interaction [36] and plastic yielding [37]. The idea of a cohesive crack model is based on the assumption that the whole crack region can be divided into two sections: one section of the crack surfaces is free of traction and the other section is subjected to a distribution of cohesive traction which is a function of surface separation. Many types of CZMs such as polynomial, exponential, bilinear, or multilinear models have been developed [38–61]. The major difference between the CZMs lies in the shape of the traction-separation response and the parameters used to describe that shape. On the experimental side, various types of experiments have been conducted to obtain CZMs, such as using adhesively bonded joints [62–66].

The fundamental idea of CZMs is that before the physical macrocrack is formed, the two adherends are held together by traction within a fracture processing zone in the adhesive layer. The cohesive stresses vary according to the relative separation of the two adherends. In mechanics, the relationship between the traction and separation is defined as the interface cohesive law for the fracture process. An obvious advantage of CZMs is that it can be and has been incorporated in commercial finite element software packages such as ANSYS and ABACUS. Also, it avoids the stress singularity, which is seen in linear elastic fracture mechanics (LEFM), and considers the physical fracture process in a more realistic manner [62]. The advantage of LEFM lies in its obvious simplicity and decent accuracy, especially for relatively brittle materials and interfaces. Within the framework of LEFM, the remote loadings can be correlated with the critical conditions (crack growth) by a local parameter, the stress intensity factor (SIF), or a global parameter, the strain energy release rate (ERR) [66]. Despite the huge success of LEFM, linear elastic fracture mechanics relies on the existence of a crack at the interface and on the assumption of small-scale nonlinearity beyond the crack tip. If any of these conditions are violated, alternative approaches are required. Specifically, with the increased use of modern toughened adhesives, the plastic zones associated with cohesive fracture along the adhesive interlayer, in many cases, could be comparable to or even larger than the thickness of the adherends [66]. Therefore, the tendency is that more and more CZMs have been used in place of classical LEFM. Consequently, if CZMs can be obtained during healing efficiency experiments, the results can be readily used in a structural analysis consisting of the self-healing materials. Also, fracture toughness is an intermediate result in the process of obtaining CZMs. Therefore, using specimens that can determine both fracture toughness and CZMs is "one stone for two birds." It is also worth noting that CZMs can be directly incorporated in analytical models to derive theoretical solutions for some types of specimens or structures with relatively simple geometries [57–61,67–70]. In the following, we will discuss three types of specimens and test procedures to determine the fracture toughness and CZMs, from which healing efficiency can be readily determined.

8.2.1 Healing Efficiency for a Double Cantilever Beam (DCB) Specimen

A typical DCB specimen is shown in Figure 8.3, which is based on an adhesively bonded joint configuration. There are two ways to prepare the test specimens for two different purposes. One is to test the healing efficiency of the material as a thin film. In this case, we just need to use the material as an adhesive and use metals such as steel or aluminum as adherends [26]. We have found that the fracture toughness obtained is a function of the adhesive thickness [66]. For instance, as the adhesive thickness increases, the interfacial fracture strength reduces and the toughness increases. Also, it is found that this tendency tends to stabilize as the adhesive thickness increases [66]. However, since the healing efficiency is defined as the ratio of the toughness of the healed specimen over that of the virgin specimen, and also the re-fracture path usually follows the previous path because the healing efficiency is generally smaller than 100%, it is believed that the adhesive thickness would not significantly affect the toughness ratio. Because thermoplastic particles will be used as the healing agent, it is believed that the adhesive thickness should be several times larger than the particle diameter to avoid the effect of the particle on the fabrication and uniformity of the adhesive layer.

The second approach is to investigate the healing efficiency of the bulk material. In this case, the material will be fabricated into a uniform beam with a mylar sheet having a certain length and thickness, which is inserted at the center of the beam before curing. The mylar sheet should have a piecewise thickness with a thicker portion representing the initial crack between the two adherends and the thinner portion representing the initial crack within the adhesive layer. In such a way, the initial crack configuration after removing the mylar sheet from the cured beam will be the same as that shown in Figure 8.3. With such treatment, the adherends and adhesive layer can be clearly defined, which is the same as that of the adhesively bonded joint. The adhesive layer thickness is equal to the thickness of the mylar sheet in the thicker portion.

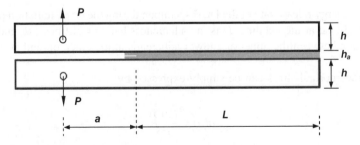

Figure 8.3 Schematic of the DCB specimen. *Source:* [66] Reproduced with permission from Elsevier

Again, as discussed above, while the thickness of the adhesive layer significantly affects the toughness, it should have an insignificant effect on the toughness ratio of the virgin specimen and the healed specimen, that is, the healing efficiency.

8.2.1.1 Theoretical Background

This section is primarily based on the work by Ji *et al.* [66]. The basic idea is that during testing of DCB specimens, we can obtain global load and displacement, as well as the rotation angle of the beam. Using a charge-coupled device (CCD) camera or any other microscopic displacement measurement device, we can measure the local crack tip opening. The cohesive law is the relationship between the local traction and the local separation. Obviously, because we cannot measure the crack tip stress accurately, we need to turn to the global test results. The famous *J*-integral can serve as a bridge to link the local and global test results. The idea is to first obtain the *J*-integral from the global test results. After that, we can obtain the local stress by using the *J*-integral and the local separation. Once the local stress is obtained, the traction–separation law, or the lump-sum cohesive law, is determined.

Now consider a typical double cantilever beam (DCB) specimen as shown in Figure 8.3, in which *a* is the initial crack length, h_a is the adhesive layer thickness, *h* is the thickness of the adherend, *P* is the peel load at the loadline, and *L* is the adhesive layer length. It is assumed that the adherends are linearly elastic during the entire fracture test process. Theoretically, the adhesive material must be nonlinearly elastic during the test in order to satisfy the condition for the *J*-integral exactly. However, for a monotonic loading process (no unloading occurs), the cohesive separation as well as plastic dissipation in the adhesive layer might still be considered by the well-known path independent integral or the Rice *J*-integral as follows [71]:

$$J = \int_{\Gamma} \left(W \, dy - T \frac{\partial u}{\partial x} ds \right) \tag{8.8}$$

where $W(x, y)$ is the strain energy density, *x* and *y* are the coordinate directions, $T = n\sigma$ is the traction vector, *n* is the normal to the curve or path Γ, σ is the Cauchy stress, and *u* is the displacement vector.

Based on the concept of energetic force and classical Euler beam theory, Andersson and Stigh [72] derived a closed-form expression of the *J*-integral for the DCB specimen as follows:

$$J = \frac{(Pa)^2}{D} + P\theta_0 = P\theta_P \tag{8.9}$$

where *J* is the energy release rate of the DCB specimen during the crack initiation process, *P* is the global peel load at the loadline, *D* is the adherend's bending stiffness, θ_0 and θ_P are the relative rotation between the upper and lower adherends at the crack tip and at the loadline, respectively.

The interface normal stress can be simply expressed by

$$\sigma(\delta) = \frac{\partial J(\delta)}{\partial \delta} \tag{8.10}$$

in which δ is the crack tip opening or separation. Clearly, $\sigma(\delta)$ is the cohesive law.

It is worth noting that the transverse shear effects may become significant for the energy-release rate when there is a large-scale fracture zone for conditions such as laminated composite joints. However, for the bonded steel joint with a slender configuration, which will be used in determining the healing efficiency as a thin film, the deviation is small. Even when the DCB specimen is prepared as a beam made of the self-healing material, as long as the slenderness ratio is sufficiently large, the deviation should also be small. Therefore, the classical beam theory can be adopted as shown in Equation (8.9).

As expected, the energy release rate J and lump-sum cohesive law can be experimentally determined if the crack tip separation δ, the loadline rotation θ_P of the adherends, and the global peel load P are simultaneously recorded during the fracture test. It is noted that this interface constitutive relationship is the equivalent interface cohesive law, not necessarily the intrinsic cohesive law. This is because, in addition to the intrinsic cohesive separation, possible plastic deformation in the adhesive layer contributes to the entire normal separation between the two adherends during the fracture test. Of course, with the decrease of the adhesive thickness, it is expected that this equivalent interface cohesive law will finally approach the intrinsic cohesive law [66].

A typical nonlinear equivalent cohesive law is shown in Figure 8.4. This typical nonlinear separation–traction law has three segments: (a) the elastic stage when the normal interfacial separation $\delta \leq \delta_o$, where the normal interfacial stress σ increases with separation until the maximum interfacial stress σ_{max} (interfacial strength) is reached; (b) the softening stage when $\delta_o \leq \delta \leq \delta_f$, where the normal traction σ decreases with separation δ; and (c) the complete debonding stage, where there is no interfacial stress when $\delta \geq \delta_f$.

The fracture toughness, K_{IC}, for the Mode I fracture test under the plane stress condition can be obtained from [23]

$$K_{Ic} = \sqrt{J_c E} \qquad (8.11)$$

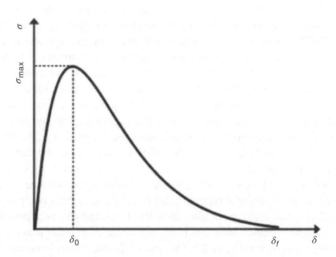

Figure 8.4 A typical nonlinear interfacial traction–separation law. *Source:* [66] Reproduced with permission from Elsevier

where J_c is the critical energy release rate and E is the modulus of elasticity of the adhesive layer.

Once the critical fracture toughness is found, the healing efficiency is defined as [22,23,27]

$$\eta = \frac{K_{\text{IChealed}}}{K_{\text{ICvirgin}}} \tag{8.12}$$

where K_{IChealed} is the fracture toughness of the healed specimen and K_{ICvirgin} is the fracture toughness of the virgin specimen.

It is believed that the above formulations are applicable to both cases – either using self-healing material as a thin film in an adhesively bonded joint or using self-healing material as both the adherends and the adhesive layer.

8.2.1.2 Application

As an example, Li, Ji, and Ouyang [26] used the above formulations to evaluate the healing efficiency of a thermosetting polymer adhesive embedded with a thermoplastic healing agent using the DCB configuration. The adhesive, LOCTITE Hysol 9460, is a modified structural epoxy adhesive. The mix ratio of resin and hardener is 1:1 by weight. According to the manufacturer, its elastic modulus, tensile strength, and elongation at break are 2.76 GPa, 30.3 MPa, and 3.5%, respectively. A thermoplastic polymer identified as a copolyester (CP) from Abifor Inc., Switzerland (with particle size $\leq 80\,\mu\text{m}$, density 1.3 g/cm^3, melting range 114–124 °C, and bonding temperature range 125–150 °C), was used as a healing agent. It is noted that the melting temperature of this thermoplastic may be too high for practical applications. Other thermoplastics such as polycaprolactone or polylactic acid, which has a melting temperature around 50–80 °C, may be investigated in future studies. A continuous glass woven fabric reinforced polymer laminate was used to fabricate the 3.1 mm thick, 25.4 mm wide, and 254.0 mm long adherends of DCB specimens. The adherend was made of 17 layers of glass woven fabric (0° and 90° cross-ply) with epoxy resin. The in-plane flexural strength is 379.2 MPa and the in-plane flexural modulus of elasticity is 18.6 GPa.

Two types of DCB specimens were prepared with the actual average adhesive thicknesses of 0.6 mm. One was regular DCB specimens without thermoplastic particles in the adhesive and the other is DCB specimens with 10 wt% thermoplastic particles in the adhesive. Clearly, this adhesive layer thickness is about 7.5 times larger than the diameter of the thermoplastic particles and thus the adhesive layer can be treated as a uniform mixture. A very thin mylar tape with thickness of 0.045 mm was carefully inserted from the edge of the adhesive layer by 2 mm at the middle height of the adhesive layer immediately after the application of the adhesive layer, in order to create a sharp initial crack within the adhesive layer. Since the distance between the loadline and the edge of the adhesive layer is 78 mm, the total initial crack length $a = 80$ mm.

A hole is drilled in each of the two tool-grade steel blocks. These two steel blocks are sanded, polished and then carefully bonded on to the surface of the laminated composite adherends at one end of the DCB specimen by super glue. In order to maintain a coaxial peel force during the testing, two self-aligned, free-rotating ball pins were designed and fabricated using tool-grade steel, as schematically shown in Figure 8.5. One end of the ball pin with threads was mated with the prefabricated threads within the holes in the steel blocks bonded on the surface of the DCB specimen, and the other end was connected to the MTS machine.

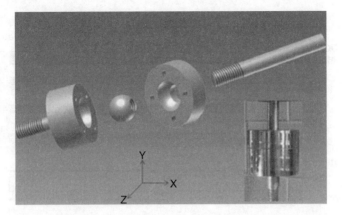

Figure 8.5 Schematic of self-aligned free-rotating ball joint. *Source:* [66] Reproduced with permission from Elsevier

Li, Ji, and Ouyang [26] used the MTS 810 machine to conduct the peel fracture test and collect the loading force data "P" and the displacement "Δ" of the DCB specimens at the loadline (see Figure 8.6). The fracture test was conducted under the displacement controlled mode. The loading rate was set as 0.6 mm/min and the data collecting frequency was 1 Hz. In order to measure the rotating angle "θ_p" of the adherends during the peel test, two digital inclinometers and sensors were attached at the free end of the adherends to collect the data during the test, as shown in Figure 8.6. The accuracy of the inclinometer is 0.01°, the test range is from −70° to 70°, and the data acquisition frequency is 1 Hz. The Sony XCD-CR90 high

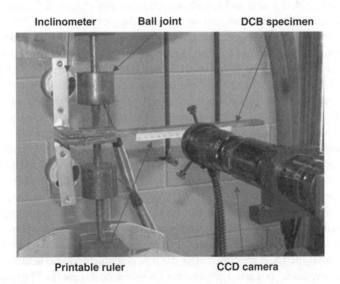

Figure 8.6 DCB specimen attached with an inclinometer during the peel test. *Source:* [26] Reproduced with permission from Elsevier

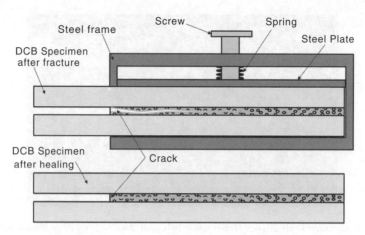

Figure 8.7   Schematic of the DCB specimen healing process by manually applying the crack-closure force. *Source:* [26] Reproduced with permission from Elsevier

resolution CCD camera with a resolution of $3.7 \times 3.7\,\mu\text{m}/\text{pixel}$ was used in this experiment to capture the crack tip opening. The position of the camera was adjusted to be perpendicular to the side of the DCB specimen and the deformation images of the DCB specimen during the test were shot with the focus on the adhesive layer, as shown in Figure 8.6. The camera shooting rate was 1 Hz. The collected images were input to an image processing toolkit, *Image J*, to post-analyze the recorded images and thus obtain the local separation at the crack tip "δ".

Because the adhesive layer does not have a shape memory capability, the first step of the close-then-heal (CTH) scheme must be done manually. The damaged DCB specimens were put into a steel frame with a screw and spring, as schematically shown in Figure 8.7, to apply a compressive force to the joint. This corresponded to the first step in the CTH scheme, that is, close the crack in the adhesive layer. The amount of compression force applied to the specimen with the screw and spring was 75 N. After the crack of the specimen was closed, the entire setup was placed into an oven at 150 °C for about 1 hour and then naturally cooled down to room temperature. This completed the second step in the CTH scheme. It is noted that the first step can be completed by other means, such as shape memory alloy (SMA) or shape memory polymer (SMP) Z-pins trained by pre-tension. For example, Li and Shojaei found that the recovery stress of polyurethane fibers after cold-drawing programming can reach 16 MPa [73]. Therefore, in order to create 75 N of compressive force, the cross-sectional area of the SMP Z-pins needs to be $75/16 = 4.68\,\text{mm}^2$. Because the SMP filament diameter used by Li and Shojaei was 0.04 mm, the number of filaments required is 3726. Usually a fiber bundle consists of 400 filaments. Therefore, $3726/400 = 9.3$, or 10 SMP fiber bundle Z-pins, each with a cross-sectional area of about $0.468\,\text{mm}^2$, should be sufficient. Once the shape memory effect is triggered by heating, the Z-pins will autonomously apply the compressive force to the DCB specimens. As a preliminary study, however, this idea was not evaluated in this work.

Figure 8.8 shows the DCB specimen after the first cracking and first healing. It is clear that the original macrocrack is closed after the close-then-heal process. The high resolution microscopy CCD camera was used to observe the local deformation of the specimen during the Mode I fracture test. Figure 8.9 (a) to (e) presents the crack initiation and propagation of the

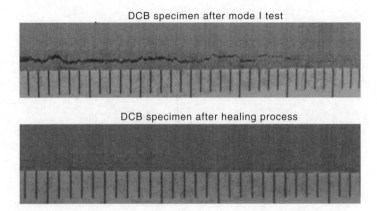

Figure 8.8 DCB specimen before and after the healing process. *Source:* [26] Reproduced with permission from Elsevier

self-healing and control DCB specimens during the fracture test. In Figure 8.9 (a), a clear initial crack is seen at the center of the adhesive layer. In Figure 8.9 (b), the initial crack starts opening and propagating. Comparing Figure 8.9 (c) and (d), it is clear that the control specimen without the healing agent has a clean crack, while the self-healing specimen shows that the healing

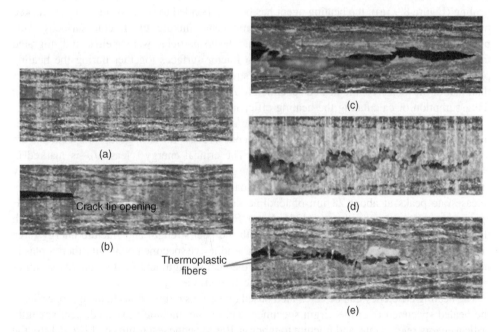

Figure 8.9 Microscopy images during the Mode I test: (a) virgin specimen; (b) crack initiation; (c) crack propagation of a regular DCB specimen without a healing agent; (d) crack propagation of a DCB specimen with a thermoplastic healing agent; and (e) crack propagation of a DCB specimen with a healing agent after the first healing. *Source:* [26] Reproduced with permission from Elsevier

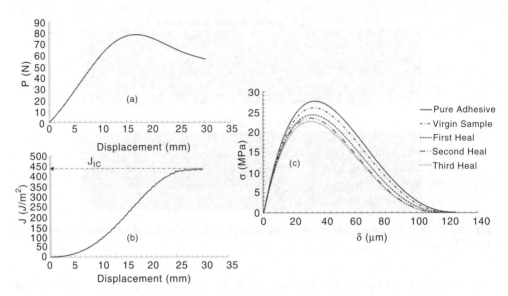

Figure 8.10 (a) Typical load versus displacement curve, (b) energy release rate versus displacement curve, and (c) cohesive laws. *Source:* [26] Reproduced with permission from Elsevier

agent is still attached to the fractur surface, blocking the view of the camera. After the first healing (Figure 8.9 (e)), the healing agent has been re-bonded to the two sides of the cracked matrix, as demonstrated by the fibrous thermoplastic linking the fractur surfaces. This phenomenon indicates that some of the thermoplastic particles were melted and migrated to the fracture surface and then glued the two fractur surfaces together during the healing process. As discussed previously, because the strength of the thermoplastic particle is lower than that of the adhesive, the healed specimen re-fractured along the previous path, satisfying the assumption of calculating the healing efficiency.

The typical load–displacement at the loadline and energy release rate – loadline displacement for the virgin specimens, and the derived cohesive law for each group of specimens are shown in Figure 8.10 (a) to (c), respectively. The critical energy release J_{IC} is marked in Figure 8.10 (b). From Figure 8.10 (a) and (b), the global load peaks before the fracture energy crests. The global load crests at about 17 mm of loadline displacement, while the critical energy release rate peaks at about 28 mm of loadline displacement. This is typical of other DCB specimens with a tough adhesive [74]. From Figure 8.10 (c), the virgin specimen (with thermoplastic particles) sees a reduction in both the peel strength and fracture energy (area enclosed by the σ–δ curve) as compared to the pure adhesive specimen (without a thermoplastic healing agent). As the healing cycle increases, the peel strength and fracture energy continuously decrease, indicating a reduction in the healing efficiency.

The healing efficiency can be calculated based on various ratios of mechanical properties of the healed specimen over the virgin specimen. These may include peak load, peel strength, critical energy release rate, and fracture toughness. Based on the test results by Li *et al.* [26], the healing efficiencies are summarized in Table 8.1. When using Mode I fracture toughness, Equation (8.12), to calculate the healing efficiency, it is assumed that the elastic modulus of the thermoplastic healing agent is 1.5 GPa. From Table 8.1, two observations can be made. One is

Table 8.1 Healing efficiency using various mechanical parameters (%)

Healing parameter	After first healing	After second healing	After third healing
Peak load (P_c)	94.19	91.04	88.22
Peel strength (σ_c)	93.62	90.42	87.23
Critical energy release rate (J_{IC})	91.34	85.62	82.46
Mode I fracture toughness (K_{IC})	67.33	63.11	60.77

that the healing efficiency reduces as the fracture–healing cycle increases. This is consistent with any self-healing system because no healing system can last forever. The other is that different parameters yield considerably different healing efficiencies. The peak load yields the highest healing efficiency and the Mode I fracture toughness gives the lowest healing efficiency. From the safety point of view for evaluating healing efficiency, toughness is a better indicator because it will lead to a conservative design. Also, this result suggests that when we make comparisons of healing efficiencies from different healing schemes, we must first make sure that they use the same criteria for comparison. Otherwise, the comparison is meaningless.

It is noted that, when using an adhesively bonded joint configuration to evaluate the healing efficiency of self-healing materials, the fracture may be driven by energy (toughness) or by stress, or by both, depending on the scale of the crack processing zone or cohesive length scale, that is, the length of plasticity zone in the adhesive layer ahead of the physical crack tip. When the cohesive length scale is small, there is only a small-scale yielding ahead of the physical crack tip; hence, the fracture is brittle and there is a stress singularity at the physical crack tip based on linear elasticity. As a result, the fracture is driven by energy, that is, fracture toughness is a better criterion to evaluate the fracture strength. If the plasticity zone is large, the fracture is ductile and yielding occurs along the crack processing zone. Therefore, the fracture is driven by the cohesive stress and thus the stress criterion is better in determining the fracture strength. For the case in between the stress driven and toughness driven criteria, both cohesive strength and toughness are critical.

8.2.2 Healing Efficiency for an End-Notched Flexure (ENF) Specimen

In order to evaluate the healing efficiency for the Mode II fracture test, end-notched flexure (ENF) specimens can be used (see a schematic in Figure 8.11). Recently, Ouyang and Li [57] derived a formulation of a Mode II energy release rate for the ENF joints bonded with dissimilar materials. When the adhesively bonded joint becomes a standard ENF specimen, that is, the two adherends have identical thickness and are made of identical material, we have [57]

$$J_{II}(\delta_0) = \int_0^{\delta_0} \tau(\delta)\,\mathrm{d}\delta = \frac{\dfrac{1}{2}\left(\dfrac{h}{D}\right)^2 Q_T^2 a^2 + \dfrac{hQ_T}{2D}\delta_0}{\dfrac{2}{A} + \dfrac{h^2}{2D}} \tag{8.13}$$

where J_{II} is the energy release rate of the ENF specimen during the crack initiation process, Q_T is the reaction force per unit width at the support near the notched end, a is the initial crack

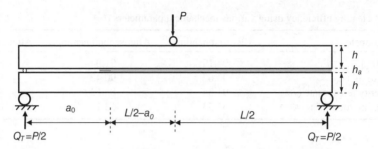

Figure 8.11 Schematic of the ENF specimen. *Source:* [63] Reproduced with permission from Springer

length, h is the thickness of the adherend, and δ_0 is the slip between the upper adherend and the lower adherend at the initial crack tip. A and D are the axial stiffness and the bending stiffness per unit width of the adherends, respectively, which can be determined for isotropic adherends as follows:

$$A = \frac{Eh}{1 - v^2}; \quad D = \frac{Eh^3}{12(1 - v^2)} \tag{8.14}$$

in which E is the modulus of elasticity and v is Poisson's ratio of the adherends, respectively.

Similar to the Mode I fracture test, the energy release rate J_{II} can be experimentally determined as a function of the crack tip slip δ_0 and the global shear force Q_T. Once the experimental J_{II}–δ_0 curves are obtained according to Equation (8.13), the Mode II interfacial traction–separation law $\tau = \tau(\delta_0)$ can be experimentally determined as follows:

$$\tau(\delta_0) = \frac{\partial J_{\mathrm{II}}(\delta_0)}{\partial \delta_0} \tag{8.15}$$

Again, once the critical energy release rate J_{IIC} is determined, the Mode II fracture toughness under the plane stress condition, K_{IIC}, can be found through the critical energy release rate as

$$K_{\mathrm{IIC}} = \sqrt{J_{\mathrm{IIC}}E} \tag{8.16}$$

Similarly, the healing efficiency under the Mode II fracture can be defined as

$$\eta = \frac{K_{\mathrm{IIChealed}}}{K_{\mathrm{IICvirgin}}} \tag{8.17}$$

where $K_{\mathrm{IIChealed}}$ is the fracture toughness of the healed specimen and $K_{\mathrm{IICvirgin}}$ is the fracture toughness of the virgin specimen, both under the Mode II fracture test.

As an example, Ji, Ouyang, and Li [63] prepared ENF specimens and conducted the Mode II fracture test. The adhesive and steel adherends used were the same as those used in Section 8.2.1.2. In Figure 8.11, $a_0 = 101.6$ mm, $L = 457.2$ mm, $h = 9.53$ mm, and $h_0 = 0.1$ mm, 0.2 mm, 0.4 mm, 0.6 mm, 0.8 mm, and 1.0 mm in order to investigate the effect of adhesive

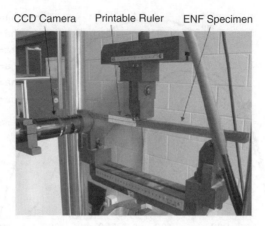

Figure 8.12 ENF specimen during the bending test. *Source:* [63] Reproduced with permission from Springer

layer thickness on the cohesive laws. Figure 8.12 shows the ENF specimen under the Mode II fracture test and the associated instrumentation. Figure 8.13 shows a typical energy release rate change with the local crack tip slip. In Figure 8.13, we marked two special energy release rate values. One is J_0, which corresponds to the peak bending load, and J_C, which indicates the start of crack propagation. It is seen that the energy release rate continues to increase after the peak bending load is reached, echoing the nature of the toughened adhesive. In Figure 8.14, the effect of adhesive thickness on the lump-sum equivalent cohesive law is demonstrated. Clearly, the shear strength (peak shear stress) and energy release rate increase as the adhesive layer thickness increases. However, as discussed in the Mode I fracture test, this effect would not affect the healing efficiency if this specimen is used to evaluate the Mode II healing efficiency,

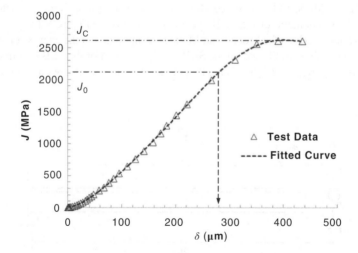

Figure 8.13 A typical relationship between the energy release rate J and the local crack tip slip δ for Group 1 specimen ($h_a = 0.1$ mm). *Source:* [63] Reproduced with permission from Springer

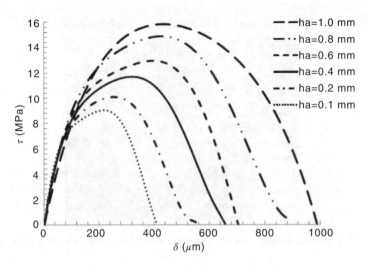

Figure 8.14 Typical shapes of the interfacial traction–separation laws at different adhesive layer thicknesses h_a. *Source:* [63] Reproduced with permission from Springer

simply because the efficiency is defined as the ratio of the healed specimen and the virgin specimen and re-cracking of the healed specimen usually follows the previous path.

8.2.3 Healing Efficiency for a Single-Lag Bending (SLB) Specimen

Engineering structures made of self-healing materials are usually subjected to complex stress conditions. As for the fracture mode, both Mode I and Mode II, and Mixed Mode I&II can be encountered. Therefore, it is desired to evaluate the healing efficiency under the mixed fracture mode. The Mixed Mode I&II facture condition can be created in a single-leg bending (SLB) specimen [64]. A schematic of the SLB specimen is shown in Figure 8.15.

Ji *et al.* [64] have derived the Mode I and Mode II components of the energy release rates and the equivalent cohesive laws based on the SLB configuration. For the SLB specimen shown in Figure 8.15, Ji, Ouyang, and Li [64] found that the Mode I energy release rate component J_I can

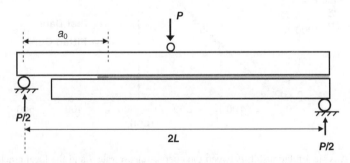

Figure 8.15 Schematic of the SLB specimen. *Source:* [64] Reproduced with permission from Elsevier

be expressed as follows:

$$J_I(w_0) = \int_0^{w_0} \sigma(w)dw = \frac{P}{4}\theta_P \tag{8.18}$$

where P is the mid-span bending load, θ_P is the relative rotation between the two adherends at the loadline, and w_0 is the local normal separation between the two adherends at the crack tip. Note that in a real SLB test, it is not convenient to measure the rotation of the bottom adherend at the loadline due to the presence of the support, and nor at the crack tip. One may choose any location between the crack tip and support because the rotations for the bottom adherend within this region are identical [64].

The Mode II energy release rate component J_{II} turns out to be the same as the ENF specimen in Equation (8.13) [57]. With the energy release rate in hand, we can find the cohesive law of both the Mode I and Mode II components as follows [64]:

$$\sigma(w_0) = \frac{\partial J_I(w_0)}{\partial w_0} = \frac{\partial\left\{\frac{1}{4}P\theta_P\right\}}{\partial w_0} \tag{8.19}$$

$$\tau(\delta_0) = \frac{\partial J_{II}(\delta_0)}{\partial \delta_0} = \frac{\partial\left\{\dfrac{\frac{1}{2}\left(\dfrac{ha}{2D}\dfrac{P}{2}\right)^2 + \dfrac{h}{2D}\dfrac{P}{2}\delta_0}{\dfrac{2}{A} + \dfrac{h^2}{2D}}\right\}}{\partial \delta_0} \tag{8.20}$$

From Equations (8.19) and (8.20), it is seen that once the crack tip opening w_0, crack tip slip δ_0, loading force P, and the rotation of the two adherends are simultaneously recorded, the interface shear stress $\tau(\delta_0)$ and interface normal stress $\sigma(w_0)$ can be determined experimentally.

In general, the relationship between the fracture toughness and energy release rate $K_{IC} \sim J_{IC}$ in Equation (8.11) and $K_{IIC} \sim J_{IIC}$ in Equation (8.16) hold in the SLB specimens. Therefore, when using the SLB specimen to evaluate the healing efficiency of self-healing materials, we can use Equations (8.12) and (8.17) to evaluate the Mode I healing efficiency and the Mode II healing efficiency, respectively.

Of course, an SLB specimen provides more information than the pure Mode I and Mode II fractures, that is, the mixed fracture mode. Therefore, we need to define a new parameter to evaluate the mixed mode healing efficiency. Because the energy release rate is a scalar, we can define the total critical energy release rate, J_C, during the mixed mode fracture process as

$$J_C = J_{IC} + J_{IIC} \tag{8.21}$$

Here we assume that the two fracture modes occur simultaneously. This assumption can be justified because, as long as one fracture mode starts propagating, the other mode cannot hold the adhesive in place. Otherwise, the crack cannot propagate, regardless of the fracture mode.

With Equation (8.21), we can define the mixed mode fracture toughness, K_C, using the G-criterion as [75]

$$K_C = \sqrt{K_{IC}^2 + K_{IIC}^2}$$ (8.22)

Once the fracture toughness is obtained, we can define the healing efficiency of self-healing materials from the mixed mode fracture resistance as

$$\eta = \frac{K_{Chealed}}{K_{Cvirgin}}$$ (8.23)

where $K_{Chealed}$ and $K_{Cvirgin}$ are the mixed mode fracture toughness of the healed specimen and the virgin specimen, respectively.

Similarly, the mixed mode cohesive law can also be constructed based on the two fracture modes. Due to the vectorial nature of the normal and tangential separations, we can define the overall crack tip separation as [64]

$$\varphi = \sqrt{w_0^2 + \delta_0^2}$$ (8.24)

The overall mixed mode equivalent cohesive law can be obtained as follows:

$$T(\varphi) = \frac{\partial J}{\partial \varphi}$$ (8.25)

where T is the overall traction or normalized traction across the crack tip and $J = J_I + J_{II}$ is the total energy release rate.

Using the same raw materials and adhesive as the ENF specimen, Ji, Ouyang, and Li [64] prepared SLB specimens. The top steel adherend has dimensions of 6.35 mm thick, 25.40 mm wide, and 304.80 mm long. The bottom steel adherend has the same width and thickness as the top adherend, but with a shorter length of 279.40 mm. Immediately after the application of the adhesive layer, a very thin mylar tape with a thickness of 0.035 mm was carefully inserted from the edge of the adhesive layer by 2 mm at the middle height of the adhesive layer to create a sharp initial crack. Since the distance between the loadline and the edge of the adhesive layer is 18 mm, the total initial crack length a_0 is 20 mm. Again, the fracture test was conducted under the displacement controlled mode. The loading rate was set as 1 μm/s and the data collecting frequency was 1 Hz. The test setup with instrumentation is shown in Figure 8.16.

Figure 8.17 shows the comparison of the overall mixed mode cohesive law and the two component cohesive laws. It is noted that the area enclosed by the T–ϕ curve is a sum of the areas enclosed by the σ–w curve and τ–δ curve, which is a direct validation of Equation (8.21).

Finally, Figure 8.18 shows the overall mixed mode traction–separation laws with various adhesive thicknesses. It is seen that the area under the T–ϕ curves increase as the thickness of the adhesive layer increases, which suggests that the energy release rate increases, as also does the fracture toughness. It is interesting to note that the peak traction is almost constant as the adhesive layer thickness increases. This is understandable because the peak traction represents the strength of the material under the complex stress condition.

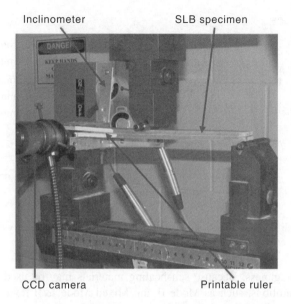

Figure 8.16 SLB test setup with instrumentation. *Source:* [64] Reproduced with permission from Elsevier

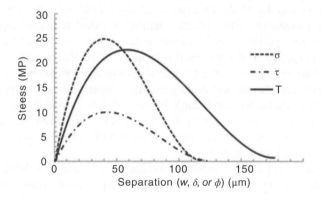

Figure 8.17 Cohesive law of the overall mixed mode fracture as compared with the interfacial traction–normal separation law and interfacial traction–slip law. *Source:* [64] Reproduced with permission from Elsevier

8.2.4 Summary

The evaluation of self-healing efficiency has been a topic of research interest since the self-healing systems were discovered. Various types of healing indicators have been used to evaluate the healing efficiency, primarily mechanical properties. It is expected that with demand of other healing properties, healing indicators other than mechanical properties are also possible, such as electrical or thermal conductivity. Of the various mechanical indicators used, fracture toughness seems popular. Actually, based on Table 8.1, fracture toughness yields the most conservative result. The current dominant toughness indicator used in the literature is

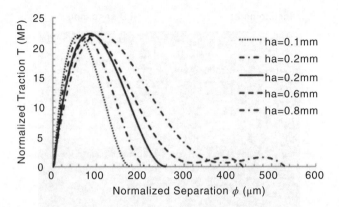

Figure 8.18 Normalized mixed traction–separation laws of an SLB specimen with different bondline thicknesses h_a. *Source:* [64] Reproduced with permission from Elsevier

Mode I fracture toughness. Because self-healing materials may be used in structures with complex stress conditions, such as Mode II and Mixed Mode I&II fracture, it is desired to evaluate the healing efficiency of the self-healing materials in terms of the Mode II fracture resistance and Mixed Mode I&II fracture resistance. To this end, we proposed two specimens; one is the ENF specimen for Mode II fracture toughness and the other is the SLB specimen for Mixed Mode I&II toughness. Closed-form equations to calculate the healing efficiency and protocols to prepare specimens and conduct tests are discussed and demonstrated with examples. One advantage of the proposed test protocols is that, in addition to the energy release rate and fracture toughness, the specimens discussed can yield lump-sum or equivalent interfacial cohesive laws for the self-healing materials, which can be incorporated in a finite element software package for conducting structural analysis and design of engineering structures made of self-healing materials. It is noted that, since different failure modes can yield different energy release rates and cohesive laws, it is expected that different failure modes may have different healing efficiencies.

The cohesive law model is not limited to evaluating self-healing efficiencies. In modern additive manufacturing, the manufacturing procedure is a process to create new interfaces. It is expected that interfacial cohesive laws will play an important role in additive manufacturing. Actually, for any material interface, such as the grain boundary in polycrystalline metals or ceramics, the boundary can be treated as an interface and the interfacial cohesive law is needed. If we treat the diffusion-formed interfacial layer during additive manufacturing or the grain boundary in polycrystalline materials as an equivalent adhesive layer, the models and test protocols for an adhesively bonded configuration may be used to determine the interfacial cohesive laws. One challenge facing the interfacial cohesive law is the adhesive thickness dependence, that is, how to determine the intrinsic cohesive law. There is a clear size effect or length scale factor involved. Atomistic modeling may find the intrinsic interfacial fracture energy and traction–separation law, but is limited to a small number of atoms. Scaling up the atomistic modeling to a continuum may be a research topic that deserves heavy investment. Some efforts have been made, such as the multiscale cohesive law model [76] and renormalization procedure [77] to address the disconnection between the atomistic and engineering cohesive descriptions, which may deserve further in-depth exploration. Finally, as discussed above, a cohesive fracture in an adhesively bonded configuration depends on yielding ahead of the physical crack tip. Therefore, there is a need to determine the characteristic yielding length scale, from which the driver for fracture, either toughness or stress or both, can be identified and higher fracture strength structures can be designed.

References

1. Zhang, M.Q. and Rong, M.Z. (2012) Theoretical consideration and modeling of self-healing polymers. *Journal of Polymer Science Part B: Polymer Physics*, **50**, 229–241.
2. Voyiadjis, G., Shojaei, A., Li, G., and Kattan, P.I. (2012) Continuum damage-healing mechanics with introduction to new healing variables. *International Journal of Damage Mechanics*, **21**, 391–414.
3. White, S.R., Sottos, N.R., Geubelle, P.H., Moore, J.S., Kessler, M.R., Sriram, S.R., Brown, E.N., and Viswanathan, S. (2001) Autonomic healing of polymer. *Nature*, **409**, 794–797.
4. Pang, J.W.C. and Bond, I.P. (2005) A hollow fibre reinforced polymer composite encompassing self-healing and enhanced damage visibility. *Composites Science and Technology*, **65**, 1791–1799.
5. Trask, R.S., Williams, G.J., and Bond, I.P. (2007) Bioinspired self-healing of advanced composite structures using hollow glass fibres. *Journal of the Royal Society Interface*, **4**, 363–371.
6. Toohey, K.S., Sottos, N.R., Lewis, J.A., Moore, J.S., and White, S.R. (2007) Self-healing materials with microvascular networks. *Nature Materials*, **6**, 581–585.
7. Chen, X., Dam, M.A., Ono, K., Mal, A., Shen, H., Nutt, S.R., Sheran, K., Wudl, F. (2002) A thermally remendable cross-linked polymeric material. *Science*, **295**, 1698–1702.
8. Varley, R., van der Zwaag, S. (2008) Towards an understanding of thermally activated self-healing of an ionomer system during ballistic penetration. *Acta Materialia*, **56**, 5737–5750.
9. Zako, M. and Takano, N. (1999) Intelligent material systems using epoxy particles to repair microcracks and delamination damage in GFRP. *Journal of Intelligent Material Systems and Structures*, **10**, 836–841.
10. Hayes, S.A., Jones, F.R., Marshiya, K., and Zhang, W. (2007) A self-healing thermosetting composite material. *Composites Part A: Applied Science and Manufacturing*, **38**, 1116–1120.
11. Li, G. and John, M. (2008) A self-healing smart syntactic foam under multiple impacts. *Composites Science and Technology*, **68**, 3337–3343.
12. John, M. and Li, G. (2010) Self-healing of sandwich structures with grid stiffened shape memory polymer syntactic foam core. *Smart Materials and Structures*, **19**, paper 075013 (12 pages).
13. Nji, J. and Li, G. (2010) A self-healing 3D woven fabric reinforced shape memory polymer composite for impact mitigation. *Smart Materials and Structures*, **19**, paper 035007 (9 pages).
14. Li, G. and Nettles, D. (2010) Thermomechanical characterization of a shape memory polymer based self-repairing syntactic foam. *Polymer*, **51**, 755–762.
15. Li, G. and Uppu, N. (2010) Shape memory polymer based self-healing syntactic foam: 3-D confined thermo-mechanical characterization. *Composites Science and Technology*, **70**, 1419–1427.
16. Nji, J. and Li, G. (2010) A biomimic shape memory polymer based self-healing particulate composite. *Polymer*, **51**, 6021–6029.
17. Nji, J. and Li, G. (2012) Damage healing ability of a shape memory polymer based particulate composite with small thermoplastic contents. *Smart Materials and Structures*, **21**, paper 025011.
18. Li, G., Meng, H., and Hu, J. (2012) Healable thermoset polymer composite embedded with stimuli-responsive fibers. *Journal of the Royal Society Interface*, **9**, 3279–3287.
19. Li, G., Ajisafe, O., and Meng, H. (2013) Effect of strain hardening of shape memory polymer fibers on healing efficiency of thermosetting polymer composites. *Polymer*, **54**, 920–928.
20. Li, G. and Zhang, P. (2013) A self-healing particulate composite reinforced with strain hardened short shape memory polymer fibers. *Polymer*, **54**, 5075–5086.
21. Raghavan, J. and Wool, R.P. (1999) Interfaces in repair, recycling, joining and manufacturing of polymers and polymer composites. *Journal of Applied Polymer Science*, **71**, 775–785.
22. Brown, E. N, Sottos, N.R., and White, S.R. (2002) Fracture testing of a self-healing polymer composite. *Experimental Mechanics*, **42**, 372–379.
23. Brown, E.N. (2011) Use of the tapered double-cantilever beam geometry for fracture toughness measurements and its application to the quantification of self-healing. *The Journal of Strain Analysis and Engineering Design*, **46**, 167–186.

24. Plaisted, T.A. and Nemat-Nasser, S. (2007) Quantitative evaluation of fracture, healing and re-healing of a reversibly cross-linked polymer. *Acta Materialia*, **55**, 5684–5696.

25. Plaisted, T.A., Amirkhizi, A.V., and Nemat-Nasser, S. (2006) Compression-induced axial crack propagation in DCDC polymer samples: experiments and modeling. *International Journal of Fracture*, **141**, 447–457.

26. Li, G., Ji, G., and Ouyang, Z. (2012) Adhesively bonded healable composite joint. *International Journal of Adhesion and Adhesives*, **35**, 59–67.

27. Wool, R.P. and O'Connor, K.M. (1981) A theory crack healing in polymers. *Journal of Applied Physics*, **52** (10), 5953–5963.

28. Voyiadjis, G., Shojaei, A., Li, G., and Kattan, P.I. (2012) A theory of anisotropic healing and damage mechanics of materials. *Proceedings of the Royal Society A:Mathematical Physical and Engineering Sciences*, **468**, 163–183.

29. Voyiadjis, G., Shojaei, A., and Li, G. (2012) A generalized coupled viscoplastic–viscodamage–viscohealing theory for glassy polymers. *International Journal of Plasticity*, **28**, 21–45.

30. Voyiadjis, G., Shojaei, A., and Li, G. (2011) A thermodynamic consistent damage and healing model for self healing materials. *International Journal of Plasticity*, **27**, 1025–1044.

31. Huang, B., Li, G., and Mohammad, L.N. (2003) Analytical modeling and experimental study of tensile strength of asphalt concrete composite at low temperatures. *Composites Part B: Engineering*, **34**, 705–714.

32. Edwards, S.F. (1967) The statistical mechanics of polymerized material. *Proceedings of the Physical Society*, **92**, 9–16.

33. de Gennes, P.G. (1971) Reptation of a polymer chain in the presence of fixed obstacles. *The Journal of Chemical Physics*, **55**, 572-579, *(1971)*

34. Flory, P.J. (1978) *Principals of Polymer Chemistry*, Cornell University Press, New York.

35. Hu, J.L. (2007) *Shape Memory Polymers and Textiles*, Woodhead Publishing Limited and CRC Press LLC, Cambridge.

36. Barenblatt, G.I. (1959) The formation of equilibrium cracks during brittle fracture. General ideas and hypothesis. Axismmetrical cracks. *Journal of Applied Mathematics and Mechanics (PMM)*, **23**, 622–636 (English translation)

37. Dugdale, D.S. (1960) Yielding of steel sheets containing slits. *Journal of the Mechanics and Physics of Solids*, **8**, 100–104.

38. Högberg, J.L. (2006) Mixed mode cohesive law. *International Journal of Fracture*, **141**, 549–559.

39. Li, S., Thouless, M.D., Waas, A.M., Schroeder, J.A., and Zavattieri, P.D. (2005) Use of a cohesive-zone model to analyze the fracture of a fiber-reinforced polymer-matrix composite. *Composites Science and Technology*, **65**, 537–549.

40. Lee, D.B., Ikeda, T. Miyazaki, N., and Choi, N.S. (2004) Effect of bond thickness on the fracture toughness of adhesive joints. *Journal of Engineering Materials and Technology*, **126**, 14–18.

41. Lorenzis, L.D. and Zavarise, G. (2009) Cohesive zone modeling of interfacial stresses in plated beams. *International Journal of Solids and Structures*, **46**, 4181–4191.

42. Zhu, Y., Liechti, K.M., and Ravi-Chandar, K. (2009) Direct extraction of rate-dependent traction–separation laws for polyurea/steel interfaces. *International Journal of Solids and Structures*, **46**, 31–51.

43. Wei, Y. and Hutchinson, J.W. (1998) Interface strength, work of adhesion and plasticity in the peel test. *International Journal of Fracture*, **93**, 315–333.

44. Wu, Z.S., Yuan, H., and Niu, H. (2002) Stress transfer and fracture propagation in different kinds of adhesive joints. *ASCE Journal of Engineering Mechanics*, **128**, 562–573.

45. Rose, J.H., Smith, J.R., and Ferrante, J. (1983) Universal features of bonding in metals. *Physical Review B*, **28**, 1835–1845.

46. Needleman, A. (1987) A continuum model for void nucleation by inclusion debonding. *ASME Journal of Applied Mechanics*, **54**, 525–531.

47. Needleman, A. (1990) An analysis of tensile decohesion along an interface. *Journal of the Mechanics and Physics of Solids*, **38**, 289–324.

48. Dourado, N., de Moura, M.F.S.F., de Morais, A.B., and Pereira, A.B. (2012) Bilinear approximations to the mode II delamination cohesive law using an inverse method. *Mechanics of Materials*, **49**, 42–50.

49. Tvergaard, V. (1990) Effect of fibre debonding in a whisker-reinforced metal. *Material Science and Engineering A*, **125**, 203–213.

50. Tvergaard, V. and Hutchinson, J.W. (1992) The relation between crack growth resistance and fracture process parameters in elastic–plastic solids. *Journal of the Mechanics and Physics of Solids*, **40**, 1377–1397.

51. Xu, X.P. and Needleman, A. (1993) Void nucleation by inclusion debonding in a crystal matrix. *Modeling and Simulation in Materials Science and Engineering*, **1**, 111–132.

52. Camacho, G.T. and Ortiz, M. (1996) Computational modeling of impact damage in brittle materials. *International Journal of Solids and Structures*, **33**, 2899–2938.
53. Hilleborg, A., Modeer, M., and Petersson, P.E. (1976) Analysis of crack formation and crack growth in concrete by means of fracture mechanics and finite elements. *Cement and Concrete Research*, **6**, 773–782.
54. Geubelle, P.H. and Baylor, J. (1998) The impact-induced delamination of laminated composites: a 2D simulation. *Composites Part B: Engineering*, **29**, 589–602.
55. Hutchinson, J.W. and Evans, A.G. (2000) Mechanics of materials: topdown approaches to fracture. *Acta Materialia*, **48**, 125–135.
56. Kanninen, M.F. (1973) An augmented double cantilever beam model for studying crack propagated and arrest. *International Journal of Fracture*, **9**, 83–92.
57. Ouyang, Z. and Li, G. (2009) Nonlinear interface shear fracture of end notched flexure specimens. *International Journal of Solids and Structures*, **46**, 2659–2668.
58. Ouyang, Z., Li, G., Ibekwe, S.I., Stubblefield, M.A., and Pang, S.S. (2010) Crack initiation process of DCB specimens based on first order shear deformation theory. *Journal of Reinforced Plastics and Composites*, **29**, 651–663.
59. Ouyang, Z. and Li, G. (2009) Local damage evolution of DCB specimens during crack initiation process: a natural boundary condition based method. *ASME Journal of Applied Mechanics*, **76**, paper 051003.
60. Ouyang, Z. and Li, G. (2009) Cohesive zone model based analytical solutions for adhesively bonded pipe joints under torsional loading. *International Journal of Solids and Structures*, **46**, 1205–1217.
61. Ouyang, Z. and Li, G. (2009) Interfacial debonding of pipe joints under torsion loads: a model for arbitrary nonlinear cohesive laws. *International Journal of Fracture*, **155**, 19–31.
62. Ji, G., Ouyang, Z., and Li, G. (2013) Effects of bondline thickness on Mode-I nonlinear interfacial fracture of laminated composites: an experimental study. *Composites Part B: Engineering*, **47**, 1–7.
63. Ji, G., Ouyang, Z., and Li, G. (2012) Local interface shear fracture of bonded steel joints with various bondline thicknesses. *Experimental Mechanics*, **52**, 481–491.
64. Ji, G., Ouyang, Z., and Li, G. (2012) On the interfacial constitutive laws of mixed mode fracture with various adhesive thicknesses. *Mechanics of Materials*, **47**, 24–32.
65. Ji, G., Ouyang, Z., and Li, G. (2011) Effects of bondline thickness on Mode-II interfacial laws of bonded laminated composite plate. *International Journal of Fracture*, **168**, 197–207.
66. Ji, G., Ouyang, Z., Li, G., Ibekwe, S.I., and Pang, S.S. (2010) Effects of adhesive thickness on global and local Mode-I interfacial fracture of bonded joints. *International Journal of Solids and Structures*, **47**, 2445–2458.
67. Ouyang, Z., Ji, G., and Li, G. (2011) On approximately realizing and characterizing pure Mode-I interface fracture between bonded dissimilar materials. *ASME Journal of Applied Mechanics*, **78**, paper 031020.
68. Klarbring, A. (1991) Derivation of a model of adhesively bonded joints by the asymptotic expansion method. *International Journal of Engineering Science*, **29**, 493–512.
69. Williams, J.G. and Hadavinia, H. (2002) Analytical solutions for cohesive zone models. *Journal of the Mechanics and Physics of Solids*, **50**, 809–825.
70. Pan, J. and Leung, C.K.Y. (2007) Debonding along the FRP–concrete interface under combined pulling/peeling effects. *Engineering Fracture Mechanics*, **74**, 132–150.
71. Rice, J.R. (1968) A path independent integral and the approximate analysis of strain concentration by notches and cracks. *Journal of Applied Mechanics*, **35**, 379–386.
72. Andersson, T. and Stigh, U. (2004) The stress–elongation relation for an adhesive layer loaded in peel using equilibrium and energetic forces. *International Journal of Solids and Structures*, **41**, 413–434.
73. Li, G. and Shojaei, A. (2012) A viscoplastic theory of shape memory polymer fibers with application to self-healing materials. *Proceedings of the Royal Society A: Mathematical Physical and Engineering Sciences*, **468**, 2319–2346.
74. Kafkalidis, M.S., Thouless, M.D., Yang, Q.D., and Ward, S.M. (2000) Deformation and fracture of adhesive layers constrained by plastically-deforming adherends. *Journal of Adhesion Science and Technology*, **14**, 1593–1607.
75. Sih, G.C. and Macdonald, B. (1974) Fracture mechanics applied to engineering problems – strain energy density fracture criterion. *Engineering Fracture Mechanics*, **6** (2), 361–386.
76. Yao, H. and Gao, H. (2007) Multi-scale cohesive laws in hierarchical materials. *International Journal of Solids and Structures*, **44**, 8177–8193.
77. Nguyen, O. and Ortiz, M. (2002) Coarse-graining and renormalization of atomistic binding relations and universal macroscopic cohesive behavior. *Journal of the Mechanics and Physics of Solids*, **50**, 1727–1741.

9

Summary and Future Perspective of Biomimetic Self-Healing Composites

Self-healing represents the next generation of technology, not only because it helps maintain the reliability and durability of engineering structures but also because it is a must for a low carbon economy and sustainability. Gradually, self-healing of engineering structures will transform from scientific dream to engineering reality. It is foreseeable that bio-inspiration will play a pivotal role in developing new and innovative self-healing systems. In the following, we will first summarize the shape memory polymer based self-healing system, which is the focus of this book, and then provide a brief outlook for further development in this field of study.

9.1 Summary of SMP Based Biomimetic Self-Healing

Self-healing based on the shape memory effect has been developed by our group towards solving a critical problem facing self-healing schemes; that is, there is a need to bring the fractured surface in close proximity or in contact before the healing mechanism takes effect. While this is not a problem at all in lab-scale specimens because the fractured parts can be brought into contact manually, unfortunately, this simple operation represents the grand challenge in real world applications. In engineering structures, the structural components are not free to move and cannot be brought in contact manually or without damaging other parts if they are forced to be in contact. Therefore, bringing the fractured parts into contact proves to be one of the largest obstacles for the use of self-healing materials in engineering applications.

Therefore, we proposed to use the shape memory effect (SME) to close a wide-opened crack first before other healing mechanisms take effect, which is the so-called close-then-heal (CTH) mechanism in our research. Bringing fractured parts into contact using SME seems a very straightforward idea. This is because shape memory is defined as the ability, if the polymer is deformed into a temporary shape, to restore its original shape upon external stimuli. Therefore, if we treat a crack or damage as a type of deformed shape, shape recovery suggests closing the crack. It sounds so simple!

Self-Healing Composites: Shape Memory Polymer-Based Structures, First Edition. Guoqiang Li.
© 2015 John Wiley & Sons, Ltd. Published 2015 by John Wiley & Sons, Ltd.

However, in real world structures, again the structural components are not free to move. Therefore, if we just use the external load caused shape memory, which is called ad hoc programming herein, we cannot guarantee that the required crack-closing force and recovery strain can be obtained. This is because some molecular level damage may occur and thus the shape recovery ratio is not 100%, which suggests that, even under the free shape recovery condition, the opened crack cannot be closed. In other words, the fractured parts cannot be brought fully into contact. Also, even though the shape recovery ratio is good enough, the recovery stress may not be able to overcome obstacles or constraints from neighboring structural components or fixed boundary conditions, which again can prevent fractured parts from marching towards each other. Therefore, crack closing by SME is more involved in science and technology than it seems to be. The SMP needs to be programmed to the proper level in order to close cracks with a certain width.

With these challenges in mind, we started with a general description of damage in various types of fiber reinforced thermosetting polymer composites in Chapter 1, which establishes the needs for self-healing. We then showed that many biological systems have a self-healing capability in Chapter 2, which provides unlimited sources for inspiration, including the CTH in this book. After that, we focused on introducing two types of programming in Chapter 3, that is, hot programming, which is above the transition temperature, and cold programming, which is below the transition temperature. Various types of programming stress conditions were also explored, including 1-D, 2-D, and 3-D. Both thermoset SMP and its syntactic foam were studied and the durability or functional stability of the shape memory capability of these smart materials under environmental attacks was evaluated. In Chapter 4, we first introduced the fundamentals of solid mechanics. We then used it to model the thermo-mechanical behavior of amorphous SMP and the SMP based syntactic foam. In Chapter 5, we considered a microphase segregated shape memory polyurethane fiber. We characterized its thermomechanical and damping properties and proposed a viscoplastic theory to model it. Particularly, we generalized the fiber as having a semicrystalline morphology, which can be easily extended to other morphologies such as pure amorphous or pure crystalline. We also investigated and explained why SMP has a good strain memory but a poor stress memory, and provided a unified definition for strain and stress fixity and recovery ratios. In Chapter 6, our focus was on SMP matrix based self-healing materials. Various types of composite structures, including notched beam, 3-D woven fabric reinforced, and grid stiffened composite structures, were evaluated for healing efficiencies. Chapter 7 focused on healing of conventional thermosetting polymers by embedding SMP fibers. SMP fibers in the form of 1-D, 2-D, and 3-D were investigated. In Chapter 8, we briefly introduced an evaluation of healing efficiency with a focus on cohesive law models using the adhesively bonded joint configuration.

9.2   Future Perspective of SMP Based Self-Healing Composites

While we believe a substantial amount of work has been done towards the SMP based self-healing system using the CTH scheme, there is considerable room left for improvement. It is still in its infancy of development. Since our goal is to develop the next generation of self-healing materials for load bearing engineering structures, we foresee that the following need to be studied further. It is noted that this is by no means comprehensive. Rather, it is for our readers of this book to have a starting point for future research and development.

9.2.1 In-Service Self-Healing

Currently, most of the studies on self-healing are based on the lab environment. True self-healing should heal damage in the in-service condition. For example, if an aircraft is damaged when it is in the air, true self-healing suggests that the damage should be healed while the aircraft is in a normal flight condition. In order to create this in-service condition, the specimens in lab-scale testing should be under various loading conditions and boundary conditions when healing is conducted. In other words, an in-service stress condition should be recreated in the specimens when healing is conducted. In a previous study by Li, Ajisafe, and Meng [1], a fixed boundary condition was used for the damaged beam specimen. Obviously, this simulated a rod under the fixed boundary condition and subjected to a 1-D longitudinal stress condition. Other boundary conditions and more complex stress conditions need to be created in order to evaluate the in-service healing capabilities.

9.2.2 Healing on Demand

Healing on demand not only suggests healing autonomously, but also indicates healing locally and timely. For the close-then-heal (CTH) scheme to become healing on demand, we need to add the sensing and heating ability to the system.

Many approaches have been investigated in the literature as damage sensors, which may include, but are not limited to, strain gages, acoustic emission, radio frequency emitter, infrared thermal tomography, eddy current, fiber optic, etc. [2,3]. In addition to sensing, heating is also needed to trigger the shape recovery process. Therefore, it is desired that the system has an integrated sensing and heating capability. In other words, we suggest that the system becomes a conductor. Many conductive fillers can be added to SMPs so that the SMPs can become a thermal and electrical conductor [4]. These fillers can be carbon nanotubes, metallic nanoparticles, carbon black particles, etc. Usually, it is believed that these fillers will not only transform the SMPs into a conductor but they may also increase the recovery stress due to the increased stiffness of the SMPs at their shape recovery temperature. Of course, the filler content must be equal to or greater than the percolation threshold, which depends on the type of fillers and the dispersion of the fillers in the SMP matrix.

Using the SMP matrix as an example, we envision that the sensing and heating system will work in the following way. First of all, the conductive SMP with a percolated conductive filler network such as carbon nanotube or carbon black serves as a sensor, which monitors the crack initiation and propagation by measuring the electrical resistance of the material. When a crack occurs, the continuity of the conductive network is damaged, leading to a jump in electrical resistance. Therefore, electrical resistance can be used as the sensing signal. Further opening and propagation of the crack lead to a further increase in electrical resistance. If the resistance reaches a threshold value, indicating a wide opening of the crack, an external electrical power source will be trigged and will apply a high electrical current to increase the material temperature. In this case, the conductive filler serves as the heater. The increase in temperature serves two purposes. On the one hand, it causes shape recovery of the SMP matrix, bringing the fractured SMP matrix closer. On the other hand, it melts the embedded thermoplastic particles, which heal the crack by the CTH scheme. As the crack heals, the damaged conductive network is partially or fully restored, and the electrical resistance of the composite will drop to a certain value, which triggers the external electrical power source to turn off. This completes one cycle of damage and healing. We envision that this cycle can be repeated.

If we prepare conductive SMP fibers by embedding conductive fillers in the SMP matrix through the melt spinning process, we envision that the SMP fiber reinforced conventional polymer matrix will also have a sensing and heating capability. The reason is that, as the crack occurs, the conductive SMP fiber will be stretched, which leads to an increase in the electrical resistance according to Ohm's law. This signal can be used as damage sensing. Again, this signal will trigger the switching-on of an external power source, leading to a temperature rise in the conductive SMP fibers and shape recovery. The constrained shrinkage of the conductive SMP fibers results in narrowing of the crack and shortening of the SMP fibers. The shortening of the SMP fibers leads to a decrease in electrical resistance, again in terms of Ohm's law. Once the electrical resistance drops to the design level, the external power source will be turned off autonomously, completing one damage–healing cycle. Again, we envision that this damage–healing cycle is repeatable, although at reducing efficiency.

9.2.3 Self-Healing by a Combination of Shape Memory and Intrinsic Self-Healing Polymers

In the CTH, we use thermoplastic particles for healing. Actually, other combinations of shape memory and healing can also be conducted, such as pairing of shape memory and a liquid healing agent. The most interesting idea may be a combination of intrinsic and extrinsic self-healing schemes. The idea may include making intrinsic self-healing polymers have a shape memory capability so that the shape memory will bring the fractured surfaces in contact and the intrinsic healing mechanisms will heal molecularly. Examples could be shape memory supramolecules, shape memory ionomers, shape memory dynamic covalent bond exchange polymers, shape memory thermally reversible covalent bond polymers, etc.

9.2.4 Manufacturing of SMP Fibers with Higher Strength and Higher Recovery Stress

9.2.4.1 Chemical Way

There is no doubt that shape memory depends first on the morphology of polymers. However, composition change also plays a key role in determining the mechanical properties, in addition to the shape memory capability. As has been demonstrated in the CTH scheme, if the recovery stress of the shape memory polyurethane fibres (SMPFs) can be further enhanced while maintaining ductility, larger cracks can be effectively closed. Therefore, there is a need to enhance the recovery stress of SMPFs further. Also, a further increase in tensile strength may make the SMPF suitable for structural reinforcement. In such a way, SMPF not only serves as a functional component but may also serve as a load bearing fiber, similar to other polymeric fibers in engineering structures. Because cold drawing has been used in our study, we need to turn to the composition for further enhancement. We prefer to focus on SMPF because its mechanical properties can be tailored in a wide range from high rigidity to flexibility. We believe that the following may be some potential ways to achieve the objectives. (1) Optimize the spinning process to reduce the fiber diameter. As the fiber becomes thinner, the defect becomes less and the molecules become more aligned along the axial direction. Consequently, the fiber becomes stiffer, stronger, and, most likely, higher in recovery stress. The spin ability of polymers is affected by the apparent shear stress, apparent shear rate, non-Newtonian index, apparent viscosity, and structural viscosity index. These parameters can be optimized so that thinner SMPFs with a higher molecular orientation can be prepared.

(2) Increase the stiffness of the polymer macromolecules. One problem with the lower recovery stress of SMPFs is the lower stiffness of the fiber at the stress recovery temperature. Therefore, increasing the fiber stiffness is preferred. Three approaches can be investigated. (a) Increase The soft segment and chain extender rigidity. Short, rigid, and a low content of polyol will improve the stiffness. A low content of polyols having low molecular weights in the hundreds can be used. Rigid chain extenders such as hydroquinone bis(2-hydroxyethyl) ether, bisphenol A, and *N*,*N*-bis (2-hydroxyethyl)-isonicotinamide can be evaluated. (b) Increase hydrogen bonds. Like those in spider silk, hydrogen bonds also play an important role in the mechanical properties of polyurethane. Buehler's research [5–8] shows that some hydrogen bonds rupture simultaneously rather than sequentially, which explains why the intrinsically weak hydrogen bonds enable strong materials. Short polyester polyols instead of polyether polyols can be used as the soft segment because polyester polyols can form hydrogen bonding with the hard segment. Diamine as a chain extender can form more hydrogen bonding by forming polyurethane/polyurea structures. (c) Increase the cross-link density. Several reports have shown that chemical cross-linking can increase the recovery stress and recovery ratios [9–11]. Unfortunately, chemical cross-linked polyurethane is not suitable for melt spinning and therefore it cannot be processed into fibers. It is envisioned that the present melt-spinning process can be modified by including a prepolymer adding system at the end of the melt-spinning process. The properties of polyurethane prepolymers can largely be modified by selecting a suitable polyol and polyisocyanate component, and by changing the molar ratio of these two reaction partners. The prepolymer at the end of the spinning process can act as a cross-linking agent to improve the mechanical strength of the fiber. At the same time, the prepolymer can decrease the polymer melt viscosity and decrease the melt elastic effect of the polyurethane. Therefore, it can increase the spin ability of the polyurethane.

9.2.4.2 Physical Way

When filaments are used as load carrying cables, they are always woven, twisted, or braided to increase the structural strength and/or deformability. For example, steel has a very low strain before yielding. However, when they are coiled into a spring, they have substantial elastic deformation. One explanation may be the considerable increase in effective length of the steel wire when it is coiled into a helical spring. Another example may be the double-spiral structure of DNA. Still another example may be the artificial muscle with the coiled carbon nanotube yarn [12–14]. Following the same line, we may also twist or coil or braid the SMPFs into wires. It is noted that twisting is also a programming process, which applies normal (length change) and shear (angle change) programming to the SMPF. Therefore, both axial and torsion recovery force can be expected. Due to the increased effective length, we envision that the overall recovery force will be increased for the SMPF to close the same crack because the constrained recovery force of the wire is transferred to the matrix through the interfacial shear stress, which is equal to integration of each differential element of the wire along the spiral length. Because of the increased effective length of the spiral fiber as compared to the straight fiber, it is expected that the helical fiber will yield a higher crack-closing force.

9.2.5 Determination of Critical Fiber Length

In self-healing with embedded short SMPFs, the fiber length plays a pivotal role in determining the crack width that can be closed. The critical fiber length has been a well-known concept in short-fiber reinforced polymer composites. To better design this type of self-healing material, we need to find a way to determine the critical fiber length. In mechanics, the crack-closing process of embedded SMPF is similar to fiber pull-out, but is not the same. (1) In fiber pull-out, an external load is applied to the fiber, while the load in the SMPF is due to the recovery stress, or body force. (2) In fiber pull-out, the fiber can be pulled out without a restriction on displacement, while the maximum displacement of the SMPF cannot be larger than the product of the crack width and the free recovery strain of the SMPF. (3) The recovery stress reduces as the recovery strain increases, that is, the recovery force continuously decreases as the crack is narrowed, while the pulling force increases in the fiber pull-out test before interfacial debonding occurs. Therefore, we need to modify the fiber pull-out model in order to evaluate the crack-closing ability of the SMPFs. A number of models have been developed to model the fiber pull-out problem [15]. Generally, there are two approaches to describe the debonding process of the fiber pull-out: one was based on fracture mechanics and modeled the debonding as a Mode II crack [16–23]; the other was based on shear lag theory and assumed that debonding initiates when the interfacial shear stress between the fiber and matrix exceeds the interfacial shear strength [24–26]. We believe that we can follow the modeling procedure by Hutchinson and Jensen [18]. Our strategy is to transform the crack-closing problem to a fiber pull-out problem. The body force due to stress recovery can be determiend by the recovery stress–strain relationship. The recovery stress can also be simulated as a thermal stress problem by assuming a negative coefficient of thermal expansion if the the fiber shrinks during recovery. Following the established procedure of solving cohesive law problems [27,28] (the equation of motion (from a free-body diagram), the constitutive law (linear elastic for both the fiber and the matrix, and bilinear for the interfacial shear), the kinematic relation (Lagrange definition of strain for the finite strain problem, that is, strain is defined by the deformation gradient), the boundary condition (clamped boundary condition for the matrix and the axial symmetry condition), and the interfacial compatibility condition (the radial stress and displacement are the same at the fiber/matrix interface)), we can solve the stress–strain distribution in the matrix piecewisely in an iterative manner. Once we know the axial strain of the matrix, integration along the axial direction will yield the surface displacement profile in the crack surface, and thus the crack width that is narrowed by the SMPF. The critical fiber length is the length that can narrow the crack to the designed level without fiber pull-out.

9.2.6 Damage–Healing Modeling

The damage–healing model has been developed to understand the damage and healing process [29–32]. While a number of models have been developed in the literature, overall the study is still in its infancy [33,34]. Currently, phenomenological models based on continuum damage mechanics, which are within the thermodynamics framework, predominate [29–32]. In these models, the damage and healing are represented by damage and healing variables, which are either scalar for isotropic damage and healing, or tensorial, for anisotropic damage and healing. Theoretically, healing is a reverse process of damage. Therefore, healing variables can be defined similarly to damage variables, such as an increase in stiffness or an increase in effective load bearing area. Of course, healing is usually time and temperature dependent. Therefore, a healing variable is a function of time and temperature. Also, healing depends on the mechanisms of healing. For intrinsic healing polymers, the fractured polymer needs to be brought into contact first. The re-establishment of a chemical bond is usually very fast. For extrinsic healing schemes, the healing efficiency depends on the healing agent. For example, while the fractured space can be fully filled in by the healing agent, the healing efficiency may be less than 100% if the healing agent has a lower strength than the matrix; on the other hand, the healing efficiency may be higher than 100% if the strength of the healing agent is higher than that of the matrix. Therefore, the classical healing variable defined in terms of the effective area needs to be revised by a factor that is defined as the quotient of the strength of the healing agent over the strength of the matrix, in order to consider the case where the healing efficiency is higher than 100%. In other words, the quotient is likely the upper bound of the healing efficiency. Furthermore, with a liquid healing agent, the recovery of strength depends on the degree of in situ polymerization; with a solid healing agent such as that in this book, the recovery of strength depends on molecular diffusion of the healing agent. Wool and O'Connor [33] have proposed a five-stage mechanism to describe the healing process by molecular diffusion. For the CTH system in this study, the molecule diffusion is enhanced by the applied pressure from the recovery force of the SMP. Therefore, molecule diffusion under external pressure needs to be further studied.

9.2.7 Development of Physics Based Constitutive Modeling of Shape Memory Polymers

In order to better design engineering structures with shape memory based self-healing capabilities, constitutive modeling, which is easily implemented in structure design, such as finite element analysis (FEA), is highly desired. Also, constitutive modeling is an important component in the hierarchical multi-length scale modeling of materials, particularly within the framework of Integrated Computational Materials Engineering (ICME), which is widely believed to be the future for integrated materials development and structure design. While considerable achievement has been made in constitutive modeling of SMPs, including thermodynamics consistent modeling [35], structural relaxation modeling [36], rheological modeling [37], molecular dynamics modeling [38], etc., one limitation of the existing constitutive modeling [39], no matter whether it is in the thermodynamic framework or the viscoelasticity framework, is the need for a large number of materials constants, which need sophisticated experimental testing and/or the curve fitting process. Therefore, there is a need to develop new models that have fewer materials constants, minimize the effort for experimental testing and curve fitting, and are suitable for the finite element modeling framework. Also, because of damage during cyclic loading and/or thermomechanical cycles, which may be further divided into physical damage such as reduction in strength and stiffness, and functional damage such as reduction in shape memory or stress memory capabilities, future constitutive modeling needs to take into account these cyclic damage effects. One potential approach may be to include statistical mechanics into the multilength and multitime scale modeling. Another fast developing modeling tool is the phase field model, which is able to quantify the microstructure evolution without explicit intervention. It is believed that this tool my be particularly useful for micro-phase segregated SMPFs because the microstructure evolution controls the memory capabilities.

9.2.8 A New Evaluation System

A shape memory polymer, as defined by its name, is characterized by its capability to memorize its shape. In different applications, various physical/mechanical properties need to be memorized, such as thermal and electrical conductivity for conducting, stress for crack closure, strain for shape, temperature for triggering memory, color for appearance, etc. However, different memory indicators correspond to different molecular level mechanisms. For example, SMPs with a perfect shape memory may have very low stress memory capability, as discussed in Chapter 5 of this book. Therefore, there is a need to understand the various memory mechanisms and define ways to evaluate them. This may need systematic and coordinated experimental testing and theoretical modeling effort.

9.2.9 Potential Applications in Civil Engineering

Shape memory polymers (SMPs), due to their unique shape memory effect, may find applications in civil engineering structures, in addition to the applications in fiber reinforced lightweight polymer composites, which are focus of this book. As an example, we would suggest some potential applications in pavement structures, which are the backbone for transportation in many countries in the world.

9.2.9.1 SMP Based Sealant

Recently, Li and Xu [40] and Li et al. [41] proposed a shape memory polymer (SMP) based syntactic foam sealant for a compression-sealed expansion joint in a bridge deck or concrete pavement. Usually, a compression-sealed sealant suffers from a couple of critical problems when the concrete wall expands at high temperature: building up of compressive stress and sealant squeezing out of the channel. By programming SMP based syntactic foam in a 2-D stress condition (compression in the horizontal or traffic direction and tension in the vertical direction) and by controlling the transition temperature of the foam below the highest temperature of the environment to be experienced, the accumulated compressive stress can be significantly reduced due to the two orders of drop in the stiffness of the foam at temperatures above its transition temperature; consequently, the concrete and the sealant will not be crushed. Simultaneously, the squeezing-out problem can also be eliminated due to the shrinkage of the foam in the vertical direction when the foam recovers (shape memory effect), which is seemingly contrary to the physics because it contracts when the temperature rises. Also, this 2-D programmed sealant can solve the problem of loss of contact between the sealant and the concrete wall when the temperature drops. For a conventional polymeric sealant, the plastic deformation accumulated at high temperatures cannot be recovered. As a result, the sealant may gradually lose contact with the concrete wall as the temperature drops, leading to leakage and gradual failure of the sealant. However, with the 2-D programmed smart sealant, compression programming in the traffic direction ensures that the plastic deformation recovers at high temperatures (the sealant tends to become wider in the traffic direction), maintaining contact with the concrete wall and minimizing the leakage problem. Further development may include an asphalt based liquid sealant and a smart sealant using two-way SMPs [42].

9.2.9.2 Asphalt Based Liquid Sealant [42]

Cracking in longitudinal, transverse, block, and alligator forms in asphalt pavement is very common [43]. These cracks are created due to thermal stress, fatigue, reflective cracking, etc. Filling of these cracks using sealant is a common practice. These cracks are usually sealed using hot-pour asphalt or cold-pour asphalt to prevent water from penetrating into the asphalt concrete layer and layers beneath. Without such treatment, water will penetrate into the layers, causing further cracks of the asphalt concrete layer and softening and pumping of layers made of soil or containing treated soil, leading to premature failure of the pavement. Because the coefficient of thermal expansion (CTE) of asphalt binder is almost an order higher than that of asphalt concrete, the asphalt sealant will debond from the cracked asphalt concrete surface when the temperature drops, leading to cracking again.

As in our previous studies [40,41], the key for the new asphalt based liquid sealant is seemingly against physics – contraction at high temperature and expansion at low temperature [42]. Inspired by the dual-shape SMP used by Li and Xu [40] and Li et al. [41], which can only have one temporary shape that can be programmed (seemingly against physics), we need to have at least two temporary shapes, one against thermal expansion in summer and the other against contraction in winter. In other words, we need to incorporate triple-shape SMP particles or fibers in liquid asphalt sealant [42]. The three or triple-shape SMPs can be block copolymers [44–47] or polymers with single broad glass transitions [48]. For triple-shape block copolymers, the SMPs will have three fundamental shapes. The first shape is the native/permanent shape (shape A) in which no programming has occurred. The second shape is a temporary shape (shape B, corresponding to programming of "block B," where block B has a relatively higher glass transition temperature (T_g) that is within a typical summer temperature range). The third shape is another temporary shape (shape C, corresponding to programming of "block C," where block C has a relatively lower glass transition temperature (T_g) that is within a typical winter temperature range). When programming such a polymer, we need to begin with high temperature tension programming of "block B". Once the temporary shape is fixed, the large tensile programmed material is machined to particles such as powders, which is conducted at temperatures lower than the transition temperature of block B to avoid its shape recovery. After that, 3-D compression programming of block C at a temperature above the transition temperature of block C but lower than the transition temperature of block B is performed on the tension programmed SMP particles. The dual programmed block copolymer particles will finally be mixed with liquid asphalt, such as asphalt emulsion or rubberized asphalt emulsion or solvent diluted asphalt, to form the smart sealant. It is noted that, although each programming will be on the entire polymer (comprising the two distinct blocks B and C), because of the different transition temperatures of the two blocks, the programming produces the maximum effect only on a particular block at the particular programming temperature. For example, when programming block C, the pre-tension in block B is largely maintained because block B is now in glassy state. It is envisioned that when such sealant is used in asphalt pavement, daytime sunshine in the winter will trigger the shape memory of block C, leading to volume expansion. As the temperature rises in the summer above the transition temperature of block B, block B is triggered and leads to contraction of the sealant, endowing the asphalt sealant with the desired properties of expansion in winter but contraction in summer [42]. Of course, determination of the prestrain levels in both blocks must consider the annual temperature changes and the service life of the sealant.

Xie [48] proves that for a dual-shape SMP with a broad glass transition, sequential programming can make this SMP have triple or multiple shapes. Therefore, for this type of polymer, we can also program them first at a higher temperature by tension, followed by another programming at a lower temperature by compression. If the two programming temperatures, which are within the broad glass transition region of the SMP, correspond to the summer and winter temperatures, respectively, it is envisioned that the programmed SMP particulate-filled asphalt sealant may behave similarly to the block copolymer SMPs [42]. It is interesting to note that Li et al. [41] show that, even for a dual-shape SMP with a narrow glass transition, sequential programming, that is, hot tension programming followed by cold compression programming, can also give the SMP a weak triple shape, suggesting the potential of its use as a sealant.

9.2.9.3 Two-Way Shape Changing Polymer Sealant [49]

Following the same line that the key requirement for a sealant is that the material must behave against physics – expansion when cooling and contraction when heating – we may find even more SMPs. For example, this may be realized by two-way SMPs. Many semicrystalline SMPs have demonstrated the two-way shape changing effect. For one-way SMPs, external thermo-mechanical programming has to be employed to deform the polymer to achieve a temporary shape [40, 41]. Two-way shape changing polymers do not require the pre-deformation. Instead, they change their shape upon stimulation. For a semicrystalline SMP to show the shape changing effect an external tensile load needs to be applied. Chung, Romo-Uribe, and Mather [50] first revealed the two-way shape changing effect on a cross-linked semicrystalline poly(cyclooctene) film to which a constant weight was applied. During cooling under a constant stress, crystallites formed in the loading direction, leading to elongation. When heated to a temperature above the melting transition of the polymer, the polymer contracted as a result of shape recovery. Therefore, it is envisioned that if such SMPs are used as a sealant, it lengthens when cooling down and shortens when heating up, which is exactly the type of behavior desired for a sealant [49].

9.2.9.4 Rutting Resistance Asphalt Concrete [51]

Rutting is a typical defect in an asphalt concrete pavement due to the inherent plastic deformation of asphalt at high temperatures, particularly in channelized traffic and/or at a stop sign where frequent acceleration/deceleration occurs [52]. The key is to reduce, eliminate, or recover plastic deformation in asphalt concrete. Various approaches have been investigated such as using a rubber modified asphalt as binder. Although the enhanced elasticity helps reduce or delay the occurrence of rutting, over the years plastic deformation still develops due to the channelized traffic load and gradual accumulation of permanent deformation. Since permanent deformation in asphalt or rubber modified asphalt cannot be ruled out, the most effective measure would be recovering the plastic deformation. In other words, the additional capacity to recover plastic deformation in asphalt concrete must be stored in order to counterbalance the gradually accumulated permanent deformation. Because SMP can recover the original shape after plastic deformation, it is envisioned that incorporation of SMP into asphalt concrete can reduce or eliminate the rutting problem. We propose that SMP can be machined into particles such as in the range of fine aggregate. A portion of fine aggregate can then be replaced by the SMP particles. In hot mix asphalt (HMA), compaction during construction at high temperature completes classical 3-D compression programming of the SMP particles, as long as the SMP particles have a higher melting or burning temperature than the mixing temperature of HMA, which is usually below 200 °C. Therefore, programming is coupled with construction and no additional programming is needed. If the transition temperature of the SMP particles is designed to coincide with the summer temperatures that the pavement will experience, the summer temperatures will trigger the shape recovery (volume expansion) of the programmed SMP particles, which helps restore the original shape of the pavement and thus eliminate rutting. Also, traffic load will serve as programming to the SMP particles, particularly in summer. Therefore, the rutting elimination ability is repeatable. Finally, warm mix asphalt can be used, instead of hot mix asphalt, which alleviates the requirement for SMPs to have higher melting temperature.

References

1. Li, G., Ajisafe, O., and Meng, H. (2013) Effect of strain hardening of shape memory polymer fibers on healing efficiency of thermosetting polymer composites. *Polymer*, **54**, 920–928.
2. Wang, Q. and Arash, B. (2014) A review on applications of carbon nanotubes and graphenes as nano-resonator sensors. *Computational Materials Science*, **80**, 350–360.
3. Li, C., Thostenson, E., and Chou, T. (2008) Sensors and actuators based on carbon nanotubes and their composites: a review. *Composites Science and Technology*, **68**, 1227–1249.
4. Behl, M. and Lendlein, A. (2007) Shape-memory polymers. *Materials Today*, **20**, 20–28.
5. Qin, Z. and Buehler, M.J. (2010) Cooperative deformation of hydrogen bonds in beta-strands and beta-sheet nanocrystals. *Physical Review E*, **82**, paper 061906.
6. Keten, S., Xu, Z., Ihle, B., and Buehler, M.J. (2010) Nanoconfinement controls stiffness, strength and mechanical toughness of beta-sheet crystals in silk. *Nature Materials*, **9**, 359–367.
7. Keten, S. and Buehler, M.J. (2008) Geometric confinement governs the rupture strength of H-bond assemblies at a critical length scale. *Nano Letters*, **8**, 743–748.
8. Keten, S. and Buehler, M.J. (2010) Nanostructure and molecular mechanics of spider dragline silk protein assemblies. *Journal of the Royal Society Interface*, **7**, 1709–1721.
9. Ortega, A.M., Yakacki, C.M., Dixon, S.A., Likos, R., Greenberg, A.R., and Gall, K. (2012) Effect of crosslinking and long-term storage on the shape-memory behavior of (meth)acrylate-based shape-memory polymers. *Soft Matter*, **8**, 3381–3392.
10. Voit, W., Ware, T., and Gall, K. (2010) Radiation crosslinked shape-memory polymers. *Polymer*, **51**, 3551–3559.
11. Yang, Z.H., Hu, J.L., Liu, Y.Q., and Yeung, L.Y. (2006) The study of crosslinked shape memory polyurethanes. *Materials Chemistry and Physics*, **98**, 368–372.
12. Lima, M.D., Li, N., de Andrade, M.J., Fang, S.L., Oh, J., Spinks, G.M., Kozlov, M.E., Haines, C.S., Suh, D., Foroughi, J., Kim, S.J., Chen, Y.S., Ware, T., Shin, M.K., Machado, L.D., Fonseca, A.F., Madden, J.D.W., Voit, W.E., Galvao, D.S., and Baughman, R.H. (2012) Electrically, chemically, and photonically powered torsional and tensile actuation of hybrid carbon nanotube yarn muscles. *Science*, **338**, 928–932.
13. Foroughi, J., Spinks, G.M., Wallace, G.G., Oh, J., Kozlov, M.E., Fang, S.L., Mirfakhrai, T., Madden, J.D.W., Shin, M.K., Kim, S.J., and Baughman, R.H. (2011) Torsional carbon nanotube artificial muscles. *Science*, **334**, 494–497.
14. Zhang, M., Atkinson, K.R., and Baughman, R.H. (2004) Multifunctional carbon nanotube yarns by downsizing an ancient technology. *Science*, **306**, 1358–1361.
15. Bowling, J. and Groves, G.W. (1979) The debonding and pull-out of ductile wires from a brittle matrix. *Journal of Materials Science*, **14**, 431–442.
16. Atkinson, C. (1982) The rod pull out problem, theory and experiment. *Journal of the Mechanics and Physics of Solids*, **30**, 97–120.
17. Mandel, J.A., Sun, W., and Said, S. (1987) Studies of the properties of the fiber–matrix interface in steel fiber reinforced mortar. *ACI Materials Journal*, **84**, 101–109.
18. Hutchinson, J.W. and Jensen, H.M. (1990) Models of fiber debonding and pullout in brittle composites with friction. *Mechanics of Materials*, **9**, 139–163.
19. Zhou, L.M., Kim, J.K., and Mai, Y.W. (1992) Interfacial debonding and fiber pull-out stresses – a new model based on the fracture-mechanics approach. *Journal of Materials Science*, **27**, 3155–3166.
20. Huang, N.C. and Liu, X.Y. (1994) Debonding and fiber pull-out in reinforced composites. *Theoretical and Applied Fracture Mechanics*, **21**, 157–176.
21. Ochiai, S. (1999) Interfacial debonding in single fibre-composite with a cracked matrix – Part 1: Debonding during cooling. *International Journal of Materials and Product Technology*, **14**, 147–166.
22. Greszczuk, L.B. (1969) Theoretical studies of the mechanics of the fiber–matrix interface in composites. *ASTM STP*, **452**, 42–58.
23. Lawrence, P. (1972) Some theoretical considerations of fiber pull-out from an elastic matrix. *Journal of Materials Science*, **7**, 1–9.
24. Gray, R.J. (1984) Analysis of the effect of embedded fibre length on fibre debonding and pull-out from an elastic matrix. *Journal of Materials Science*, **19**, 861–870.
25. Naaman, A. (1991) Fiber pullout and bond slip. I: Analytical study. *Journal of Structural Engineering*, **117**, 2769–2790.

26. Kim, J.K., Baillie, C., and Mai, Y.W. (1992) Interfacial debonding and fibre pull-out stresses. *Journal of Materials Science*, **27**, 3143–3154.

27. Ouyang, Z. and Li, G. (2009) Local damage evolution of DCB specimens during crack initiation process: a natural boundary condition based method. *ASME Journal of Applied Mechanics*, **76**, paper 051003 (8 pages).

28. Ouyang, Z., Ji, G., and Li, G. (2011) On approximately realizing and characterizing pure Mode-I interface fracture between bonded dissimilar materials. *ASME Journal of Applied Mechanics*, **78**, paper 031020 (11 pages).

29. Voyiadjis, G., Shojaei, A., Li, G., and Kattan, P.I. (2012) Continuum damage-healing mechanics with introduction to new healing variables. *International Journal of Damage Mechanics*, **21**, 391–414.

30. Voyiadjis, G., Shojaei, A., Li, G., Kattan, P.I. (2012) A theory of anisotropic healing and damage mechanics of materials. *Proceedings of the Royal Society A: Mathematical, Physical and Engineering Sciences*, **468**, 163–183.

31. Voyiadjis, G., Shojaei, A., Li, G. (2012) A generalized coupled viscoplastic–viscodamage–viscohealing theory for glassy polymers. *International Journal of Plasticity*, **28**, 21–45.

32. Voyiadjis, G., Shojaei, A., and Li, G. (2011) A thermodynamic consistent damage and healing model for self healing materials. *International Journal of Plasticity*, **27**, 1025–1044.

33. Wool, R.P. and O'Connor, K.M. (1981) A theory crack healing in polymers. *Journal of Applied Physics*, **52**, 5953–5963.

34. Darabi, M.K., Abu Al-Rub, R.K., and Little, D.N. (2012) A continuum damage mechanics framework for modeling micro-damage healing. *International Journal of Solids and Structures*, **49**, 492–513.

35. Liu, Y., Gall, K., Dunn, M.L., Greenberg, A.R., and Diani, J. (2006) Thermomechanics of shape memory polymers: uniaxial experiments and constitutive modeling. *International Journal of Plasticity*, **22**, 279–313.

36. Li, G. and Xu, W. (2011) Thermomechanical behavior of thermoset shape memory polymer programmed by cold-compression: testing and constitutive modeling. *Journal of the Mechanics and Physics of Solids*, **59**, 1231–1250.

37. Tobushi, H., Hashimoto, T., Hayashi, S., Yamada, E. (1997) Thermomechanical constitutive modeling in shape memory polymer of polyurethane series. *Journal of Intelligent Material Systems and Structures*, **8**, 711–718.

38. Diani, J. and Gall, K. (2007) Molecular dynamics simulations of the shape-memory behaviour of polyisoprene. *Smart Materials and Structures*, **16**, 1575–1583.

39. Nguyen, T.D. (2013) Modeling shape-memory behavior of polymers. *Polymer Reviews*, **53**, 130–152.

40. Li, G. and Xu, T. (2011) Thermomechanical characterization of shape memory polymer based self-healing syntactic foam sealant for expansion joint. *ASCE Journal of Transportation Engineering*, **137**, 805–814.

41. Li, G., King, A., Xu, T., and Huang, X. (2013) Behavior of thermoset shape memory polymer based syntactic foam sealant trained by hybrid two-stage programming. *ASCE Journal of Materials in Civil Engineering*, **25**, 393–402.

42. Li, G. (2013) Liquid Sealant with Thermally Adaptive Properties. US Provisional Patent Application Number 61897437.

43. Koutsopoulos, H.N. and Downey, A.B. (1993) Primitive-based classification of pavement cracking images. *ASCE Journal of Transportation Engineering*, **119**, 402–418.

44. Bellin, I., Kelch, S., Langer, R., and Lendlein, A. (2006) Polymeric triple-shape materials. *Proceedings of the National Academy of Science of USA*, **103**, 18043–18047.

45. Bellin, I., Kelch, S., and Lendlein, A. (2007) Dual-shape properties of triple-shape polymer networks with crystallizable network segments and grafted side chains. *Journal of Materials Chemistry*, **17**, 2885–2891.

46. Behl, M., Bellin, I., Kelch, S., Wagermaier, W., and Lendlein, A. (2009) One-step process for creating triple-shape capability of AB polymer networks. *Advanced Functional Materials*, **19**, 102–108.

47. Zotzmann, J., Behl, M., Feng, Y., and Lendlein, A. (2010) Copolymer networks based on poly(ω-pentadeca-lactone) and poly(ε-caprolactone) segments as a versatile triple-shape polymer system. *Advanced Functional Materials*, **20**, 3583–3594.

48. Xie, T. (2010) Tunable polymer multi-shape memory effect. *Nature*, **464**, 267–270.

49. Li, G. (2014) A smart sealant made of two-way shape memory polymer. Technology Disclosure Form (OIP number 1504). Office of Intellectual Property, Commercialization & Development, Louisiana State University.

50. Chung, T., Romo-Uribe, A., and Mather, P.T. (2008) Two-way reversible shape memory in a semicrystalline network. *Macromolecules*, **41**, 184–192.

51. Li, G. (2014) A rutting and cracking resistance asphalt concrete. Technology Disclosure Form (OIP number 1422). Office of Intellectual Property, Commercialization & Development, Louisiana State University.

52. Hu, S., Zhou, F., and Scullion, T. (2011) Development, calibration, and validation of a new M-E rutting model for HMA overlay design and analysis. *ASCE Journal of Materials in Civil Engineering*, **23**, 89–99.

Index